THE ICEBERG IN THE MIST: NORTHERN RESEARCH IN PURSUIT OF A "LITTLE ICE AGE"

A C.I.P. Catalogue record for this book is available from the Library of Congress.

ISBN 978-90-481-5644-3 ISBN 978-94-017-3352-6 (eBook)
DOI 10.1007/978-94-017-3352-6

Printed on acid-free paper

THE ICEBERG IN THE MIST: NORTHERN RESEARCH IN PURSUIT OF A "LITTLE ICE AGE"

Edited by

Astrid E. J. Ogilvie

Institute of Arctic and Alpine Research, University of Colorado, 1560 30th Street, Campus Box 450, Boulder, CO 80309-0450, U.S.A.

and

Trausti Jónsson

Department of Research and Processing, Icelandic Meteorological Office, Bústaðavegur 9, ÍS-150 Reykjavík, Iceland

Reprinted from Climatic Change
Volume 48, No. 1, 2001

Springer-Science+Business Media, B.V.

Contents

4. Synoptic Climatology

5. Human Dimensions

This special issue of Climatic Change is dedicated to:

Knud Frydendahl (Denmark)

Erik Wishman (Norway)

Páll Bergþórsson (Iceland)

GUEST EDITORIAL

This special issue of *Climatic Change* developed partly from a conference held at the University of Iceland in Reykjavík in August 1997 by EACH (European and Atlantic Climate Historians) and NABO (the North Atlantic Biocultural Organization). The conference was funded in part by a grant from the US National Science Foundation (OPP-952359) entitled: 'Historical Dimensions of Sustainability in the North Atlantic ca. 2,000BP to the present' and also in part by *Veðurstofa Íslands* (the Icelandic Meteorological Office).

The somewhat dry title of the 1997 conference was: 'Climatic and Environmental History of Northern Europe and the North Atlantic Region over the past 1000 years'. However, there was a rather more provocative sub-title: 'Was there a "Little Ice Age" in Northern Europe and the North Atlantic Region and, if so, Where and When?' This subtitle paraphrased the title of the introductory article to the special issue of *Climatic Change* on the "Medieval Warm Period" edited by Malcolm Hughes and Henry Diaz. This article was entitled: 'Was there a "Medieval Warm Period", and if so, Where and When?' Inspired by those authors, Astrid Ogilvie and Trausti Jónsson, disturbed by the often glib and *ad hoc* usage of the term the "Little Ice Age", found themselves provoked to produce a similar volume, this time focusing on what has become known as the "Little Ice Age".

This volume is dedicated to three colleagues who have contributed much to palaeoclimatology, particularly to the field of historical climatology. These are: Páll Bergþórsson, who retired as Director of the Icelandic Meteorological Office in 1993; Knud Frydendahl, who retired from the Danish Meterological Office in 1991; and Erik Wishman, who retired from his position as meteorologist with the Archaeological Museum in Stavanger, Norway, in 1996. These men all worked tirelessly and selflessly to promote our understanding of the climate of the past. They have inspired us, and others, to conduct our research in the same spirit.

We hope that, with this volume, a small step has been taken along the long road towards understanding the nature of climate change. We hope also that we have drawn attention to the fact, that, even in the sober science of climatology, our own personal viewpoints can easily colour our perceptions of climatic reality, and hence our visions of what is the signal, and what is merely the noise, in the records of past climate. In this continuing quest, we note that the recent flourishing of the emphasis on interdisciplinary and multidisciplinary approaches seems particularly fruitful. In this vein, the current volume includes research from a number of different palaeoclimatic fields. These are: the use of documentary sources, often referred to as 'historical climatology'; early instrumental records; fossil insect data; ice-core records; glacial evidence; lichenometry; synoptic climatology; and also the human dimensions of climate change. The papers presented reflect state-of-the-art knowledge, as well as thought-provoking new insights in these subject areas.

The value of the interdisciplinary approach has been illustrated recently by the development of a number of cooperative projects, and not least by the burgeoning symbiosis of the two groups mentioned above, NABO and EACH. The latter group was originally established as 'The North European Group on Historical Climatology', part of a main European group focused on the Euro-Climhist climate data base in Bern led by Christian Pfister. These groups were formed at a workshop held at the Academy of Sciences and Literature in Mainz in March 1990 on 'European climate reconstructed from documentary data: methods and results'. This workshop was funded by the European Science Foundation (ESF) within the EPC Programme (European Palaeoclimate and Man since the last Glaciation). This programme was initiated by Peter Fricker, Burkhard Frenzel, and others, in 1988. Further workshops were held, for example in Stavanger in Norway in 1992, with a focus on the Maunder Minimum period of 1675 to 1715, and in Tallinn in Estonia in 1994. The proceedings of these workshops were published mainly in the series *Paläoklimaforschung*. The first Chair of the north European climate group was Erik Wishman. At a meeting in Finland in 1995, Trausti Jónsson and Astrid Ogilvie were jointly elected the second Co-Chairs of the group. In 1997, at the above-mentioned meeting in Reykjavik, the name of the group was changed to EACH (European and Atlantic Climate Historians) conveying the sense that each member of the group had something to contribute to the whole, and to reflect the enlarged focus which now includes, not just northern Europe, but the Atlantic isles, and potentially the American seaboard as well. The group remains an informal organisation, open to all who are interested. Currently, there are some 50 members representing, among other countries: Belgium; the Czech republic; Denmark; Estonia; Finland; France; Germany; Greece; Hungary; Iceland; Italy; Latvia; Norway; Poland; Russia; Sweden; Switzerland; the UK; the USA; and China.

The NABO group was founded at a meeting at Hunter College in New York in 1992. Standing for the 'North Atlantic Biocultural Organization' *nabo* means "neighbour" in several Scandinavian languages. The acronym thus suggests a friendly and neighbourly cooperation between colleagues who mostly do not live close to each other geographically speaking, but who offer each other advice and collaboration in academic and scholarly matters. NABO is an internationally recognized non-governmental cooperative with around 450 members in 14 countries. It includes 28 major northern research centres in North America and Europe and has received funding from US, European and Canadian funding sources. Particular areas of NABO expertise are: human impacts on terrestrial and marine ecosytems; effects of climate change on cultural and natural landscapes and seascapes; inter-cultural interactions and global impacts in the North Atlantic (see also www.geo.ed.ac.uk/nabo/home.html). The Co-Chairs of the climate working group within NABO are Astrid Ogilvie and Paul Buckland, who were also Co-Founders of NABO.

The existence of this special issue owes a great deal to our fellow organizers of the 1997 Reykjavík conference: Thomas McGovern; Gísli Pálsson; Jón Haukur Ingimundarson; Andy Dugmore; and Ingibjörg Jónsdóttir. We thank them for their neighbourly collaboration and support over many years. We would also like to acknowledge our colleagues in EACH and NABO for inspiring discussions on the issue of what consitituted the "Little Ice Age", in particular, the authors contributing to this special issue, and also: John Andrews, Rudolf Brázdil; Keith Briffa; Dario Camuffo; Gaston Demarée; Matti Eronen; Thór Jakobsson; Tómas Jóhannesson; Phil Jones; Ingibjörg Jónsdóttir; John Kington; Sjöfn Kristjánsdóttir; Christian Pfister; and Tom Wigley. We thank the reviewers of this volume for their careful consideration of the individual papers, and their cheerful willingness to far exceed the call of duty. Special thanks in this regard are due to Øyvind Nordli. Sincere thanks are also owed to Stephen Schneider and Katarina Kivel of *Climatic Change* for their enthusiasm and encouragement.

The authors of the papers presented in this special issue of *Climatic Change* are drawn from both NABO and EACH. The papers themselves reflect the interdisciplinary focus of these groups as well as the wide-ranging interests within climate history of our friends and colleagues, Páll Bergþórsson, Knud Frydendahl and Erik Wishman. This special issue of *Climatic Change* is hereby respectfully and affectionately dedicated to them.

A. E. J. OGILVIE

INSTAAR,
University of Colorado,
Campus Box 450,
Boulder, Colorado 80309-0450,
U.S.A.

T. JÓNSSON

Icelandic Meteorological Office,
Bústaðavegi 9,
ÍS 150, Reykjavík,
Iceland

PREFACE

THE "LITTLE ICE AGE": LOCAL AND GLOBAL PERSPECTIVES

P. D. JONES and K. R. BRIFFA

Climatic Research Unit, University of East Anglia, Norwich, NR4 7TJ, U.K.

This volume of *Climatic Change* is devoted to the study of the climate of the last 1000 years, with a major emphasis on the last few centuries. The timespan encompasses what has been referred to as the "Little Ice Age" (Bradley, 1992). This term was originally coined by glaciologists, with reference to the most recent major glacial advance of the Holocene (Bradley and Jones, 1993). Although other such advances in different parts of the world may not have been synchronous, the term "Little Ice Age" has come to be associated with the period of a widespread foreward movement of European glaciers between about 1450 to 1850, as well as with relatively cooler temperatures. The issue of whether or not this concept is appropriate, is a major theme of many of the papers included in this volume.

The main geographical focus here is on the North Atlantic and European sectors. Much more is known about the climate history from these regions than other parts of the world. The wealth of knowledge available thus enables different types of proxy climate evidence to be intercompared, and extensive climate calibration of these data is possible using early instrumental observations which extend back to the mid-to-late eighteenth century. Another focus of this special issue is the climate of Iceland, and its climatological links with Greenland, and the western half of Europe. A provocative introductory chapter (Ogilvie and Jónsson) considers the origins and usage of the term "Little Ice Age", and also discusses whether this concept has, in fact, outlived its usefulness as an accurate description of the general climate of ca. 1600 to 1900. This paper also outlines early Icelandic and Scandinavian research on the history of climate, as well as aspects of the present-day climate of Iceland.

Several papers in this volume focus on analyses of palaeoclimatic data. Thus, the paper by J. M. Grove shows that late-thirteenth century and fourteenth-century glacier advances were of a similar magnitude as later advances that have previously been used as a definition of the "Little Ice Age" period. She also makes the point that, strictly speaking, the term "Little Ice Age" should refer only to glaciation and not to temperature. The marine core data analysed by Jennings et al., indicate accelerated environmental degradation at approximately the same time as an apparent frequency in sea-ice incidence off the coast of Iceland in the late-eighteenth century. The paper by Barlow considers the Greenland ice-core isotope record, and notes the absence of a persistent multi-century signal in the ice cores during the "traditional" "Little Ice Age" time period. Alluvial sedimentation in mediterranean Europe during the Holocene is the subject of the paper by A. T. Grove, who concludes, *inter alia,* that the climatic conditions that induced the "Little Ice Age" glacier advances also were responsible for an increase in flooding frequency and sedimentation rate during the same time period. The use of insect assemblages as a climatic indicator within the "Little Ice Age" time period is discussed in the paper by Buckland and Wagner. They emphasise the difficulty of discriminating between "natural" and human impacts. However, they conclude that there was a downturn in temperature in the late medieval period, and that it is likely that this had an impact upon insect distribution. Kirkbride and Dugmore question the use of lichenometry as an absolute dating tool, and present evidence for the method's underestimation of the age of late "Little Ice Age" landforms in southern Iceland, concluding that it is evident that many of the small outlet glaciers there were at their maximum extension in the mid-eighteenth century (or even earlier), rather than a century later.

Early instrumental series are also considered in this volume. The proliferation of meteorological instruments in Europe in the late-seventeenth and early-eighteenth centuries has provided observational time series that are generally 100 or more years longer than those available from other parts of the world. In western Europe, for example, a number of series have been developed, some reaching back to the mid-seventeenth century (e.g., Manley, 1974). Extensions of series beyond the northwestern European core of early instrumental records are of great value, as they give a more complete picture of the past than many proxy sources can provide. The paper by Jónsson and Garðarsson discusses early instrumental data from Iceland. The oldest surviving observations were made there from 1749 to 1751. From 1775 onwards further observations were taken at different times, and in different parts of the country. There are continuous measurements from 1822. The paper by Tarand and Nordli provides valuable new data for Estonia, and deals with the use of ice break-up and rye harvest dates as proxy temperature indicators back to A.D. 1500 (winter) and 1706 (spring and summer). In his paper on proxy data from farmers' diaries in Norway, Nordli demonstrates how detailed records of cereal harvesting can be used for quality checking and adjustment of early instrumental temperature records. He notes large inhomogeneities of the "classical" Trondheim series. The paper by Jacobeit et al., underlines the value of early instrumental records in the analysis of the climate of recent centuries. The main emphasis in this study, however, lies in the field of synoptic climatology, and focuses on the non-stationarity of the relationship between two zonal indices on the one hand, and the temperature in Europe on the other, during the period 1780 to 1995. A major shift in circulation-temperature relationships is identified for both winter and summer conditions.

A "human dimensions" aspect is provided by Vasey, who considers climatic and environmental impacts in Iceland. Here he discusses the buffers that were in place within the food chain to prevent hunger and malnutrition, and concludes that available buffers were large, despite the marginal nature of the farming economy.

The twelve chapters in this special issue of *Climatic Change* are written by leading authors in their fields. They bring together a comprehensive body of evidence about the period; information gleaned over many decades and from many disciplines. Currently, more proxy climate data are available from Europe and North America than other regions of the world, and these show variations in timing of the colder and warmer periods over the last 600 years. Further research in Iceland and Greenland may help to reconcile or explain the differences. Certainly, the demands of the climate change detection issue are putting pressure on palaeoclimatologists to synthesise their various data sets in order to provide, not only large regional and hemispheric timeseries of past climate, but also detailed and widespread evidence of changing spatial patterns of climate variability that can form the basis for comparisons with output from General Circulation Model integrations.

The continued use of the phrase “Little Ice Age”, as suggestive of an epoch colder than today, is forcing palaeoclimatologists to conform to a term that is likely to be inappropriate in many parts of the world. Even in the North Atlantic and European regions it is far from ideal. Recent syntheses of extensive palaeoclimatic data for the northern landmasses (e.g. Jones et al., 1998; Mann et al., 1998, 1999; and Briffa, 2000) clearly show that on this large scale the period was not ubiquitously colder than today. Some decades of the eighteenth century (particularly the 1730s) were as warm as today in many parts of Europe. The coldest phases of the period were also spatially as well as temporally variable. Within the timeframe of the last thousand years, the seventeenth century is likely to have been the coldest in Europe, while the nineteenth century seems to have been the coldest in North America. It has been suggested that the “Little Ice Age”, as traditionally defined, is manifest most clearly in the North Atlantic region. However, in their introductory paper to this volume, in noting the need to carefully analyse historical sources of sea ice, Ogilvie and Jónsson point out that one single problematic source may have helped to give rise to a belief in an early "Little Ice Age" phase in the North Atlantic. Iceland also provides an intriguing contrary picture with little evidence of sea ice in the mid-seventeenth century, and extensive sea ice in the late eighteenth/early nineteenth centuries. Clearly, further advances in the acquisition and interpretation of new data are required. More regionally focused studies, of the type emphasised in this special issue of *Climatic Change*, will help to improve our knowledge of the history of climate over this period. This knowledge is crucial to gaining a clearer understanding of the causes of climatic variations. By better defining the past we can provide clearer answers to the urgent questions posed by the climate change community in the context of global change.

References

Bradley, R. S.: 1992, 'When was the "Little Ice Age"?', in Mikami, T., (ed.), Proceedings of the International Symposium on the Little Ice Age Climate, Department of Geography, Tokyo Metropolitan University, Tokyo, 1-4.

Bradley, R. S. and Jones, P. D.: 1993, '"Little Ice Age" summer temperature variations: their nature and relevance to recent global warming trends', *Holocene* **3**, 367-376.

Briffa, K. R.: 2000, 'Annual climate variability in the Holocene: interpreting the message of ancient trees', *Quaternary Science Reviews* **19**, 87-105.

Jones, P. D., Briffa, K. R., Barnett, T. P. and Tett, S. F. B.: 1998, 'High-resolution palaeoclimatic records for the last millennium: interpretation, integration and comparison with General Circulation control run temperatures', *Holocene* **8**, 455-471.

Manley, G.: 1974, 'Central England temperature data 1659-1973', *Quart. J. Roy. Met. Soc.* **100**, 389-405.

Mann, M. E., Bradley, R. S. and Hughes, M. K.: 1998, 'Global-scale temperature patterns and climate forcing over the past six centuries', *Nature* **392**, 779-787.

Mann, M. E., Bradley, R. S. and Hughes, M. K.: 1999, 'Northern Hemisphere temperatures during the past millennium: inferences, uncertainties and limitations', *Geophys. Res. Letts.* **26**, 759-762.

"LITTLE ICE AGE" RESEARCH: A PERSPECTIVE FROM ICELAND

A. E. J. OGILVIE[1] and T. JÓNSSON[2]

[1]*INSTAAR, University of Colorado, Campus Box 450, Boulder, Colorado, 80309-0450, USA*
[2]*Icelandic Meteorological Office, Bústaðavegi 9, ÍS-150, Reykjavík, Iceland*

Abstract. The development during the nineteenth and twentieth centuries of the sciences of meteorology and climatology and their subdisciplines has made possible an ever-increasing understanding of the climate of the past. In particular, the refinement of palaeoclimatic proxy data has meant that the climate of the past thousand years has begun to be extensively studied. In the context of this research, it has often been suggested that a warm epoch occurred in much of northern Europe, the north Atlantic, and other parts of the world, from around the ninth through the fourteenth centuries, and that this was followed by a decline in temperatures culminating in a "Little Ice Age" from about 1550 to 1850 (see e.g. Lamb, 1965, 1977; Flohn, 1978). The appelations "Medieval Warm Period" and "Little Ice Age" have entered the literature and are frequently used without clear definition. More recently, however, these terms have come under closer scrutiny (see, e.g. Ogilvie, 1991, 1992; Bradley and Jones, 1992; Mikami, 1992; Briffa and Jones, 1993; Bradley and Jones, 1993; Hughes and Diaz, 1994; Jones et al., 1998; Mann et al., 1999; Crowley and Lowery, 2000). As research continues into climatic fluctuations over the last 1000 to 2000 years, a pattern is emerging which suggests a far more complex picture than early research into the history of climate suggested. In this paper, the origins of the term "Little Ice Age" are considered. Because of the emphasis on the North Atlantic in this volume, the prime focus is on research that has been undertaken in this region, with a perspective on the historiography of historical climatology in Iceland as well as on the twentieth-century climate of Iceland. The phrase "Little Ice Age" has become part of the scientific and popular thinking on the climate of the past thousand years. However, as knowledge of the climate of the Holocene continues to grow, the term now seems to cloud rather than clarify thinking on the climate of the past thousand years. It is hoped that the discussion here will encourage future researchers to focus their thinking on exactly and precisely what is meant when the term "Little Ice Age" is used.

Key Words: Climate Variability; Historical Climatology; Iceland; "Little Ice Age"; "Medieval Warm Period"; Sea Ice

1. Introduction

The term "Little Ice Age", like the "Medieval Warm Period", is a concept that has entered both the scientific and popular literature. The reality of the "Medieval Warm Period" has been extensively discussed in a special issue of *Climatic Change* edited by Hughes and Diaz (1994). In the introductory article to this issue, they state that this appelation "Medieval Warm Period" has been used to describe the time period from about the "ninth through the fourteenth centuries". This current volume of *Climatic Change* seeks to follow in the footsteps of those researchers by considering what exactly may be meant when the phrase "Little Ice Age" is used, and how useful the term really is. Although the dominant usage has come to mean a general and widespread cold period occurring some time during the past millennium, most commonly from around AD 1500 or 1550 to

Climatic Change **48**: 9–52, 2001.

1800 or 1850 (see e.g. Lamb, 1965, 1977; Flohn, 1978; Mikami, 1992) it has often not been defined clearly in the scholarly literature. There are also conflicting opinions as to when and where it may be said to have occurred. In this paper, we seek to elucidate some of the origins of the term and briefly consider prevailing viewpoints, both in the general literature and in this volume.

Prior to the era of modern meteorological data, knowledge of the climate of the past 1000 years comes from a variety of sources, including early instrumental observations, documentary historical evidence, and other high-resolution proxy climate data such as ice cores, marine and lake-core data, dendroclimatological data, phenological data, and glacial records. While we will discuss certain elements of this research, a review of all the relevant literature is clearly beyond the scope of this article. However, in the sections that follow, we outline some of the research that has been undertaken in the North Atlantic and northern European sectors, and also consider a wider perspective. In the context of this special issue of *Climatic Change* with a main focus on the climate of the North Atlantic, particularly Iceland, we also feel it to be appropriate to offer a discussion on what may be termed "the historiography of climate" in Iceland. Much pertinent research has been done on climate history in Iceland, but most of this is not generally well known outside that country. Finally, in the concluding discussion, we volunteer our own perspective.

2. Origins and Use of the Term "Little Ice Age"

The term "Little Ice Age" naturally has its origin as a reflection of a larger "Ice Age" (or "Ice Ages") when ice sheets advanced, and cold climatic regimes prevailed, before the current Holocene or "recent" period of the last 10,000 years or so. We may speculate that, as climatologists and palaeoclimatologists in the early and middle part of the twentieth century concentrated on unravelling the details of the climate of the recent past, it was tempting to compare what they perceived to be an interesting phenomenon with a similar episode (albeit of greater magnitude) in the past.

A detailed discussion on the origins and uses of the term the "Little Ice Age" may be found in the seminal work on the topic (J. M. Grove, 1988). However, it is of interest here to mention briefly some of the early uses of the concept. The term "Little Ice Age" appears to have been first introduced into the literature by Matthes, although his definition was very different from that which has passed into general usage. As Chairman of a committee on glacier fluctuations, he stated: '...accordingly it may well be said that we are living in an epoch of renewed but moderate glaciation - a "little ice-age" that already has lasted about 4000 years' (Matthes et al., 1939 p. 520). The subsequent entry of the term into common usage (and with the specific sense of referring to some variant of the period from somewhere around AD 1550 to 1850) may be attributed largely to researchers such as Lamb and Flohn. It is also of interest to note that Manley, another pioneer in the field, did not favour the term. In 1974, in one of his classic papers on the Central England temperature record, he wrote: 'We need to know more...about the onset and progress of what some have been inclined to call, not very satisfactorily, the Little Ice Age...' (Manley, 1974, p. 403). For further early discussions of "post-glacial climatic change" see also Hoyle et al., 1949.

Three main viewpoints on the definition of the nature of the term may be mentioned. Perhaps best-known is the view which equates the "Little Ice Age" with a time period during which temperatures were relatively cold. This is expressed succinctly by Lamb, who states: 'The period we are discussing has been dubbed "The Little Ice Age" because not only in Europe but in most parts of the world the extent of snow and ice on land and sea seems to have attained a maximum as great as, or in most cases, greater than, at any time since the last major ice age' (Lamb, 1977, pp. 461-462). A second viewpoint is presented by J. M. Grove. She states: 'The term "Little Ice Age" does not refer directly to climate but to the most recent period during which glaciers extended globally and remained enlarged...' (J. M. Grove, this volume). Elsewhere she states clearly that the term "Little Ice Age" should refer specifically and only to glaciation, not temperature (J. M. Grove, 1988, p. 3). A third view argues that there was, in fact, no "Little Ice Age" at all in the commonly accepted sense. This is stated most forcefully by Landsberg: '...it is a misnomer because *continental* (his italics) glaciation did not increase nor was there any sustained low global temperature...It would be far better to expurgate the scientifically misleading (even if journalistically appealing) term "Little Ice Age", because of the fact that the interval so designated was not uniformly cold in space or time' (Landsberg, 1985, p. 62).

As regards the time period covered, as Landsberg also noted, 'The literature is not even consistent about what interval is designated as the "Little Ice Age": for example, 1550-1700, eighteenth and nineteenth centuries, 1550-1850, fifteenth through nineteenth centuries' (Landsberg, 1985, footnote to p. 62). However, what may be called the orthodox or the classical viewpoint approximates closely to Flohn's suggestion: '...the cold period, often known as the Little Ice Age, which followed and reached its climax in various parts of Europe at different times between about A.D. 1550 and 1850' (Flohn and Fantechi, 1984). In many of Lamb's works his dates approximate roughly to this time, with minor variations, for example, 'the cold Little Ice Age climate of about 1550 to 1800' (1977, p. 104) and 'It is reasonable to regard the time from about 1550 to 1700 as the main phase for most parts of the world…' (1977, p. 463). Another approach suggests that the "Little Ice Age" began much earlier and extended later. Indeed, Lamb (1977, p. 449) discusses the 'climatic worsening (*Klima-Verschlechterung*) of the late Middle Ages' (AD 1200-1400) and, further: 'The course of the climatic deterioration over five centuries from AD 1200 can quite well be traced…' (Lamb, 1977, p. 451). J. M. Grove states that: 'The "Little Ice Age" was well under way during the thirteenth and fourteenth centuries' (Grove, this volume). It is interesting to note that, with this definition, the start of the "Little Ice Age" encroaches quite significantly on the "Medieval Warm Period" time period of the 'ninth through the fourteenth centuries' as discussed by Hughes and Diaz (1994).

Although the "Little Ice Age", both in the Matthesian and Lambian sense, is a mid-twentieth century construction, discussion on climatic changes in the North Atlantic region during historical time had already started by the early nineteenth century. An example of this is the survey given in 1824 by the Swedish scholar and diplomat, Ehrenheim, on the variability of climate. This discusses both "modern" (i.e. late-eighteenth, early-nineteenth century) variations as well as historical ones. There is no definite conclusion in the work, but he notes that severe winters with heavy ice conditions, both in Scandinavia, and elsewhere in Europe, seem to be getting progressively fewer since the late Middle Ages. He also suggested that the summers have at the same time become cooler, although the more "recent" development is complex, as there is evidence for a cooling in Scandinavia during instrumental measurement time (i.e. from the beginning of the instrumental records in the eighteenth century to his time of writing in the third decade of the nineteenth century). For Ehrenheim, the crux of the matter is variability. According to his ideas, the occurrence of a cold period indicates that a warm period is imminent, or in progress elsewhere. By stretching the point, one can find both a "Little Ice Age" of sorts and a "Medieval Warm Period" within his system, but these do not occur at the same time everywhere. In more modern terminology, Ehrenheim is emphasizing "continentality" as the important variable of climate.

By the end of the nineteenth century, other climatologists were already debating the decreased continentality of the European climate during the preceding nineteenth century and the increased maritime influence (Ekholm, 1899). Enhanced continentality eventually became an important attribute of the "Little Ice Age" (Schove, in Hoyle et al., 1949) and, more recently, by e.g., Jönsson and Holmqvist (1995). Lately, perceived changes in the continental and maritime condition, and the evolution towards a more maritime climate, include the North Atlantic/Arctic Oscillation (NAO/AO) as a player in the "Little Ice Age" drama (Barlow, this volume, and references cited therein).

3. Current Issues and Perspectives

3.1. INTRODUCTION

Most current researchers would probably agree that a perceived end to the "Little Ice Age" in the latter part of the nineteenth, or early part of the twentieth century, is uncontroversial. The reasons for this are at least twofold. One is that there are many more climate data available for this period than for earlier periods, and it is therefore, relatively speaking, easier to determine the nature of climatic change around this time. Another is the perspective afforded by the increase in global temperatures. Since the late nineteenth century, the Earth has warmed by about 0.6°C (see, e.g., Jones, et al., 1999). At least part of this warming has been attributed to human activities, specifically, the emissions of carbon dioxide and other trace gases. The Intergovernmental Panel on Climate Change (IPCC) states: 'The balance of evidence suggests a discernible human influence on global climate' (Houghton et al., 1996, p. 4). Another perspective is presented by Kreutz et al., (1997), however. From their measurements on sea-salt deposition in ice cores from both Antarctica and Greenland they suggest that if an onset of the "Little Ice Age" can be characterised by the increase in deposition about AD 1400 these conditions are still persisting, and they conclude that '...in terms of polar atmospheric circulation, we suggest it is possible that conditions common during the LIA, may have persisted into the 20th century and may still persist' (Kreutz et al., 1997, p. 1296).

Outside the immediate climate history community, the "Little Ice Age" is generally perceived as a disastrous event that occurred suddenly, at a certain time in the past. However, specialists have always been aware of the difficulty of defining exactly the onset of the "Little Ice Age" and, as stated above, even a cursory glance at the literature reveals that many dates within a 400 year period (1200–1600) are mentioned either as the onset of the "Little Ice Age" or some manifestation of it. It even seems that every climate related change during this particular period, gradual or not, is tied to the "Little Ice Age". Thus the term has become a self re-enforcing one. The recent proliferation of detailed proxy records will in the coming years undoubtedly elucidate the "onset problem" and favour a more focused discussion. This new era is already evident in the literature e.g., in Briffa (2000), Mann et al. (1999) and Kreutz et al. (1997).

3.2. DATA SETS AND APPROACHES

3.2.1. *Paleoclimatic Data Sets*

Different locations in the Atlantic Arctic region are represented by different proxy climate indicators as not all areas possess the same type of records. Thus, for example, long ice-core records (of e.g., stable isotopes, dust, chemical composition) are available from a number of sites in Greenland (see, e.g., Dansgaard, 1964; Dansgaard et al., 1975; Johnsen et al., 1989; Mayewski et al., 1993; Barlow, 1994, this volume; White et al., 1997; Kreutz et al., 1997). Accumulation has also been studied in the ice cores (e.g. Friedman et al., 1995; Cuffey and Clow, 1997). Iceland is well known for its excellent historical and early instrumental records (Ogilvie, 1991, 1992; Jónsson and Garðarsson, this volume). Historical and early instrumental records are also found in Scandinavia (see e.g. Jönsson, 1998; Selsing, 1998; Moberg and Bergström, 1997). Dahl-Jensen et al. (1998) report on temperatures in the Greenland boreholes. Huang et al. (2000) give a survey of borehole temperatures around the world, but only a few originate in the North Atlantic region. Long tree-ring series are available from a number of sites in Scandinavia, and also North America (Bartholin and Karlén, 1983; Briffa et al., 1990, 1992; Briffa and Jones, 1994; Jacoby et al., 1999; Briffa, 2000). Glacier variations have been documented by, for example, Karlén, 1982; J. M. Grove, 1988, this volume; Grove and Switzur, 1994; Guðmundsson, 1997; Geirsdóttir et al., 2000). Marine data from *foraminifera* and sediment cores give valuable proxy climate information for the past 1000 to 2000 years, and may also act as a proxy sea-ice indicator when these data are lacking (Jennings and Weiner, 1996; Jennings et al., this volume). Lake-core records are available from, for example, Iceland (see, e.g., Doner, 1999; Harðardóttir et al., in press) and Baffin (Doner and Overpeck, 1995; Hughen et al., 2000). Syntheses of a selected number of different proxy records have been made by, for example, Lamb (1977); Williams and Wigley (1983); Bradley and Jones (1993); Briffa and Jones (1993); Overpeck et al. (1997); Jones et al. (1998); Mann et al. (1998, 1999) and Crowley (2000). Drawing on some of the above-mentioned research, a few of the varying viewpoints regarding the climate of the last millennium are outlined below.

3.2.2. *Recent "Little Ice Age" perspectives*

As mentioned above, the most commonly stated orthodoxy regarding the time frame of the "Little Ice Age" has been to confine it to some time from around AD 1550 to 1850. An early attempt to rigorously analyse palaeoclimatic data sets noted: '...the "Little Ice Age" of the last few centuries, a "Medieval Warm Period" around the 12th century AD., and an earlier cold period some time between the 8th and 10th centuries' (Williams and Wigley, 1983). Recent research seems to be finding less and less evidence for the concept of the "Medieval Warm Period" (see e.g., Hughes and Diaz, 1994; Ogilvie and Farmer, 1997. However, see also Dahl-Jensen et al., 1998; Pfister et al., 1998). Thus, for example, Jones et al., (1998, p. 464) state: '...we can only concur with Hughes and Diaz (1994) that there is little evidence for the "Medieval Warm Period"...'. In the composite northern latitude tree-ring data series for the last two millenia (Briffa, 2000) one might use the label "Medieval Warm Period" for a temperature maximum that is evident during c. 980 to ca. 1100. It is very doubtful, however, if such a short period fits the idea of that concept. The general perception of the "Little Ice Age" concept is also challenged by Briffa's tree-ring dataset, as the post-1200 period is composed of both cold and warm periods, and definite intervals of multi-century deviation are difficult to find. The longest lasting negative deviations during the last two millennia as suggested by Briffa's data set is in the thirteenth and early fourteenth centuries (the "Medieval Cold Period") and the coldest years in the series seem to occur during ca. 534 to ca. 550.

In 1998, Jones et al., concluded that there is clear evidence for: '...cooler centuries between 1500 and 1900, particularly the seventeenth and nineteenth centuries which are indicative of a two phase "Little Ice Age"...' (1998, p. 469). However, a wider perspective is also noted: 'The new series...are challenging the concepts embodied in the terms "Medieval Warm Period" and "Little Ice Age" (Jones et al., 2001).

Based on analyses of a variety of palaeoclimatic data, a further interpretation of the temperature changes during the last 1000 years is put forward in Mann et al. (1999). The changes are interpreted as a cooling trend (0.028°K per century) during the whole period, with an abrubt end about 100 years ago. It is suggested that 'this cooling is possibly related to astronomical forcing' (Mann et al., 1999, p. 762). The late eleventh, late twelfth and late fourteenth centuries rival *mean* (their italics) twentieth century temperature levels. In this view, an indistinct starting time of the "Little Ice Age" is not a problem; it does not start, it creeps upon us. They suggest further that '...cooling following the fourteenth century could be viewed as the initial onset of the Little Ice Age *sensu lato*' (Mann et al., 1999, p. 762). The paper by Crowley (2000) challenges the idea of relating the AD 1000 to 1800 cooling to "astronomical forcing". Instead, the low-frequency variability of temperature during the last millennium is in large part attributed to volcanic forcing, changes in solar irradiance and greenhouse gas variability. Working with the GISP2 and GRIP ice-core records, Barlow (1994, this volume) found that the isotopic signals do not support the concept of a centuries-long low temperature time period representative of the so-called "Little Ice Age", and suggests that the climate record was rather variable in the 1500s and 1600s (GISP2 and GRIP records) and into the 1700s (GISP2). On a century time scale the lowest temperature time period during the last 1000 years may have been the 1300s (Dansgaard, et al., 1975: Barlow et al., 1997; Barlow, this volume).

In addition to the uncontroversial fact that glaciers advanced and retreated during the last millennium, analysis of borehole temperatures from bedrock and ice sheets present the most forceful 'evidence' for a cooler period preceding the twentieth century. The idea of using subtle variations in bedrock temperature gradients as manifested in boreholes in order to make inferences about climate history has existed for some time, at least since the 1920s (see e.g., references in Beck, 1992). The use of the method for detecting specifically the "Little Ice Age"and a preceding "Little Climatic Optimum" (i.e. a "Medieval Warm Period") was compellingly introduced by Cermák (1971). A brief history of the subject and more recent developments are given by Beck (1992) and by Beltrami and Mareschal (1995). Recently Huang et al. (2000) have analyzed 616 borehole profiles from all continents except Antarctica. Almost 80 per cent of these 'show a net warming over the past five centuries' (p. 757) and they suggest that the cumulative temperature change during the last 500 years exceeds 'recent estimates from conventional climate proxies' (p.756). The borehole approach has also been used for analysis of past Greenland ice surface temperatures (Clow et al., 1996; Cuffey and Clow, 1997; Dahl-Jensen, et al., 1998). Dahl-Jensen et al. (1998) resolve a "Little Ice Age" (their own label) with two distinct minima (at 1550 and 1850) and a preceding period of warmer temperatures as well.

The problems regarding the use of borehole temperatures for estimation of ground surface temperature history have been discussed in detail in the literature (see e.g. Beck, 1992; Clow, 1992; Beltrami and Mareschal, 1995; Huang et al., 1996). Clow et al. (1996) discuss the particular problems encountered in both the measurement and interpretation of ice-core borehole temperatures. A discussion of most of these problems is out of the scope of this paper, as are the differences between results obtained by the miscellaneous analysis methods.

Cermák (1971) already stresses the 'uncertain relation between surface ground temperature and the mean air temperature' and considers this a 'great obstacle' that 'requires a quantitative assessment' (Cermák, 1971, p. 17). Later studies support this cautionary point of view and show how changes in land use (deforestation, agriculture) can have a significant influence on the local ground surface temperature histories that diverge greatly from the mean air temperature changes. Here it suffices to mention the study by Skinner and Majorowicz (1999), for northwestern America, and Sebagenzi et al. (1992) for Zaire in Africa. The detailed Skinner-Majorowicz study reveals large ground surface temperature changes, independent of climatic change, at the time of deforestation, and conversion of prairie grassland to an agricultural area. The Sebagenzi et al., paper regarding Zaire is interesting in our context because the large warming that is detected (3-4°C) is almost exclusively attributed to deforestation and other surface changes. Climate changes are deemed to be 'a smaller term which could not be resolved by analyses of the available data' (p. 216). It is thus important that boreholes for a climate analysis are carefully selected and that surface ground temperature changes are not automatically attributed to climate change.

The temporal resolution of the borehole temperature deteriorates very rapidly backwards in time as a consequence of the heat diffusion process. Information about surface temperature variations slowly diffuses downwards, but at the same time the amplitude of the variations attenuates. Shen et al. (1996) apply two different analysis methods at three different measurement noise levels on two synthetic 1000-year-long data series (years 1000 to 2000 in their notation). One of these is a saw-tooth signal with an amplitude of 2°C and a period of 400 years, the other a single ramp signal starting at 1800, with a 3°C increase in temperature until 2000. At realistic noise levels (10m°K) analysis of the two very different histories render a strikingly similar result before 1800 (pp. 51 and 52).

There is a qualitative difference between the ice-core borehole data and data recovered from bedrock boreholes. In the case of the ice, the ice itself "propagates" downwards to greater depths as snow accumulates on top of it. It thus carries information about its original temperature downwards. At the same time, the ice continuously exchanges heat with the layers above and below by diffusion. The diffusion gradually smears out the original gradients (Clow et al., 1996). Eventually, at greater depths, the geothermal gradient also dominates the ice holes, just as in the bedrock ones (clearly seen in Dahl-Jensen et al., 1998, p. 269). The difference in the amplitude of the temperature variations between the

Dye and Grip core holes (Dahl-Jensen et al., 1998) is of some concern. One should also note that the past temperature estimates made by the ice-borehole temperature measurements are independent of the now "traditional" stable isotope methods.

It is very difficult to doubt the borehole records as indicators of real, albeit not fully scaled, temperature changes, but it is easy for the non-specialist to overlook the importance of resolution. There is a problem of presentation, as the resolution of one end of a bore hole time plot is much better than at the other. This gives the decidedly incorrect impression that the variability in the past was much smaller than near the present. Drawing progressively wider uncertainty (or standard error) envelopes towards the past is not appropriate either, as the early entries are multi-century values and are not very uncertain in the technical sense of the word. A possible solution would be to restrict the plot to a few points, progressively more sparse towards the past, not connected by a continuous line.

Although the borehole data present convincing evidence of a warming of the northern and mid-latitude continents during the last 200 years or so, care must be taken in the interpretation of its association with mean air temperature changes. The tropical data show that other explanations can at least in some cases also be valid. This is especially the case in areas subject to large changes in land use during the last 200 years.

3.2.3. *Changing viewpoints*

Advances in the study of palaeoclimatic proxy records have led to increased questioning regarding the accepted wisdom of the reality of the "Little Ice Age" and its companion, the "Medieval Warm Period". Thus, as gaps in the knowledge of the past thousand years (and further back in time) are increasingly filled, it has become clear that there is considerable spatial variability (Hughes and Diaz, 1994; Mann et al., 1999) and that percieved cold periods are invariably punctuated by milder episodes and vice versa (Ogilvie, 1992; Ogilvie and Farmer, 1997; Mann et al., 1999). A further perspective is noted by Jones et al. (1998). This latter paper focuses on the coherence of proxy data series used for temperature reconstructions. Little coherence is found between the proxies on the 30-50 year timescale. This result is also valid for those that are geographically quite close and 'there does not appear to be anything approaching a pattern of coherent near-global temperature change implied by the results' (p. 467). There are some strong teleconnections in the 0.03 cpy (cycles pr. year, 33-year period), but these are seriously degraded by a small frequency change down to 0.02 cpy (50-year period). Individual data series can thus not be expected to reveal anything about the structure, onset, or an end of a hemispheric (or global) "Little Ice Age". Jones et al. (1999) conclude that there is evidence that the seventeenth century was the coldest of the period since 1400, being 'on the average only 0.5° – 0.8°C below the 1961-1990 base' (p. 190). Also, as Overpeck et al. (1997) have noted in a recent paper on the Arctic climate of the past four centuries: 'interannual to century-scale Arctic climate variability is the norm' (p. 1254). Kreutz et al. (1997) sum up much current research: 'Although it was a globally distributed event, the LIA was not a 500-year period of global cooling. High-resolution tree-ring records from several areas have suggested that although there is substantial decadal-scale variability related to temperature changes over the last 2000 years, no distinct LIA signal is recorded. Indeed it is possible that the LIA was not simply a cooling everywhere but instead a period of both warm and cold anomalies that varied in importance geographically' (p. 1294).

4. The Historiography of Climate in Iceland

4.1. INTRODUCTION

Perhaps because Icelanders have long been aware of living near the climatic margin for an economy based on animal husbandry and agriculture (Ogilvie, 2000), their records of meteorological events tend to be full and varied and extend far into the past. In particular, historical records exist of the incidence of sea ice off the coasts of Iceland. Although many early writers commented on the weather and climate, and specifically on the sea ice, our discussion will focus on the principal research done in this area . This includes work by: Hannes Finnsson (first published 1796, see the 1970 edition of his work); Þorvaldur Thoroddsen (1892-1904, 1908-1922, 1914, 1916-1917); Jón Eyþórsson (1926, 1949, 1950, 1965, 1966); Lauge Koch (1945); Sigurður Thórarinsson (1956a b, 1974) and Páll Bergþórsson (1967, 1969a b, 1985). As a comprehensive survey of research into historical climatology in Iceland has not been published previously (however, see Ogilvie, 1982) it is appropriate to include it here as a context for our general discussion of research on the climate of the past millennium.

In this discussion of historical climate in Iceland, three separate viewpoints regarding climatic change will be alluded to. One may be termed the "uniformitarian" view. As this implies, it suggests a fairly constant climate, with no variation to speak of. A second may be called the "deterioration" view. This involves a transition from a relatively favourable early climate to an unfavourable later climate. This has several variants. A third may be termed the "relative" view. This emphasises that climate is constantly changing, and suggests that, even if there were relatively long cold or warm time periods they nevertheless encompass much annual-to-decadal variability.

Specific research on past climate may be said to have begun in Iceland with the publication in the late-eighteenth century of a scholarly treatise on the effects of dearth years and severe seasons. This work, *Mannfækkun af Hallærum á Íslandi*, ("Loss of Life as a Result of Dearth Years") by Hannes Finnsson was first compiled in 1793, and subsequently published in 1796 and 1970. (The edition from 1970 is used here). Finnsson defined a dearth year as one in which there was famine and deaths from hunger all over the country. He does not consider the possibility that the climate may have varied during the period of his survey (from early medieval times to the late-eighteenth century) but he does broach the subject of whether dearth years have increased or decreased. He suggests that it is difficult to make comparisons between the number of deaths from famine in the eighteenth century and earlier times because of the lack of data for the earlier period (a notably modern perspective). However, he suggests that there were fewer dearth years before AD 1280 than after, and that the main reason for this was that trade was more favourable for Iceland then. Finnsson's work is of great value and interest; it is one of the earliest attempts to investigate past climate and climate-society interactions written in Iceland, or anywhere else, and is a most scholarly and well-written work. Although he did include some works which more recent research has shown to be unreliable (Vilmundarson, 1969, 1972; Ogilvie, 1984) his careful attention to the documentation of his sources, for example, is in the best tradition of rigorous scholarship.

Other minor works on the general subject of the effects of severe seasons continued to be written in the nineteenth century. However, these are of little interest here. They consist mainly of lists of severe years, with little or no discussion, and most of them are no more than extracts from Finnsson's work. There were no new developments in the field in Iceland until the early decades of the twentieth century, when the concept of a changing climate, and the consequences of this for societies, was first suggested and subsequently discussed.

4.2. THE WORK OF ÞORVALDUR THORODDSEN

The major author on the past climate of Iceland is arguably Þorvaldur Thoroddsen (1855-1921). His photograph is shown in Figure 1. Thoroddsen's primary works on the subject of climate are the compilation *Árferði á Íslandi í þúsund ár* ('The Climate of Iceland Through one Thousand Years'), published in the same volume as the monograph, *Hafís við strendur Íslands* ('Sea Ice off Iceland's Coasts') in 1916-17. These works are in Icelandic and have never been translated into any other language. (Quotations from these given below have been translated by the authors.) Thoroddsen's compilations bring together much information on many different kinds of weather phenomena, from the time when Iceland was first settled in the late ninth century, to the early 1900s. They have been extensively used by researchers interested in the past climate of Iceland.

It is clear from his published work, that Thoroddsen adhered to the view that the climate of Iceland did not change markedly from Iceland's settlement to the time when he was writing at the end of the nineteenth and beginning of the twentieth centuries (Thoroddsen, 1914, 1916-17). He suggested that changes in the fortunes of the population were more related to the unwillingness of the inhabitants to admit to the facts of natural variability and their failure to behave accordingly, rather than being caused by long, protracted periods of hostile weather. As noted above, this is the "uniformitarian" view of climatic change.

Figure 1. Þorvaldur Thoroddsen (1855-1921). From Thoroddsen (1922-1923).

> Everything regarding weather and climate is everywhere very irregular, fickle and unstable...In all ages average and good years have been much more numerous than the bad. But it is a depressing truth that most common men, in spite of centuries of penance, seem unable to learn foresight and prudence (Thoroddsen, 1916-17, p. 350).

However, almost at the end of his commentary, after a discussion of the shortage of data prior to AD 1600, there is possibly an indication of a new perspective:

> The last three centuries can in all likelihood be compared and a close scrutiny reveals that the difference is not large. But it is quite certain that the climatic conditions of the nineteenth century, in spite of periods of hardships, have been considerably better than the conditions of the seventeenth and the eighteenth centuries (Thoroddsen, 1916-17, p. 354).

In view of the effort that Thoroddsen put into the compiling of *Árferði á Íslandi,* and the fact that it has been used so extensively by later authors on climatic variations in Iceland, it is ironic that he was convinced that the climate of Iceland had not changed significantly in the past. This he states categorically:

> It can be said with certainty that since Iceland was settled no significant changes in the climate or weather have occurred. Sagas and annals show quite clearly that good and bad years, in the past as now, have alternated with long and short intervals. Then, as now, the sea ice came to the coasts, the glaciers were the same, the desert areas the same and the vegetation the same (Thoroddsen, 1908-1922, Vol. II, p. 371).

However, it seems that it is possibly the definition of "climatic change" that is at issue here for he is ready to concede that '...good and bad years...have alternated with long and short intervals'. Another interesting point is that it has been suggested that, during the early centuries of settlement, the comparatively favourable climate of Iceland was similar to that of the early twentieth century. Presuming this to be true, Thoroddsen's remark that 'the climatic conditions have, on the whole, been the same (during the settlement period) as during the present (1914, p. 205) would be correct at least as far as this particular period was concerned.

4.3. A VICTORY FOR THE UNIFORMITARIANS?

In the early years of the twentieth century, Thoroddsen entered a heated debate on the climate of medieval Scandindavia. In 1913, the Swedish oceanographer, Otto Pettersson, published a paper entitled *Klimatförändringar i historisk och förhistorisk tid* (Pettersson, 1913). This was published in English the following year as 'Climatic Variations in Historic and Prehistoric time' (Pettersson, 1914). Here Pettersson asserted that not only does climate vary with time, but also that an economic decline in the late Middle Ages in northwestern Europe (especially Scandinavia) coincided with a period of severe climate and that this was an important causal factor for the decline. This corresponds to the "deterioration" viewpoint mentioned above. As Pettersson puts it:

> All these (climatic) events are recorded in ancient chronicles which also depict the social and economic state of the comunities, which were greatly influenced by these violent climatic variations and their consequences: famine and disease (Pettersson, 1914).

Pettersson noted that the first inhabitants of Iceland were able to cultivate grain and that this later became impossible. He also concluded that from the thirteenth century onwards, Iceland began to be blocked by drift ice and that the amount and duration of the ice was far greater than at the present time (i.e. 1914). From these observations he sought to refute the belief held by '...most geographers of our time' who 'take it for granted that the climate of Iceland has not altered in historic time' (Pettersson, 1914).

In contrast to many early (and some later) writers on climatic change, Pettersson suggested a physical explanation for the changes; a tidal theory (Pettersson, 1914). According to this view, climatic changes are driven by an 1800-year variation in the tidal forces of the moon and sun. At the tidal maxima the oceanic circulation is more vigorous, and the climate of the North Atlantic region gets colder. The Pettersson theory was later heavily critisised by Nansen (1925) as being full of inconsistencies and lack of oceanographic rigor. However, the 1800-year cycle has recently been revived (Keeling and Whorf, 2000) as a possible cause of changes evident in deep sea cores in the North Atlantic region (Bond et al., 1997). Whether this development will rehabilitate Pettersson's viewpoint remains to be seen.

As suggested above, Thoroddsen was probably one of the geographers criticised by Pettersson. However, he did believe that the climate, whether it varied or not, was of great importance for the people of Iceland. He wrote, for example:

> The sea ice...has been the most important causal factor in the dearth-years, price-rises and famines, and has done more harm to the Icelandic population than all the volcanic eruptions and earthquakes (Thoroddsen, 1914, p. 205).

In 1915, a Norwegian historian, Edvard Bull, entered the debate. He agreed with Pettersson that the climate had changed. In order to try and prove this, he listed a number of severe years within the period 1291 to 1392, and claimed that this period stood out as having a far more unfavourable climate than the periods immediately preceding or following it. Bull also called for a more precise analysis, and suggested that the most important question in the debate was not whether the Icelandic climate in early medieval times was essentially more favourable than that of the present day (c. 1915), but whether a definite climatic decline could be observed at a particular time. He went on to comment that a response to this change, and therefore its importance historically, depended not only on the magnitude of the change, but also on how quickly it occurred, and on the country's distance from the limit for successful settlement.

Bull must have been the first to suggest that because of Iceland's marginal location, only minor climatic variations needed to occur in order to make conditions difficult for the inhabitants, and that climate must therefore be of great importance for the history of such a country. He also raised another extremely important point: the need to use accurate and reliable sources for the reconstruction of climate (see e.g., Bell and Ogilvie, 1978). He pointed out that many of the sources used by Thoroddsen did not come into this category (Bull, 1915).

The debate on climate change in the early part of the twentieth century in fact interested a number of Scandinavian researchers (see e.g. Hildebrandsson, 1915; Norlind, 1915). It is not possible to mention all of these; however, it is of interest to note the research done by the Dane, Speerschneider. His main work was *Isforholdene i de Danske Farvande i ældre og nyere Tid, Aarene 690-1860* ('The State of the Ice in Danish Waters in Former and Present Times, AD 690 to 1860'). He was writing around the same time as Thoroddsen, Pettersson and Bull, and was another devotee of the "uniformitarian" viewpoint. In particular, he challenged the notion of changes in ice severity (at least around Denmark). In the English summary to his above-mentioned work (Speerschneider, 1915, p. 141) he categorically stated: 'There is no reason to believe in any marked difference in the amount of ice in Danish waters during the winters of former periods and of the present day'.

In the early 1920s, the Norwegian scientist and explorer, Fridtjof Nansen, took up the debate. He was well acquainted with medieval Scandinavian literature, in particular with what may be termed early geographical texts (see e.g., Ogilvie, 1991). One of these may be mentioned here. This is the *Konungs skuggsjá* ('The King's Mirror') believed to have been written in Norway around 1250 (Anonymous, 1917; Ogilvie, 1988). In this work, ice conditions around Greenland are described very clearly and accurately (see e.g. Lehn and Schroeder, 1979). Nansen was also very familiar with the ice conditions in polar waters due to his extensive travels in these regions. Because of his knowledge of the early texts such as 'The King's Mirror' (which to him seemed to describe the ice conditions as they were in his own day) and his own experiences as an explorer, Nansen found it it difficult to see where the idea of a milder climate in medieval times had originated. He therefore speculated that this concept was due to a misunderstanding caused by the names of the two Norse settlements in Greenland, *Eystribyggð*, the "Eastern Settlement" and *Vestribyggð*, the "Western Settlement". (In reality, both settlements were located in southwest Greenland, with the "Western Settlement" having the more northerly location.) It was thought (Nansen assures us) that the "East" component in the name *Eystribyggð* referred to the geographical area east of Cape Farewell. This area is now infested with sea ice in almost every year, but, according to this argument, could not have been during the Norse settlement period. *Ergo*, the climate must have been milder during that time. However, (Nansen continues) as the Eastern Settlement is really on the western side of southern Greenland, and as the climate there is now (for Nansen "now" is the early twentieth century) capable of supporting both cattle raising, and some cereal growing, there is, according to him, no need to invoke a complicating deterioration theory for this region. In a similar manner, Nansen also opposed the idea of a late medieval deterioration of the climate of Scandinavia and the Baltic (Nansen, 1925) and thus must be seen as siding with the uniformitarians.

The discussion was soon carried further. The first formally trained Icelandic meteorologist, Jón Eyþórsson, published a paper in 1926 with the title *Um loftslagsbreytingar* ("On Climatic Change"). In it he neatly summarizes the then recent debate and tends to agree with the conclusions of Thoroddsen and Nansen. The main emphasis of the paper is, however, a systematic refutation of the existence of what later became known as the "Medieval Warm Period" and its alleged benefit for the colonisation of Greenland.

The conclusions of Nansen and Eyþórsson were firmly based on the meteorological and oceanographical knowledge of the time. As noted above, one of their main arguments was that reliable medieval sources described ice and oceanographic conditions exactly as they are today. This conclusion is still difficult to dismiss. Regarding "land-based" arguments for a warm medieval period, they allude to human-induced environmental degradation complicated by socio-economic developments as the major forcing factor, rather than climate. They also emphasise the error of assigning a weight to the medieval *Sagas of Icelanders* as a reliable source of climate data (see Ogilvie, 1991). Both authors do confess to a belief in the reality of a climatic optimum in the early part of the Holocene, and they look to the sun or to orbital changes as the ultimate agents of climatic variability on all timescales. At the same time, they entirely dismiss the moon-sun tidal cyclicity theories of Pettersson. It is often forgotten that the main cyclic orbital changes had been calculated, and their influence on climate were already being vigorously debated in the nineteenth century. A useful overview of pre-Milankovitch knowledge on orbital changes and climate is given by Ekholm (1899, p. 381ff.).

Nansen and Eyþórsson thus found Thoroddsen's uniformitarian view the more rational one, and their long and well-argued papers could possibly have been the final blow to the "deteriorarians" if the climate itself had not intervened, so to speak. The warming that had started before the turn of the century accelerated in the 1920s, and by 1950, the reality of large climatic fluctuations between colder and warmer periods was no longer open to dispute.

4.4. A CHANGE OF CLIMATE

That a considerable warming was in progress had, in fact, become evident before 1940 (Callendar, 1938). Lauge Koch, the Danish geologist, and author of the major work on the Arctic sea ice, *The East Greenland Ice*, published in 1945, put forward the interesting idea that Thoroddsen's adherence to the view that climate does not change may have been partly due to the climatic regime that prevailed during his lifetime.

> Thoroddsen wrote his first survey of the ice conditions in Iceland immediately after the two heavy ice years 1881 and 1882. The only period with a mild climate to which he gave attention was the fourteen-year period 1840 to 1854. In 1916 to 1917 he treated the Icelandic ice conditions for the last time. In the intermediate years an amelioration of the climate did not occur until after Thoroddsen's death and consequently he never himself experienced a real and protracted amelioration of the climate like that known from the last two decades. (Here Koch means the 1920s and 1930s.) Koch (1945, pp. 210-211).

In his later papers (1949, 1950) Eyþórsson admitted that there had been some shift in the climate during the preceding thirty years. As the founder of the "Icelandic Glaciological Society" he was very well aware of the rapid recession of most of the Icelandic glaciers, and he began his 1950 paper with the statement that: 'Every middle aged and old person, living his life in Iceland, agrees that there is less snow in the winter now and that the climate is milder than it was before the turn of the century and in the beginning of the present one' (p. 67). (See also Thórarinsson, 1943). At the end of the paper, Eyþórsson cautiously states: 'We do not know if this temperature increase will continue or decline. An additional warming is very unlikely and during the last 10 years the temperature seems to have stagnated or even started on a downward route' (p. 84).

With fifty years of hindsight, it is interesting now to be able to confirm that Eyþórsson was certainly correct in his view of the climate of the first half of the twentieth century. The twentieth-century maximum warming was already then a thing of the past, but the temperature did not revert to pre-1925 levels until 1965. In one of his last papers on climate (1965) we witness a partial withdrawal of the earlier view on the "stationarity" of the Icelandic climate of historical times:

> As my experience and research on climate, sea ice and glaciers has grown, the more convinced I become, that the period 1600-1900 (or 1920) was a distinctively cold and hard one and therefore can from our point of vantage

be labelled "The Lesser Ice Age" or "The Final Ice Age" (Eyþórsson, 1965, p. 6).

Eyþórsson's final text on climate history is a short commentary on a list of mild winters compiled from Thoroddsen's work of 1916-1917 (Eyþórsson, 1966). It is tempting to view this choice of subject as an allusion to the possibility of "hidden" warm periods spanning many decades. After all, the most recent warm period (prior to 1966) only lasted 40 years.

As late as the 1950s, the Icelandic geologist, Sigurður Thórarinsson, found it necessary to point out that the debate on medieval climate had gone on so long that, during the time that had elapsed, the object of comparison, the climate of the present day, had changed to a large extent (Thórarinsson, 1956a). It seems that, by 1960, there was a general agreement on the overall shape of a curve showing temperature conditions in Iceland since settlement in the late ninth century. According to this view, it was generally warm until the thirteenth century. Then the cold set in, and lasted to 1920, with some amelioration in the fifteenth and sixteenth centuries. The cold period around 1300 is fairly well documented in written sources (see Ogilvie, 1991). Palaeoglaciological work (see e.g., Guðmundsson, 1997) also indicates major glacial advances at approximately that time. However, does this indicate that the "Little Ice Age" started as early as AD 1250 in Iceland? Was this late medieval cold period a part of so-called "Little Ice Age" 'proper'?

Questions concerning the actual rather than relative temperature changes had not been answered either. However, in a series of papers written before 1970, Páll Bergþórsson sought to address these (Bergþórsson, 1967, 1969a b). His effort was mainly concentrated on two issues: i) the determination of the actual temperature range since the settlement and; ii) the effect of temperature changes on agriculture and economy (Bergþórsson, 1985). His approach to the temperature determination is described in a paper published in English in 1969 (Bergþórsson, 1969a). Here he draws attention to the issue of a diffuse beginning of a "Little Ice Age" versus an abrupt end. In an Icelandic version of the same paper he is more explicit: 'It is thus most likely that a nearly continuous cold period stretched from the late twelfth century until the early present century' (Bergþórsson, 1969b, p. 343).

Central to Bergþórsson's approach, is a comparison of temperatures as measured at two Icelandic stations (Stykkishólmur at the west coast, from 1845 to 1969, and Teigarhorn at the east coast, from 1874 to 1969) and the sea-ice incidence during the same period. During the first part of this period (1846 to 1919) Bergþórsson finds a good correlation (-0.69) between the annual temperatures and sea ice. During the latter part (1920 to 1969) the correlation is not as good (-0.39). He then assumes that the first mentioned relationship is valid during the period 1591 to 1845 with the *ad hoc* premise that prior to 1780 "light ice" months are under-reported and 'that one has to increase the apparent incidence of ice'

(1969a, p. 94). He also asserts that due to the better reporting of ice after the Icelandic Meteorological Office was established in 1920, it 'was therefore inevitable to reduce the observed ice amount of the last decades considerably' (p. 95). Although Bergþórsson is probably qualitatively correct regarding an under-reporting of "light ice" during the seventeenth and early-eighteenth centuries, the quantification of this problem is open ended. As he himself concedes 'it must also be admitted that this fact reduces considerably the confidence of these ice records' (p. 95). For the pre-sixteenth century estimation, Bergþórsson relied on a count of "severe years" made first by Finnsson (1796, 1970) and then by Thoroddsen (1916-1917). This he correlates with the sea-ice incidence (r=0.69) during the 1591 to 1940 period. Very cold years were not always heavy ice-years, but the correlation is fairly high, and therefore Bergþórsson's assumption is probably justified.

By the method outlined above, Bergþórsson estimated 30-year temperature running means in Iceland (Stykkishólmur and Teigarhorn average) as ranging from 4.5°C in the early tenth and twelfth centuries, down to slightly below 3.0°C at the end of the thirteenth century, and in the late-eighteenth and late-nineteenth centuries. The corresponding 1931-1960 average of these stations is 4.3°C, and the 1961-1990 mean 3.6°C. According to Bergþórsson, the temperature depression of the "Little Ice Age" is thus about 0.5°C compared to the present values. However, this is about 1.2°C relative to the 1931-1960 average.

The graphs from the Bergþórsson (1969a) paper have been widely quoted, most recently in, for example, Broecker et al. (1999) and Crowley and Lowery (2000). It is disconcerting that researchers who have used Bergþórsson's ice index tend to ignore the caveats that Bergþórsson himself recognises.

4.5. HISTORICAL SOURCES, SEA ICE AND CLIMATE

The history of the uses of the term "Little Ice Age" has, like most histories, many strands. One is how it has been defined by previous researchers, another is how accurate they were in their conclusions. Just as a detailed survey of the former is beyond the scope of this article, so is any detailed analysis of the latter. However, a few points may be made here. Many of the early discussions on the "Little Ice Age" were based on historical records. Examples are the work by Lamb, Flohn, and, indeed, the Icelanders we have been discussing, from Finnsson and Thoroddsen onwards. One flaw in many early reconstructions is the failure to carefully analyse the historical sources in order to ensure their reliability. This has been pointed out by comparatively recent research (see e. g. Bell and Ogilvie, 1978). However, it was also emphasised by Bull (see above) and by Lauge Koch.

Koch's reconstruction of past ice incidence is largely based on the compilation by Thoroddsen, described above, particularly the section on sea ice. However, Koch is critical of some of Thoroddsen's methods. For example, Koch points out that Thoroddsen frequently assumed the presence of sea ice not only on the basis of direct descriptions of ice but also on indirect evidence and this was 'sometimes on a very vague basis'. As an example, Koch refers to the report that 35 ships were lost in 1118. Thoroddsen suggests that, in all likelihood, these ships were crushed by sea ice. As Koch puts it, 'this does not seem very convincing' Koch, (1945, pp. 204-205).

Koch also gives an extensive review of the literature on the past climate of Iceland, including the papers described above by Eyþórsson, Pettersson and Bull, and shows himself to be aware of many of the problems involved in using documentary sources in the reconstructions of past climate, or sea-ice incidence. For example, he postulates an argument by Nansen that records of ice incidence only reflect the number of the ice records in the literature, not the quantity of ice. As Koch points out, according to this argument the amount of ice recorded would apparently increase as records became fuller. This is not always the case, however, and Koch notes the great variations in ice conditions from about 1780 onwards, when records of ice had become more or less complete (Koch, 1945, p. 248).

As stated, in more recent times, temperature and ice incidence for Iceland during the period AD 940 to 1969 have been estimated (see above) by Páll Bergþórsson. He also based his analyses on the work of Thoroddsen. When considering sea-ice history the issue comes up again as to how the ice is estimated when descriptions are oblique or lacking. It has been suggested that the absence of ice cannot simply be assumed during times when there is little evidence due to lack of data (Vilmundarson, 1969, 1972; Ogilvie, 1984, 1992). As Koch also pointed out (see above) Thoroddsen frequently assumed the presence of ice on rather flimsy evidence. However, it may not be appropriate to assume the presence of ice on indirect evidence. In discussing this issue, Bergþórsson gives an example. He says that in 1758 'the annals tell us there was no drift ice at the coast, and that this had hardly occurred in memorable time' and that 'even if no ice is mentioned in the annals for 1751, 1752 and 1753, it is therefore not justifiable to consider them ice-free' (1969a, p. 94). Upon examining the sources more closely, this assumption would seem to be incorrect. Firstly, the phrase upon which the evidence for ice in the years mentioned hinges, 'this had hardly occurred in memorable time', is a sort of stock-phrase used frequently in sources such as the annals, and it is questionable how much reliance may be placed upon it. Secondly, if we look closely at the sources for the other years: in fact sea ice was recorded in 1751. In a contemporary source, *Höskuldstaðaannáll*, it is stated: 'Ice came in at góa (c. third week in February) and filled up all of Húnafjord...The ice lay for a long time' (*Annálar* IV, p. 490). However, sea ice was not recorded for 1752, 1753, 1754 or 1755. The description of ice given in Thoroddsen (1916-17, p. 156) for 1755 is an error for 1756. That there should not be ice seems surprising in view of the cold weather described during these years, but it seems too much of a coincidence that none of the reliable contemporary sources consulted for these years record ice. A further discussion of ice conditions during 1755 and 1756 may be found in Ogilvie (1984) and also in Vilmundarson (1969, 1972) in reference to an unpublished paper (Ólafsson, 1967) on the severe years 1751-1757, in which the compounded errors regarding 1755 are pointed out.

What is most striking, however, is not Thoroddsen's omission of sources but the fact that he consulted as many as he did. That many of these sources (most of the later annals, for example) were in difficult-to-read manuscript form, makes Thoroddsen's work even more laudable; it is unfortunate that he was unable to benefit from the critical work done by the editors of the annals when these works were later published. In the final instance, Thoroddsen's compilation must be valued for the pioneering work that it is.

The importance of carefully analysing historical sources may also be seen in regard to a much broader issue than the accuracy of an account for one particular year. Lamb (and others) have concluded that the old sailing route to Greenland had been abandoned by about AD 1342 because of increasing sea ice, and hence, it has been inferred that the "Little Ice Age" in the North Atlantic set in around this time (Lamb, 1977, p. 6). The source for this description of a changed sailing route is a work entitled *Grænlandslýsing Ívars Bárðarsonar* ('The Description of Greenland according to Ívar Bárðarson'). This is a problematic source (Ogilvie, 1998) and, although the greater part of the work may well be reliable, the mention of sea ice is almost certainly not part of the original account, but a later interpolation. It is not known when the insert on sea ice was written or, indeed, if it is accurate. The information from this difficult source has passed further to other researchers, creating, almost single-handedly, an early "Little Ice Age" in the North Atlantic. Thus, Porter (1981, p. 99) referring to the suggestion of Dansgaard et al. (1975) that a 'comparatively cold Little Ice Age' began in the twelfth century at the site of the Crête ice core, continues: 'At about that time sea ice began to expand around Greenland, and its southern limit moved southward, possibly forcing abandonment of traditional sailing routes between Iceland and the Norse colonies in Greenland...The Little Ice Age may therefore have begun at least two to three centuries earlier in the North Atlantic than is generally inferred'. Of course, Porter's discussion is here lifted out of context, and it should be noted that he does also discuss other evidence, such as glacial advances. That other, reliable, sources regarding sea-ice conditions in the fourteenth century may exist is a separate issue (Ogilvie, 1991, 1998).

4.6. CURRENT RESEARCH ON THE PAST CLIMATE OF ICELAND

The very name of "Iceland" conjures up a country living in a permanent ice age - whether minor or major. However, Iceland's climate history over the past 1000 years is notable for its great variability. As research into historical and early instrumental records continues, this history has been increasingly refined. As stated above, in his work, *The East Greenland Ice*, Lauge Koch presented a record of sea-ice variations back to c. AD 865. Comparing the curves that he produced to 'lowlands and mountain chains', Koch concluded that: 'a lowland occurred in the saga time, a low mountain chain from 1200-1400, and from 1600-1900 actual alps' (Koch, 1945, p. 260). This is similar to the broad conclusions of Eyþórsson, mentioned above. However, as our knowledge of the past climate of Iceland has increased, the complexity of the situation has become more evident, and this picture is no longer valid. Perhaps the most striking aspect of the climatic history of Iceland is the great variability that may be seen. (However, see the discussion below on the topic of variability). Documentary and early instrumental climate data for Iceland also highlight some interesting anomalies. Current knowledge regarding the past climate of Iceland to ca. AD 1900 is outlined briefly below.

Documentary historical climate records exist for Iceland from medieval times, but they do not become prolific until the seventeenth century. Research on these records has been undertaken in recent years by, for example: Ogilvie (1982; 1984; 1991; 1992; 1993; 1996); Jónsson (1998); Guðmundsson (1997); Kristjánsdóttir, 1998; Jónsson and Garðarsson (this volume). Although it has frequently been assumed that the climate of Iceland was relatively favourable during settlement times and shortly thereafter (from c. AD 871 onwards) there are no contemporary documentary sources for this early period. However, other types of proxy evidence do suggest that the climate of that time may have been similar to the warmest part of the twentieth century in Iceland (see, e.g., Bergþórsson, 1997, p. 175-177; Ogilvie, et al., 2000). It is also possible to draw inferences on the climate of the time from our knowledge of the glacial record in Iceland. Of particular interest in this regard, is the siting of farms, prosperous in settlement times, in areas which were later rendered uninhabitable by encroaching glaciers (Ogilvie, 1984, p. 140). However, relatively cold periods are suggested during ca. 1180 to 1210, and again during the 1280s and 1290s. The fourteenth century was extremely variable. Some evidence exists for a mild climate between 1430 and 1560 (Ogilvie, 1991). (See Figure 2 for a visual presentation of these data). Furthermore, for much of this latter period, documentary sources are lacking. For the last part of the sixteenth century, sources indicate a comparatively harsh climate, extending, with some variability, to around 1630, when a particularly severe period seems to have prevailed until around 1640. Relatively mild conditions then lasted to around 1680. The late-seventeenth century was severe. The early decades of the 1700s were mild in contrast to the very cold 1690s, 1740s and 1750s. The 1760s and 1770s show a return to a milder regime. The 1780s are likely to have been the coldest decade of the century, with the harsh conditions compounded by volcanic activity (Ogilvie, 1986). The 1810s, 1830s, 1860s and 1880s were also comparatively cold. Temperatures appear to have been fairly mild in the middle of the nineteenth century, but colder in the latter part (Ogilvie, 1992; Jónsson and Garðarson, this volume).

As initially noted by Bergþórsson (1969a b) the Arctic sea ice is another important climate proxy indicator (see Figures 2 and 3a b). The few documentary

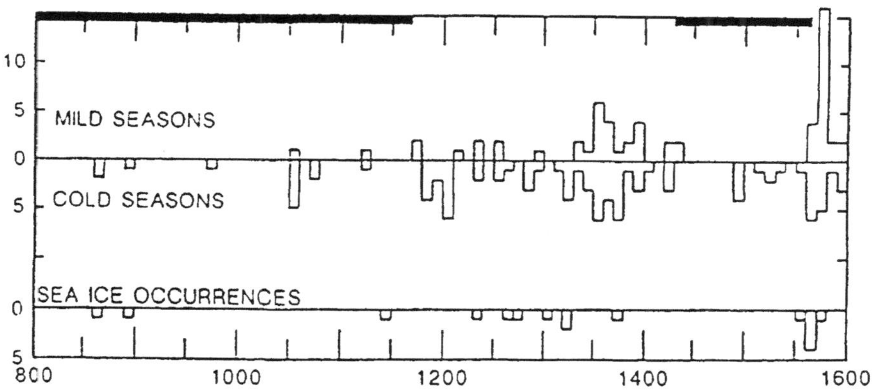

Figure 2. Mild and cold seasons in Iceland c. AD 865 to 1598, showing the number of cold or mild seasons per decade. Occurrences of sea ice are also shown; specifically, the number of years per decade in which sea ice was mentioned. The few instances of sea ice recorded at this time, undoubtedly reflect poor data coverage rather than lack of ice. The heavy lines at the top of the diagram indicate periods of very poor data (from Ogilvie, 1991).

descriptions of sea ice that are available before AD 1600 are shown in Figure 2. As may be seen, severe ice-years are documented in 1145, 1233, 1261, 1275, 1306, 1320, 1321, 1374, and then not until 1552, 1564, 1566, 1567, 1568 and 1572. The lack of descriptions of sea ice for much of this time frame must be due largely to the paucity of data, particularly in the period ca. 1430 to ca. 1560. The occurrence of severe sea-ice years in the 1590s is also implied by the sources (Ogilvie, 1991). From AD 1600, the sea-ice data are sufficient to be able to construct indices (see Figure 3a b). While sea ice was extensive in the early years of the seventeenth century, the years 1640 to 1680 seem to have been relatively ice free. No sea ice at all was reported during the 1650s. This coincides with the period when the documentary evidence suggests that mainly mild temperatures predominated. This mild period is in contrast to the marked coldness of the mid seventeenth century in Europe. The decades with most ice present during the last 500

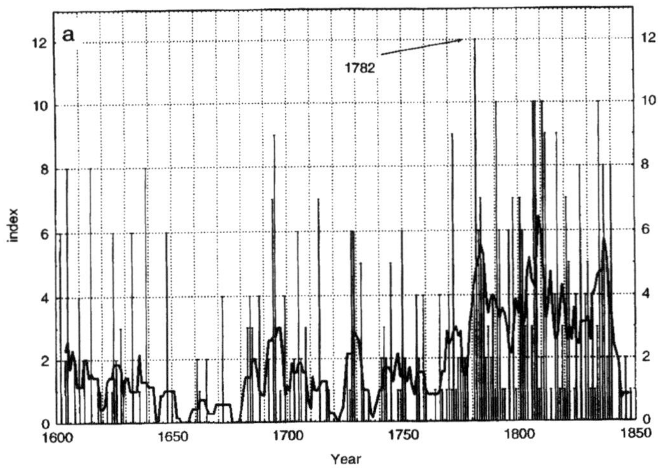

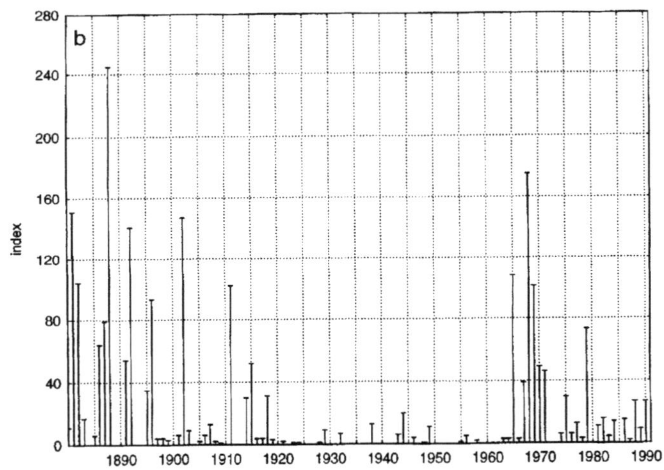

Figure 3 (a) Sea-ice index for Iceland AD 1601-1850 based on contemporary documentary sources (discussed in Ogilvie, 1992). The data are shown as individual years and also as 7-year running means. The occurrence of sea ice off the north, northwest, east and south coasts has here been quantified in order to provide an index reflecting both geographical extent and duration. (It may be noted, however, that sea ice rarely reaches the south coast; and, due to the prevailing ocean currents, virtually never occurs off the western coasts.) One "sea-ice point" was awarded for each of the above 4 regions which reported the presence of ice in any given year. A further point was accrued for each season in which ice was reported. Sea ice occurs off Iceland mainly during winter, spring and summer. Autumn ice is rare. The period 1851-1879 is in preparation. *(b)* The Koch ice-index 1881–1990. The figure is based on a table in Wallevik and Sigurjónsson (1998, p. 7–9, see also Koch, 1945). The index is a function of the "residence time" of sea ice near the coast, and a measure of the length of the coast affected each year. Due to much more intensive observations after 1965 some inhomogeneity is to be expected in the series. The "Ice Years" of the late 1960s are outstanding and the difference between the first 3 decades in the figure and the following 5 is clearly seen.

years were probably the 1780s, the 1810s and the 1830s. From 1840 to 1855 there was little ice off the Icelandic coasts. From that time to the end of the century there was frequent ice again, although the incidence does not seem to have been as heavy as in the earlier part of the century. Further clusters of sea-ice years occurred again from ca. 1864 to 1872. The 1880s contained several very severe sea-ice years. Some sea-ice years occurred in the 1890s, but less than in the 1880s. From 1903 onwards sea-ice incidence falls off dramatically. In comparison with the entire period AD 1600 to 1900, sea ice off Iceland has occurred less frequently in the twentieth century (Jónsdóttir and Ogilvie, 1997; Jónsson and Garðarsson, this volume).

5. The "Little Ice Age" and the Icelandic Temperature Record

5.1. TEMPERATURES IN WESTERN ICELAND 1830-1999

Figure 4 shows the annual temperature in Stykkishólmur in Iceland during the last 170 years, with the thick line representing 10-year running means (for station history see Sigfúsdóttir, 1969). The temperatures during the 1830-1845 period are proxies calculated from measurements made in Reykjavík. Both of the stations were in operation with a similar set of measurements during the 1845-1854 period, which was used as a basis for the calculation (Jónsson, 1989; Jónsson and Garðarsson, this volume).

The period of continuous measurements can be divided into a few sub-periods or episodes. The mid-nineteenth century warm period is clearly seen, ending abruptly in 1859. The next warming did not start until the 1920s, but closer scrutiny of the graph reveals that, after 1892 extremely cold years do not occur, so it is convenient to split the long 1859 to 1925 cold period into two, the first covering 1859 to 1892, and the second from 1893 to 1925. The 1859 to 1892 period is distinguished by large variability not seen in other periods. The coldest years are all characterized by heavy sea ice during the late winter, extending through the spring and into the summer. According to anecdotal evidence, older farmers in the mid-twentieth century reportedly said that by 1893, 'the worst was over' (Sigfúsdóttir, 1969, p. 75). It is worth noting that the two cold periods as defined above, together more or less coincide with the lifetime of Þorvaldur Thoroddsen (1855-1921). The warming in the 1920s started in the winter season, but by 1927 the warm period proper had started. It culminated in the 1930s and 1940s but came to an abrupt end in 1965 when the sea ice returned after being absent for most of the post-1920 period. In Iceland, the period 1965 to 1971 is always referred to as the "IceYears". With the exception of 1979, the ice has only made a few very short visits after 1971, but cool years persisted until the mid-1980s. Later, there has been a slight amelioration.

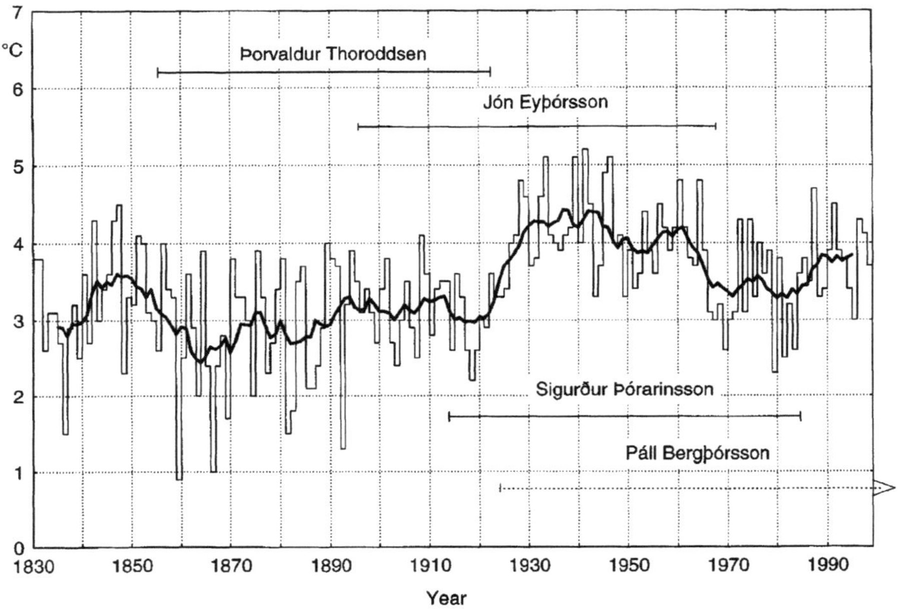

Figure 4. The Stykkishólmur temperature series 1830–1999. Individual bars are drawn for every year. The thick line represents the 10-year running means. The thin arrows show the lifelines of 4 eminent Icelandic climate researchers.

The changes in temperature from one period to another are quite abrupt, but there seems to be an underlying trend of about 0.6°C over the period as a whole. The great twentieth-century warm period is well defined in the record. As outlined in section 4 above, it certainly enhanced the image of a cold and dark colonial age, flanked by much more benevolent warm periods of freedom and liberty. The concept was also reinforced greatly by the "discovery" of a similar period in Europe, if not the whole world. After the abrupt cooling of the 1960s, the view back to a well defined "Little Ice Age" in Iceland has not been as clear as before. The warm period lasted about 40 years. It is now natural to question the uniqueness of this interval during historical time in Iceland. If one adheres to the position that all warm periods were both very short and insignificant in Iceland within the alleged "Little Ice Age" period, it seems to imply that temperature variability on the scale of two decades or more was suppressed during this period, compared to the changes of the last century or so. There is no reason to assume that this is the case during any part of the last millennium.

As mentioned above, the year-to-year variability was particularly large during the coldest part of the instrumental temperature record (1859 to 1892). It is worth noting that, if heavy ice years are removed from the record and replaced by "ordinary" cold years, the 1859 to 1892 temperature would have been about the same as it was during the following 1893 to 1925 period. The next two sections focus on sea-ice influence.

5.2. ENHANCED YEAR-TO-YEAR VARIABILITY DURING THE COLDEST PERIODS

All the extreme cold "excursions" in the Stykkishólmur temperature record were accompanied by the presence of sea ice and its characteristic temperature anomaly patterns. When the ice is present, instead of being an island, Iceland for a short time period and, in a dramatic manner, becomes more like a promontory of a Greenland-Arctic frozen continent. In an "ordinary" cold winter month the negative anomalies are largest inland, but smaller at the coast, also in the North. When there is heavy sea ice at the coast, the coastal anomalies are also large. The roughly circular (i.e., the shape of the country) anomaly pattern of the ordinary cold winter month is replaced by a large north/south anomaly gradient. The impact of the anomalies that accompany the sea ice is largest in the northern and eastern coastal districts, larger than at Stykkishólmur. It is irrefutable that the sea ice was much more frequent at the coasts of Iceland in the eighteenth and nineteenth centuries than later (see Figure 3a b). Did the sea ice just enhance the effects of the "Little Ice Age" in Iceland, or is there no "Little Ice Age" signal there without the sea ice?

In a "normal" March, the temperature difference between the islands of Vestmannaeyjar off the south coast on the one hand, and Grímsey, a solitary island off the northern coast, on the other, is about 2.5 to 3.0°C. Figure 5 shows a time series plot of this temperature difference (gradient) in March 1878 to 1999. As clearly seen, the late 1960s are highly abnormal. These are the "Ice Years" mentioned above. The break in the time axis signifies a major relocation of the Vestmannaeyjar station in 1921, which involved a change in height above sea level of almost 110 m. A correction, -0.6°C, of the Vestmannaeyjar values prior to the relocation, is applied to the series as plotted. However, a simple comparison of values before and after the break is open to doubt. All the same, it is tempting to interpret the apparent trend throughout the period as a real one. The extreme gradient of March 1881 is not an error. The sea-ice cover was extensive. It is worth pointing out that the next two pre-1921 extreme differences occurred in the heavy ice months of March 1888 and 1902 (see Figure 3b).

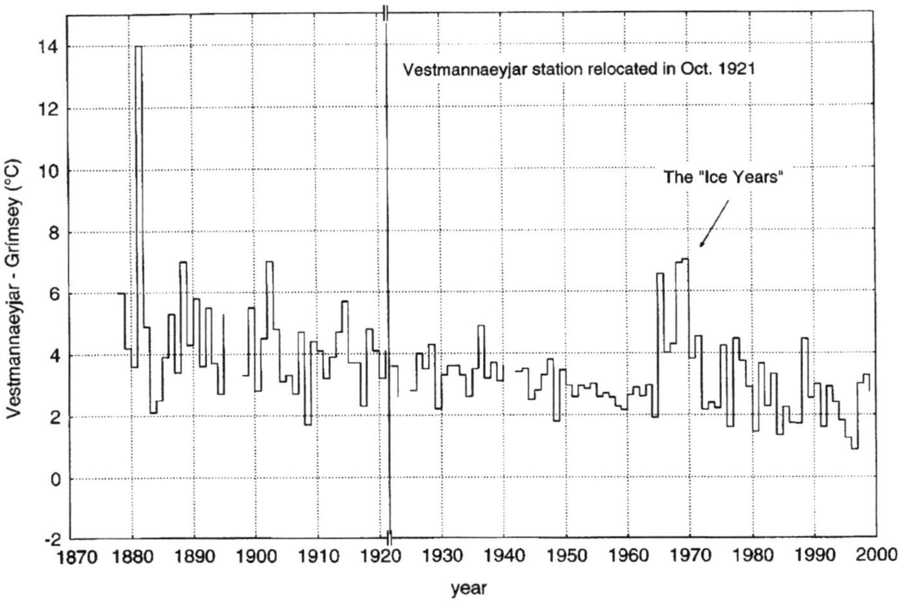

Figure 5. Vestmannaeyjar / Grímsey temperature gradient in March, 1878–1999. See text for explanation.

5.3. VERTICAL STABILITY AND THE "LITTLE ICE AGE"

The vertical stability of the atmosphere over the relatively warm ocean surrounding Iceland is at a minimum during the winter. During severe ice-years the stability at the northern and eastern coasts increases as the ice covers ocean surface areas that are usually open, and operate as heat sources. In this situation, the frequency of cold inversions (in the broad sense of the word) over the affected areas increases. The information exchange beween the near-surface layer and higher levels decreases. The convective activity associated with cold outbreaks over an open ocean diminishes in ice-affected areas, relative to the normal situation. "Normal" circulation/temperature relationships "fail". An example is seen in Figure 6, a scatterplot showing the Vestmanneyjar/Grímsey gradient introduced above, plotted against the difference of the average 500/1000hPa thickness along 70°N and 60°N between 10° and 30°W. (Thickness values from the NCEP dataset; Kalnay et al., 1996). As seen, the larger thickness gradients are usually ac-

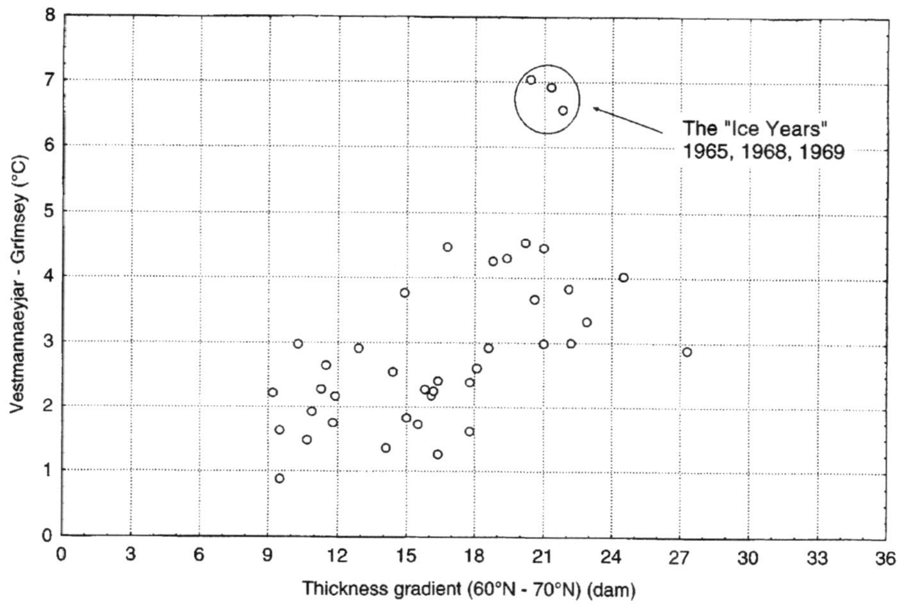

Figure 6. Vestmannaeyjar / Grímsey temperature gradient (vertical axis). NCEP 500/1000hPa thickness gradient between 60°N and 70°N averaged over the longitude interval 10°W to 30°W. Time period: 1958–1998. See text for explanation.

companied by a relatively large Vestmannaeyjar/Grímsey gradient. The group far outside the "cloud" consists of the three heavy-ice March months of 1965, 1968 and 1969. The surface gradient is in this case 2-4°C larger than expected during a month with a similar thicknes gradient during an open ocean situation.

This implies that it is not to be expected that the Summit area of Greenland (3000 m a.s.l.) gets the same message about the Icelandic climate during an ice maximum as at a minimum. The presence of the ice thus enhances cold anomalies in the lowest layers of the troposphere relative to the upper ones, at least during the 1960s ice episode. Figure 5 seems to imply that the southern/northern Iceland temperature gradient was influenced in a similar manner during the ice years prior to 1920, and these earlier occurrences were also characterised by the relative intensification of low-level negative temperature anomalies.

It is possible to split cold signals in the Iceland/southern Greenland area into two components. On the one hand, there is a deep tropospheric cooling, involving the whole troposphere down to the surface. On the other, is a low-level (shallow) cold anomaly associated with "anomalous" extensions of the sea ice or low oceanic surface temperatures. It is highly likely that both components are operating. The weight of each component is open to speculation. The sea-ice enhanced negative anomalies present in the Icelandic temperature record seem to indicate that the low-level component is considerable in that particular area. If this is the case, one would expect a smaller "Little Ice Age" signal in the highest level ice cores than at sea level, as the latter experiences both types of cold anomalies, while the former only one. One should also expect a "non-stationarity" of relationships between general tropospheric variables (circulation) and surface temperature records if this two-type-anomaly hypothesis is valid. The hypothesis is probably testable, but in the next few years it is more likely that the evidence will remain circumstantial rather than conclusive. It is appropriate at this point to recall the amplitude difference of the Dye and GRIP borehole temperatures mentioned above (Dahl-Jensen et al., 1998). This difference has not been explained in a satisfactory manner, but it supports the notion of the low-level conceptual model introduced above.

6. Perspectives from the Present Volume

The papers in this special issue of *Climatic Change* are drawn from a number of different subject areas within paleoclimatic research. What we have categorized as: "Palaeoclimatic Data Analyses" are the papers by: J. M. Grove; A. T. Grove; Jennings et al.,; Barlow; Buckland and Wagner; and Kirkbride and Dugmore. Papers focusing on "Early Instrumental Observations" are by: Jónsson and Garðarsson; Nordli; and Tarand and Nordli. The paper of Jacobeit et al., comes under the general heading of "Synoptic Climatology". A "Human Dimensions" section is represented by the paper by Vasey. Some of these studies face the question of the meaning of the term "Little Ice Age" directly. Others focus on issues other than a specific definition of the term, and allude to it in varying terms, from clear and precise to oblique.

The paper by J. M. Grove 'The initiation of the "Little Ice Age" In Regions Round the North Atlantic' clearly shows that the late-thirteenth century, and fourteenth-century glacier advances, were of a similar magnitude as later advances that have earlier been used as a definition of the "Little Ice Age" period. She is also quite specific on the definition of the "Little Ice Age" stating: 'The term "Little Ice Age" does not refer directly to climate, but to the most recent period during which glaciers extended globally, and remained enlarged, while their fronts fluctuated about more forward positions'. She also notes that the timing attributed to the "Little Ice Age" 'has varied considerably from author to author, perhaps reflecting real regional differences. However, this lack of agreement could be due in part to the uneven distribution and character of the evidence available, to the dating techniques used, and their resolution, and possibly to differing degrees of effort devoted to unravelling glacial history.' As regards the time frame of the "Little Ice Age", she concludes: 'It is evidently reasonable to view the whole period between the start of expansion in the thirteenth century and the great retreat of the late nineteenth and twentieth centuries as one complex "Little Ice Age", with each century-scale fluctuation itself made up of smaller scale oscillations...'. Furthermore, 'In all the regions around the North Atlantic, physical evidence of pre-sixteenth century glacial advances has been identified...General glacial advances took place in the North Atlantic region in the thirteenth and fourteenth centuries' (J. M. Grove, this volume).

The paper by Barlow 'The Time Period AD 1400-1980 in Central Greenland Ice Cores in Relation to the North Atlantic Sector' considers the Greenland isotope record and the absence of a persistent multi-century signal in the ice cores during the 'traditional' "Little Ice Age" time period. This apparent discrepancy between the signals from Greenland and Europe is mainly attributed to the large weight of the NAO governed Europe/Greenland temperature seesaw in the synoptic climate of the area. The paper shows a fairly good connection with temperature variations in Iceland on one hand, and the Greenland isotope record on the other.

The results of the Jennings et al., paper 'Oceanographic Change and Terrestrial Human Impacts in a Post AD 1400 Sediment Record from the Southwest Iceland Shelf' throws new light on a point that has already been made by Ogilvie (1992). A scrutiny of Icelandic historical sources seems to indicate that the sea ice visited the coasts of Iceland more frequently in the eighteenth than in the seventeenth century and there was indeed a difference between the two centuries. As stated above, the general view has been that the period 1600-1900 was more or less uniformly bad in Iceland, except for a few short and insignificant warm spells. It is of course not unlikely that the increased frequency of sea ice is at least partly a consequence of increased data density. However, the core data analysed by Jennings et al., indeed indicate accelerated environmental degradation at approximately the same time as the apparent sea-ice frequency increased. As the authors clearly point out, however, it is difficult to separate entirely the climatic effect on the environment from the impact of humans. Nevertheless, the potential of shallow shelf cores as a tool in late Holocene research is very clear.

In their paper 'Can Lichenometry be used to Date the "Little Ice Age" Glacial Maximum in Iceland?' Kirkbride and Dugmore question the use of lichenometry as an absolute dating tool. They present evidence for the method's underestimation of the age of late "Little Ice Age" landforms in southern Iceland, and conclude that it is evident that many of the small outlet glaciers there were at their maximum "Little Ice Age" extension in the mid-eighteenth century (or even earlier), rather than a century later. The evidence thus supports the notion of a complicated glaciation history of the last few centuries in Iceland.

The use of ice break-up and other data, cereal growing, and changes in the insect fauna, as proxies for climatic changes, are the subjects of the papers by Tarand and Nordli, Nordli, and Buckland and Wagner, respectively. The paper by

Tarand and Nordli, 'The Tallinn Temperature Series Reconstructed Back Half a Millennium By Use of Proxy Data' deals with the use of three different climate proxies: i) ice break-up in the port of Tallinn as a proxy for mean winter temperature; ii) ice break-up on the rivers in northern Estonia as a proxy for the beginning of spring; and iii) the first day of the rye harvest as a proxy for spring and summer temperatures. On the basis of these, the mean winter temperature could be extended back to AD 1500, and the spring and summer temperature to 1706. From the late-eighteenth century, the records document an overall warming trend in both seasons. In his paper, 'Reconstruction of Nineteenth Century Summer Temperatures in Norway by Proxy Data from Farmers' Diaries', Nordli demonstrates how good records of cereal harvesting can be used for quality checking and adjustment of early instrumental temperature records. 'Huge inhomogeneities of the "classical" Trondheim series were detected' (quotation from the abstract). The potential of using such records for the estimation of temperatures prior to the advent of instrumentation is emphasised. Discussing the period from the early to the late-nineteenth century, Nordli concludes that 'the summer temperature during the last part of the "Little Ice Age" was low compared with the climate in the twentieth century. The use of insect assemblages as a climatic indicator within the "Little Ice Age" time period is discussed in the paper by Buckland and Wagner, 'Is There an Insect Signal for the "Little Ice Age"?' The difficulty of discriminating between "natural" causes and human impact is emphasized by the authors, and they point out that a site that has not been perturbed by human interference throughout the last two millennia has yet to be found. Buckland and Wagner conclude that, in spite of this, 'the downturn in temperature in the late medieval period is likely to have had a 'significant impact upon insect distribution' and that: 'To the question posed as the title to this paper...the answer has to be a guarded, "Yes, probably" '.

Alluvial sedimentation in Mediterranean Europe during the Holocene is the subject of the paper by A. T. Grove, 'The "Little Ice Age" and its Geomorphological Consequences in Mediterranean Europe'. Concurring with J. M. Grove, the author states that: 'The "Little Ice Age" is best defined as being the interval within the second millennium A.D. during which glaciers extended further down their valleys than in the twentieth century.' According to A. T. Grove, much previous research has concluded that rapid sedimentation in river valleys and deltas in medieval and early modern times may be attributed to soil erosion. This erosion, in turn, is the result of human activity, particularly in the form of agricultural expansion and deforestation. However, A. T. Grove argues that: 'climatic extremes, especially a greater frequency of deluges, possibly associated with "Little Ice Age" advances, were more important than is usually recognized.' He thus concludes that the climatic conditions that induced the "Little Ice Age" glacier advances also were responsible for an increase in flooding frequency and sedimentation rate during the same time period. Similar earlier events could explain the formation of the so-called "younger fill" fluvial terraces in the area.

Noting that there 'appears to be a link between glacier advances and some types of extreme weather close by in the Mediterranean' he concludes that if climatic variations are responsible then: 'Rarely does it seem necessary to invoke forest clearance, the spread of agriculture, or the abandonment of land to explain the sedimentary features'.

So-called "non-stationarity" is emerging as one of the more acute problems facing climate researchers. It is becoming clear that relationships between climatic variables that have appeared to be both robust and simple in recent decades are not so strong in the long run. The paper of Jacobeit et al., 'Zonal indices for Europe 1780-1995 and running correlations with temperature' focuses on the non-stationarity of the relationship between two zonal indices on the one hand, and the temperature in Europe on the other, during the period 1780 to 1995. A major shift in circulation-temperature relationships is identified for both winter and summer conditions. It is suggested that the period 1850 to 1870 indicates a transition period in which circulation modes change from the "Little Ice Age" to the present climate of Europe. The paper underlines the value of early instrumental records in the analysis of the climate of recent centuries and the quest for its origins.

The paper by Jónsson and Garðarsson concerns 'Early instrumental meteorological observations in Iceland'. A substantial part of these early data has been digitized, and the Icelandic instrumental record is thereby extended back to the 1770s. This article focuses on the available sources rather than a climate analysis; however, the authors note that the eighteenth and nineteenth centuries appear to have been difficult climatically, with lower temperatures and heavier sea-ice incidence than experienced in this area in the twentieth century. Milder periods were found, though. Furthermore, it seems likely that the seasonal variation of temperature was somewhat larger than the present day, and the temperature gradient between southern and northern Iceland during winter was also larger than now. This research provides a solid basis for a further close study and comparison of both documentary and instrumental data from the period around 1770 onwards (Jónsson and Ogilvie, in preparation).

The final paper in this special issue is by Vasey, and focuses on the "human dimensions" aspects of climatic and environmental impacts in Iceland. Entitled 'A quantitative assessment of buffers among temperature variations, livestock, and the human population of Iceland, 1784 to 1900' the paper gives a modern analysis of issues that occupied Hannes Finnsson and other early researchers. Rather than focus purely on potential impacts of climate on the populace, Vasey discusses the buffers that were in place within the food chain to prevent hunger and malnutrition. He concludes that available buffers were, in fact, large, in spite of the marginal nature of the farming economy, and the relative poverty of the inhabitants.

7. Conclusions

Knowledge concerning climatic variability during the Holocene is rapidly advancing. As J. M. Grove and others note, it has become clear that the "Little Ice Age" (in its strict "glacial" sense) was not a unique event. Several phases of glacial expansion during the Holocene comparable in scale with those noted during the last millennium have been traced in both the Swiss and Austrian Alps (J. M. Grove, 1996, this volume, and references cited therein). Discussing a broader viewpoint, encompassing also a temperature signal, J. M. Grove's suggestion is reinforced by Kreutz et al., (1997, and references cited therein) who state that it now appears that 'several LIA-type events have occurred throughout the Holocene and that relatively minor forcings may be responsible for these events' (p. 1294). From another type of (marine-based) palaeoclimate proxy, evidence is accumulating for a semi-regular variation in the deposition of debris carried by ice into areas south of the modern ice edge (Bond et al., 1997). The so-called Ice-rafted-debris-cycle (IRD) is typically 1500 years long. It is interesting that this period is close to the Pettersson 1800-year tidal cycle mentioned above (Pettersson, 1914; Keeling and Whorf, 2000). Evidence for the linkage of the IRD to subtleties in the global thermohaline circulation is compelling (Broecker et al., 1999). Bond et al., (1997) suggest that the "Little Ice Age" is 'the most recent cold phase in the series of millennial-scale fluctuations' (p. 1264).

One of the grave difficulties regarding the "Little Ice Age" concept is that it has been (and is being) used to describe not just local, but also regional and hemispheric conditions. On the hemispheric scale, it is evident (see e.g. Jones et al., 1998; Mann et al., 1999; Briffa, 2000; and Crowley, 2000) that the temperature during the centuries preceding the twentieth were slightly lower than earlier in the period. However, there is no consensus on the nature of the cooling; it can be interpreted either as a slow "ramp" cooling or a period consisting of a few steps, each characterized by more uniform conditions. These steps are so small, however, that their significance is easily contested. On the regional scale, the overall picture tends to dissolve in details of variability. Although on this scale, it is possible in many cases to label certain cold periods as the starting dates of a "Little Ice Age", different proxy records tend to support different dates at different places (even at close locations). For these reasons, the tendency to implicate the "Little Ice Age" in any apparent perturbation in a climate or proxy variable during the last millennium, has, in the view of the present authors, been taken to the extreme, not least because it confers a one-dimensional view of climate variability. It is not to be denied that the appellation "Little Ice Age" is an evocative one, far more so certainly than, for example, "Ice rafting debris episode" (see above) and, because of this, if for no other reason, it may well continue to be used.

Regarding the North Atlantic region, the greater part of the evidence continues to support the notion that the general climate during ca. 1250 to ca. 1900 was at least slightly colder than the twentieth century overall. Conditions in this part of the world have, as we have seen, considerably contributed to the emergence and maintenance of the "Little Ice Age" concept. As described above, the large early-twentieth century warming greatly facilitated the acceptance of the concept in Iceland as well as in the Nordic region. With this perspective, the "Little Ice Age" had surely ended. However, it could be argued that it then set in again. The intensity of the late-twentieth century cold period in the Iceland/southern Greenland area seems to be generally under-appreciated. Just as the early-twentieth century warm period had the effect of focusing the perception of a "Little Ice Age" in Iceland, the subsequent cold period had the effect of blurring it. As background to this point, it may be noted, that the decadal-scale temperature variability during the instrumental era has been large in Iceland and Greenland. Certainly, the concept of an uninterrupted cold period stretches credibility, as this would imply that the decadal variance was suppressed during this time. However, as mentioned above (section 5.1), the year-to-year variability in Iceland was largest in the coldest part of the instrumental record. This merely serves to underline the need to carefully consider the meaning of "variability" as an attribute of climate (see e.g., Wigley and Raper, 1995).

The recent millennial temperature reconstructions (see e.g., Jones et al., 1998; Mann et al., 1999; Briffa, 2000; and Crowley, 2000) all show a marked warming during the last 150 to 200 years. The evidence for a change in radiative forcing due to accumulation of greenhouse gases is persuasive. The astronomical forcing proposed by Mann et al. (1999); the thermohaline variability suggested by Broecker et al. (1999) and the somewhat more complex model of Crowley do not contradict, but rather complement each other, and, along with the other multi-proxy reconstructions mentioned above, would appear to be ushering in a new era in historical climatology. Nevertheless, there is no room for complacency, as a number of issues still require attention. The most serious of these concerns the association between the instrumental temperature record and the proxy indicators (see e.g. Wigley and Raper, 1995). It is unfortunate that potentially disruptive (in the records) anthropogenic influences (for example, the land-use changes noted by Skinner and Majorewicz, 1999; Crowley, 2000; Mann, 2000; and the indirect fertilization of trees noted in Briffa, 2000) commences at about the same time as systematic temperature measurements. During the last 20 years or so, great effort has been expended on purging similar effects (urban warming, screen changes etc.) from the instrumental temperature record (see Jones et al., 1999, and references cited therein). The proxy world is clearly ripe for an analogous sanitising action. Non-stationarities of relationships suggest that there is a missing piece in the jigsaw (see chapter 5 above). It is also worth noting that when multiproxies on the large scale have been calibrated to instrumentally-measured temperatures, they have, on the whole, mainly been calibrated to rising temperatures; a cooling might show a different response.

The tale of the fall of "uniformitarianism" as outlined in section 4 above is a cautionary one, illustrating as it does how the view on the present tends to influence the perception of the past. At the present moment, a unanimous definition of the "Little Ice Age" remains as elusive as ever. The elegant solution of J. M. Grove (the "Little Ice Age" as an exclusively glacial phenomenon) is no longer a practical option. The term has already escaped its confinement. As a name for a relatively small general hemispheric scale temperature excursion it is marginally acceptable. Used in the plural as in a "Little Ice Age" event (Wigley and Raper, 1995) it may also be acceptable. However, as an indiscriminate label of every cold, dry, wet or stormy period during the last 800 years it is to be avoided. The preference of the present authors would be to see it disappear from use.

It is certainly the case that recent work has brought us closer to a partial solution, namely, finding linkages between, or common causes for, the numerous phenomena that are seen as evidence for a "Little Ice Age". An appropriate strategy for future research is a continuing quest for the discovery of further historical sources as well as the addition and elaboration of new proxy data in the coming years. At the same time, climate researchers should continue to seek to chart the climate of the past thousand years with a fresh approach rather than attempting to fit their findings into the convenient straightjacket of those hackneyed labels, the "Medieval Warm Period" and the "Little Ice Age".

To return, in concluding, to our Icelandic perspective; it may be noted again, that the task of assessing the climate signal in the environmental history of Iceland is particularly problematic. The impact of the settlement of Iceland and subsequent resource depletion easily dwarfs most signals of a climatic origin, except for changes in the glaciers. A discussion of the impact of climate on the population of Iceland has been beyond the scope of this article. However, it is appropriate to recall again the words of Þorvaldur Thoroddsen: 'The sea ice...has done more harm to the Icelandic population than all the volcanic eruptions and earthquakes' (Thoroddsen, 1914, p. 205) and to remember the effects of a sometimes difficult climate on the populace. The picture shown in Figure 7 is a sketch by Fridtjof Nansen made in the year 1882 showing a fine example of "pancake" ice. In Icelandic terminology, although a similar word is used to describe this type of ice formation, a very different term also refers to it as "plates of hunger". In seeking to unravel the meaning of the term "Little Ice Age", the present authors question the use of the term when it is used without precise definition; however, this is not to deny the reality of the numerous oscillations in climate over the past thousand years, or the difficulties often experienced by the Icelandic populace, if not during long "ice ages" but, most certainly, during "ice years".

Íslummur („hungur-diskar"). 18. apríl 1882.

Figure 7. "Hunger plates" (pancake ice). A drawing made by Fridtjof Nansen on 18 April, 1882, during his first visit to the Arctic. The drawing is from the book on his travels during the year 1882 (Nansen, 1967).

Acknowledgements

This special issue of *Climatic Change* developed from a conference held in Reykjavik in August 1997. This was funded by a grant from the US National Science Foundation (OPP-952359). Grant OPP-9726510 is also acknowledged. We would like to thank our colleagues for helpful discussions on the thorny problem of what consitituted the "Little Ice Age", in particular, the authors contributing to this special issue, and also Tómas Jóhannesson, Þór Jakobsson, Sjöfn Kristjánsdóttir, Ingibjörg Jónsdóttir, Keith Briffa and Christian Pfister. Funding for research which contributed to these discussions has also been provided by the EU Fourth framework program ADVICE project (ENV4-CT95-0129), *Vísindasjóður* (the Icelandic Science Foundation) and *Veðurstofa Íslands* (the Icelandic Meteorological Office). This paper was formally reviewed by the self-confessed Malcolm Hughes and Anders Moberg, and informally by Phil Jones, Gaston Demarée and Ingi Sigurðsson. We thank these colleagues for their most helpful and constructive comments.

References

Annálar 1400-1800, I – VII: 1922 - 1998, Hið íslenska bókmenntafélag, Reykjavík.

Anonymous: 1917, 'The King's Mirror' (Speculum Regale – Konungs skuggsjá), translated by Larson, M., in *Scandinavian Monographs* **3**, The American Scandinavian Foundation, New York

Barlow, L. K.: 1994, *Evaluation of seasonal to decadal scale deuterium and deuterium excess signals, GISP2 ice core, Summit, Greenland, A.D. 1270-1985*, Unpublished Ph.D. dissertation, University of Colorado, Boulder.

Barlow, L. K.: 2001, 'The Time Period AD 1400-1800 in Central Greenland Ice Cores in Relation to the North Atlantic Sector', this volume.

Barlow, L. K., Sadler, J. P., Ogilvie, A. E. J., Buckland, P. C., Amorosi, T., Ingimundarson, J. H., Skidmore, P., Dugmore A. J., and McGovern, T. H.: 1997, 'Interdisciplinary investigations of the end of the Norse Western Settlement in Greenland', *Holocene* **7**, 489-499.

Bartholin, T. S. and Karlén, W.: 1983, 'Dendrokronologi i Lapland AD 436-1981', *Dendrokronologiska Sällskapet, Meddelanden* **5**, 3-16.

Beck, A. E.: 1992, 'Inferring past climate change from subsurface temperature profiles: some problems and methods', *Global and Planetery Change* **6**, 73-80.

Bell, W. T. and Ogilvie, A. E. J.:1978, 'Weather Compilations as a Source of Data for the Reconstruction of European Climate During the Medieval Period', *Climatic Change* **1**, 331-348.

Beltrami, H. and Mareschal J.-C.: 1995, 'Resolution of ground temperature histories inverted from borehole temperature data', *Global and Planetary Change* **11**, 57-70

Bergþórsson, P.: 1967, 'Kuldaskeið um 1300?' (A cold period at 1300?), *Veðrið* **12**, 55-58

Bergþórsson, P.: 1969a, 'An estimate of drift ice and temperature in Iceland in 1000 years', *Jökull* **19**, 95–101.

Bergþórsson, P.: 1969b, 'Hafís og hitastig á liðnum öldum', (Sea ice and temperatures in earlier centuries) in Einarsson, M. Á. (ed.), *Hafísinn*, Almenna bókafélagið, Reykjavík 1969, pp. 333-345.

Bergþórsson, P.: 1985, 'Sensitivity of Icelandic Agriculture to Climatic Variations', *Climatic Change* **7**, 111–127.

Bergþórsson, P.: 1997, *Vínlandsgátan* (The Vinland mystery), Mál og menning, Reyjavík.

Bond, G., Showers, W., Cheseby, M., Lotti, R., Almasi, P., deMenocal, P., Priore, P., Cullen, H., Hajdas, I., and Bonani, G.: 1997, 'A pervasive millennial-scale cycle in North Atlantic Holocene and glacial climates', *Science* **278**, 1257-1266.

Bradley, R. S. and Jones, P.D.: 1992, 'Climatic variations over the last 500 years', in Bradley, R. S. and Jones P. D. (eds), *Climate since A.D. 1500*, London, Routledge, pp. 649-665

Bradley, R. S. and Jones, P. D.: 1993, ''Little Ice Age' summer temperature variations: their nature and relevance to recent global warming trends', *Holocene* **3** 367-376.

Briffa, K. R.: 2000, 'Annual climate variability in the Holocene: interpreting the message of ancient trees', *Quarternary Science Reviews* **19**, 87-105

Briffa, K. R., Bartholin, T. S., Eckstein, D., Jones, P. D., Karlén, W., Schweingruber, F. H., and Zetterberg, P.: 1990, 'A 1,400-year tree-ring record of summer temperatures in Fennoscandia', *Nature* **346**, 434-439.

Briffa, K. R., Jones, P. D., Bartholin, T. S., Eckstein, D., Schweingruber, F. H., Karlén, W., and Zetterberg, P.: 1992, 'Fennoscandian summers from AD 500: temperature changes on short and long timescales', *Climate Dynamics* **7**, 111-119.

Briffa, K. R. and Jones, P. D.: 1993, 'Global surface air temperature variations during the twentieth century: Part 2, Implications for large-scale high-frequency palaeoclimatic studies', *Holocene* **3**, 77-88.

Briffa, K. R. and Jones, P. D.: 1994, 'Summer temperatures across northern North America: Regional reconstructions from 1760 using tree-ring densities', *Journal of Geophysical Research* **99**, 25,835-25,844.

Broecker, W., Sutherland, S., and Peng, T.-H.: 1999, 'A possible 20th-century slowdown of Southern Ocean deep water formation' *Science* **286**, 1132-1135.

Buckland, P. C. and Wagner, P. E.: 2001, 'Is There an Insect Signal for the "Little Ice Age?', this volume.

Bull, E.: 1915, 'Islands klima i Oldtiden' (Medieval Climate of Iceland), *Geografisk Tidskrift* **23**, 1-5.

Callendar, G. S.: 1938, 'The artificial production of carbon dioxide and its influence on temperature', *Quarterly Journal of the Royal Meteorological Society* **64**, 223-240.

Cermák, V.: 1971, 'Underground temperature and inferred climatic temperature of the past millennium', *Palaeogeography, Palaeoclimatology, Palaeoecology* **10**, 1-19.

Clow, G. D.: 1992, 'The extent of temporal smearing in surface-temperature histories derived from borehole temperature measurements', *Global and Planetary Change* **6**, 81-100.

Clow, G. D., Saltus, R. W., and Waddington, E. D.: 1996, 'A new high-precision borehole-temperature logging system used at GISP2, Greenland, and Taylor Dome, Antarctica', *Journal of Glaciology* **42**, 576-584.

Crowley, T. J.: 2000, 'Causes of climate changes over the past 1000 years', *Science* **289**, 270-277.

Crowley, T. J. and Lowery, T. S.: 2000, 'How warm was the Medieval Warm Period?', *Ambio*, **29**, 51-54.

Cuffey, K. M. and Clow, G. D.: 1997, 'Temperature, accumulation, and ice sheet elevation in central Greenland through the last deglacial transition', *Journal of Geophysical Research* **102**, 26383-26396.

Dahl-Jensen, D., Mosegaard, K., Gundestrup, N., Clow, G. D., Johnsen, S. J., Hansen, A. W., and Balling, N.: 1998, 'Past temperatures directly from the Greenland ice sheet', *Science* **282**, 268-271.

Dansgaard, W.: 1964, 'Stable isotopes in precipitation', *Tellus* **16**, 436-438.

Dansgaard, W., Johnsen, S. J., Reeh, N., Gundestrup, N., Clausen, H. B., and Hammer, C. U.: 1975, 'Climatic changes, Norsemen and modern man', *Nature* **225**, 24-28.

Doner, L. A.: 1999, 'Late-Holocene paleoclimate records from northwest Iceland lake sediments', Abstracts with programs, **31**, *Geological Society of America, 31st Annual Meeting*, Denver, Oct. 25-28, 1999, p. A 315.

Doner, L. A. and Overpeck, J. T.: 1995, 'Comparsion of late-Holocene pollen records from Ogac lake with other Baffin Island records: is there a Little Ice Age signal?', Abstracts, *25 th International Arctic Workshop*, 16-18 March, 1995, Université Laval, Quebec City, Quebec, pp. 30-31.

Ehrenheim, F. V. von: 1824, *Om Climaternes rörlighet,* Kongl. Vetenskaps Academien, Stockholm, 1824.

Ekholm, N.: 1899, 'Om klimatets ändringar i geologisk och historisk tid samt deras orsaker', *Ymer* **1899**, 353-403.

Eyþórsson, J.: 1926, 'Um loftslagsbreytingar á Íslandi og Grænlandi síðan á landnámsöld' (On climatic changes in Iceland and Greenland since settlement times), *Skírnir* **100**, 113-133.

Eyþórsson, J.: 1949, 'Temperature variations in Iceland' in Mannerfelt, C. M. (ed.), *Glaciers and climate: geophysical and geomorphological essays dedicated to Hans W:son Ahlmann*, Svenska sällskapet för antropologi och geografi, Stockholm, pp. 36-55.

Eyþórsson, J.: 1950, 'Hitafarsbreytingar á Íslandi' (Temperature variations in Iceland), *Náttúrufræðingurinn*, **20**, 67-85.

Eyþórsson, J.: 1965, 'Hafís og ísrek' (Sea ice and sea-ice drift), *Veðrið* **10**, 22-29.

Eyþórsson, J.: 1966, 'Mildir vetur á Íslandi' (Mild winters in Iceland), *Veðrið* **11**, 3-8.

Finnsson, H.: 1796, 'Um Mannfæckun af Hallærum á Íslandi' (Depopulation and Famine in Iceland), *Rit Lærdómslistafélags* **14**, 30-226.

Finnsson, H.: 1970, *Mannfækkun af hallærum*, Eyþórsson, J. and Nordal, J. (eds), (A second annotated ed. of Finnsson (1796)), Almenna bókafélagið, Reykjavík

Flohn, H.: 1978, 'Die Zukunft unseres Klimas: Fakten und Probleme', *Promet* **8**, 1-21.

Flohn, H. and Fantechi, R. (eds): 1984, *The climate of Europe: Past, present and future*, D. Reidel Publishing Company, Dordrecht, Boston, Lancaster

Friedmann, A., Moore, J. C., Thorsteinsson, Th., Kipfstuhl, J., and Fischer, H.: 1995, 'A 1200 year record of accumulation from northern Greenland', *Annals of Glaciology* **21**, 19-25.

Geirsdóttir, Á., Harðardóttir, J., and Andrews, J. T.: 2000, 'Late-Holocene terrestrial glacial history of Miki and I. C. Jacobsen Fjords, East Greenland', *Holocene* **10**, 123-134.

Grove, A. T., 2001, 'The "Little Ice Age" and its Geomorphological Consequences in Mediterrean Europe', this volume.

Grove, J. M.: 1988, *The Little Ice Age*, Methuen, London and New York.

Grove, J. M.: 1996, 'Little Ice Ages', an entry in Schneider, S. (ed.): *Encyclopedia of climate and weather* vol. 2, Oxford University Press, New York, Oxford, 1996, pp. 475-478.

Grove, J. M.: 2001, 'The Initiation of the "Little Ice Age" in Regions around the North Atlantic', this volume.

Grove, J. M. and Switzur, R.: 1994, 'Glacial Geological Evidence for the Medieval Warm Period', *Climatic Change* **26**, 143-169.

Guðmundsson, H. J.: 1997, 'A review of the Holocene environmental history of Iceland', *Quaternary Science Reviews* **16**, 81-92.

Harðardóttir, J., Geirsdóttir, Á., and Thórdarson, Th.: in press, 'Tephra layers in a sediment core from Lake Hestvatn, S-Iceland: implications for evaluating sedimentation processes and environmental impact of tephra fall deposits in th surrounding watershed', in White, J. and Riggs, N., (eds), *Sedimentologists (IAS) Special Publication,* Blackwell, Oxford.

Hildebrandsson, H. H.: 1915, 'Sur le prédendu changement du climat Européen en temps historique', *Nova acta regiae societatis scientarium upsaliensis*, Ser. IV, vol. 4. N. 5.

Houghton, J. T., Meira Filho, L. G., Callander, B. A., Harris. N., Kattenberg, A., and Maskell, K. (eds): 1996, *Climate Change 1995: The Science of Climate Change*, Cambridge University Press, Cambridge.

Hoyle, F., Godwin, H., Manley, G., and Brooks, C. E. P.: 1949, 'Discussions – Post-glacial climatic change', *Quarterly Journal of the Royal Meteorological Society* **75**, 161-185.

Huang, S., Shen, P. Y., and Pollack, H. N.: 1996, 'Deriving century-long trends of surface temperature change from borehole temperatures', *Geophysical Research Letters* **23**, 257-260.

Huang, S., Pollack, H. N., and Shen ,P.: 2000, 'Temperature trends over the past five centuries reconstructed from borehole temperatures', *Nature* **403**, 756-758.

Hughen, K. A., Overpeck, J. T., and Anderson, R. F.: 2000, 'Recent warming in a 500-year paleotemperature record from varved sediments of Upper Soper Lake, Baffin Island', *Holocene* **10**, 9-19.

Hughes, M. K., and Diaz, H. F.: 1994, 'Was there a "Medieval Warm Period", and if so, Where and When?', *Climatic Change* **26**, 109-142.

Jacobeit, J., Jönsson, P., Bärring, L., Beck, C., and Ekström, M.: 2001, 'Zonal Indices for Europe 1780-1995 and Correlation with Temperature', this volume.

Jacoby, G. C., Workman, K. W., and D'Arrigo, R. D.:1999, 'Laki eruption of 1783, tree rings and disaster for northwest Alaska Inuit', *Quaternary Science Reviews* **18**, 1365-1371.

Jennings, A. E. and Weiner, N. J.: 1996, 'Environmental change in eastern Greenland during the last 1300 years: evidence from foraminifera and lithofacies in Nansen Fjord, 68°N', *Holocene* **6**, 179-191.

Jennings, A. E., Hagen, S., Harðardóttir, J., Stein, R., Ogilvie, A. E. J., and Jónsdóttir, I.: 2001, 'Oceanographic Change and Terrestrial Human Impacts in a Post AD 1400 Sediment Record from the Southwest Iceland Shelf', this volume.

Johnsen, S. J., Dansgaard, W. D., and White, J. W. C.: 1989, 'The origin of Arctic precipitation under present and glacial conditions', *Tellus* **41B**, 452-468.

Jones, P. D., Briffa, K. R., Barnett, T. P., and Tett, S. F. B.: 1998, 'High-resolution palaeoclimatic records for the last millennium: interpretation, integration and comparison with general circulation model control-run temperatures', *Holocene* **8**, 455-471.

Jones, P. D., New, M., Parker, D. E., Martin, S., and Rigor, I. G.: 1999, 'Surface air temperature and its changes over the past 150 Years', *Reviews of Geophysics* **37**, 173-199.

Jones, P. D., Ogilvie, A. E. J., Davies, T. D., and Briffa, K. R.: 2001, 'Climate and climate impacts: the last 1000 years: developments in research', in Jones, P. D., Ogilvie, A. E. J., Davies, T. D. and Briffa, K. R. (eds), *Climate and climate impacts: the last 1000 years*, Kluwer Academic/Plenum Publishing, in press.

Jónsdóttir, I. and Ogilvie, A. E. J.: 1997, 'Sea ice off the coasts of Iceland in historical times', in Lewkowicz A. G. and Kokel, S. V. (eds) *Program and Abstracts*, 27th Arctic Workshop, 27 February-2 March, 1997. University of Ottawa, Canada, pp. 124-125.

Jönsson, P. and Holmquist, B.: 1995, 'Wind direction in southern Sweden 1740-1992: Variation and correlation with temperature and zonality', *Theoretical and Applied Climatology* **51**, 183-198.

Jönsson, P.: 1998, 'Meteorological observations from Lund, southern Sweden, since 1740: wind direction', in Frenzel, B. (ed). *Documentary Climatic Evidence for 1750-1850 and the 14th century. Paläoklimaforschung* **23**, *Special Issue: ESF Project "European Palaeoclimate and Man"* **15**, European Science Foundation, Strasbourg, Akademie der Wissenschaften und der Literatur, Mainz, pp. 121-135.

Jónsson, T.: 1989, *Afturábak frá Stykkishólmi: Veðurathuganir Jóns Þorsteinssonar landlæknis í Nesi og í Reykjavík* (The observations of Jón Thorsteinsson in Nes and Reykjavík 1820-1854, and their relation to the Stykkishólmur series), Icelandic Meteorological Office report, Reykjavík.

Jónsson, T.: 1998, 'Reconstructing the temperature in Iceland from early instrumental observations: data availability and a status report', in Frenzel, B. (ed) *Documentary Climatic Evidence for 1750-1850 and the 14th century. Paläoklimaforschung* **23**, *Special Issue: ESF Project "European Palaeoclimate and Man"* **15**, European Science Foundation, Strasbourg, Akademie der Wissenschaften und der Literatur, Mainz, pp. 87-98.

Jónsson, T. and Garðarsson, H.: 2001, 'Early Instrumental Meteorological Observations in Iceland', this volume.

Jónsson, T. and Ogilvie A. E. J.: in preparation, 'Temperature variations in Iceland 1775 – 2000'.

Kalnay, E., Kanamitsu, M., Kistler, R., Collins, W., Deaven, D., Gandin, L., Iredell, M.,Saha, S., White, G., Woollen, J., Zhu, Y., Leetmaa, A., Reynolds, R., Chelliah, M., Ebisuzaki, W., Higgins, W., Janowiak, J., Mo, K. C., Ropelewski, C., Wang, J., Jenne, R., and Joseph, D.: 1996, 'NCEP/NCAR 40-year reanalysis project', *Bulletin of the American Meteorological Society* **77**, 437-471.

Karlén, W.: 1982, 'Holocene glacier fluctuations in Scandinavia', in *Holocene Glaciers* (ed.) Karlén, W., *Striae*, (Special issue), Uppsala **18**, 26-34.

Keeling, C. D. and Whorf, T. P.: 2000, 'The 1800-year oceanic tidal cycle: a possible cause of rapid climate change', *PNAS* **97**, 3814-3819.

Kirkbride, M. P. and Dugmore, A. J.: 2001, 'Can Lichenometry be used to date the "Little Ice Age" Glacial Maximum in Iceland?', this volume.

Koch, L.: 1945, 'The East Greenland Ice', *Meddelelser om Grønland* **130** (3), Kommisionen for Videnskabelige Undersøgelser i Grønland, Copenhagen.

Kreutz, K. J., Mayewski, P. A., Meeker, L. D., Twickler, M. S., Whitlow, S. I., and Pittalwala, I. I.: 1997, 'Bipolar changes in atmospheric circulation during the Little Ice Age', *Science* **277**, 1294–1296.

Kristjánsdóttir, S.: 1998. 'Sources of climatic information in the manuscript collection of the National Library in Iceland', in *The Second Climate and History Conference, 7-11 September, 1998*, Abstract Volume, Climatic Research Unit, Norwich, UK, p. 60.

Lamb, H. H.: 1965, 'The early medieval warm epoch and its sequel', *Palaeogeography, Palaeoclimatology, Palaeoecology* **1**, 13-37.

Lamb, H. H.: 1977, '*Climate: Present, Past and the Future, Vol. 2, Climatic History and the Future*', Methuen, London.

Landsberg, H. E.; 1985, 'Historic weather data and early meteorological observations', in Hecht, A.D. (ed.) *Palaeoclimatic Analysis and Modelling*, New York, Wiley, pp. 27-69.

Lehn, W. H. and Schroeder, I. S.: 1979, 'Polar mirages as aids to Norse navigation', *Polarforschung* **49**, 173-187.

Manley, G.; 1974, 'Central England temperatures: monthly means 1659 to 1973', *Quart. J. R. Met. Soc.* **100**, 389-405.

Mann, M. E.: 2000, 'Lessons for a new millennium', *Science* **289**, 253-254.

Mann, M. E., Bradley, R. S., and Hughes, M. K.: 1998, 'Global-scale temperature patterns and climate forcing over the past six centuries', *Nature* **392**, 779-787.

Mann, M. E., Bradley, R. S., and Hughes, M. K.: 1999, 'Northern Hemisphere temperature during the past millennium: inferences, uncertainties and limitations', *Geophysical Research Letters* **26**, 759-763.

Matthes, F. E., Reid. H. F., Hobbs, W. H., Church, J. E., Martin, L., Osgood Field Jr., Wm, Trager, E. A., Parker, G. L., Kehrlein, O., Phillips, K. N., Cooper, W. S., FitzGerald, G., and Gould, L. M.: 1939, 'Report of committee on glaciers, April 1939', *Transactions, AGU* **20**, 518-523.

Mayewski, P. A., Meeker, L. D., Morrison, M. C., Twickler, M. S., Whitlow, S. I., Ferland, K. K., Meese, D. A., Legrand, M. R., and Steffensen, J. P.: 1993, 'Greenland ice core "signal" characteristics: an expanded view of climatic change', *Journal of Geophysical Research* **98 D** 12839-12847.

Mikami, T. (ed.): 1992, *Proceedings of the International symposium on the Little Ice Age climate. Tokyo, Japan 1991*, Editorial Committee of LIAC Proceedings, Department of Geography, Tokyo Metropolitan University.

Moberg, A. and Bergström, H.: 1997, 'Homogenization of Swedish temperature data. Part III: The long temperature records from Stockholm and Uppsala', *International Journal of Climatology* **17**, 667-699.

Nansen, F.: 1925, 'Klimat-vekslinger i Nordens Historie', *Avhandlinger utgitt av Det norske videnskaps-akademi, 1. Matematisk-naturvidneskapelig klasse 1925* **3**, 1-63.

Nansen, F.: 1967, *Hjá selum og hvítabjörnum* (Icelandic translation of Nansen, F.: 1924, 'Blandt sel og björn'), Ísafoldarprentsmiðja, Reykjavík.

Nordli, P. Ø.: 2001, 'Reconstruction of Norwegian Nineteenth Century Summer Temperatures by Proxy Data from Farmer's Diaries', this volume.

Norlind, A.: 1915, 'Till frågan om det historiska klimatet, särskilt i Nord- och Mellaneuropa', *Ymer* **1915**, 83-97.

Ogilvie, A. E. J.: 1982, *Climate and society in Iceland from the medieval period to the late eighteenth century,* A doctoral dissertation, Climatic Research Unit, School of Environmental Sciences, Norwich.

Ogilvie, A. E. J.: 1984 'The Past Climate and Sea-ice Record from Iceland. Part 1: Data to A. D. 1780', *Climatic Change* **6**, 131-152.

Ogilvie, A. E. J., 1986: 'The climate of Iceland, 1701-1784', *Jökull* **36**, 57-73.

Ogilvie, A. E. J.: 1991, 'Climatic changes in Iceland A. D. c. 865 to 1598', in *The Norse of the North Atlantic* (Presented by Bigelow, G. F.), *Acta Archaeologica* **61** 1990, 233-251.

Ogilvie, A. E. J.: 1992, 'Documentary evidence for changes in the climate of Iceland, A. D. 1500 to 1800', in Bradley, R. S. and Jones, P. D. (eds), *Climate Since A.D. 1500,* Routledge, London and New York, pp. 92-117.

Ogilvie, A. E. J.: 1993, 'The climate of Scandinavia in medieval times', in Pulsiano, P., and Wolf, K. with Acker, P. and Fry, D. K. (eds), *Medieval Scandinavia: An Encyclopedia. Garland Encyclopedias of the Middle Ages,* Garland Publishing, New York and London, pp. 94-96.

Ogilvie, A. E. J.: 1996, 'Sea-ice conditions off the coasts of Iceland A.D. 1601-1850 with special reference to part of the Maunder Minimum period (1675-1715)', *AmS-Varia* **25**, Archaeological Museum of Stavanger, Norway, 9-12.

Ogilvie, A. E. J.: 1998, 'Historical accounts of weather events and related matters in Iceland and Greenland, A.D. c. 1250 to 1430', in Frenzel, B. (ed.), *Documentary Climatic Evidence for 1750-1850 and the 14th century, Paläoklimaforschung* **23**, *Special Issue: ESF Project "Euro-*

pean Palaeoclimate and Man" **15**, European Science Foundation,Strasbourg, Akademie der Wissenschaften und der Literatur, Mainz, pp. 25-43.

Ogilvie, A. E. J.: 2000, 'Climate and farming in northern Iceland, ca. 1700-1850', in Sigurðsson, I. and Skaptason, J. (eds), *Aspects of Arctic and Sub-Arctic History. Proceedings of the International Congress on the History of the Arctic and Sub-Arctic Region, Reykjavík 18-21 June 1998*, University of Iceland Press, Reykjavík, in press.

Ogilvie, A. E. J. and Farmer, G.: 1997, 'Documenting the medieval climate', in Hulme, M. and Barrow, E. (eds), *Climates of the British Isles*, Routledge, London and New York, pp. 112-133.

Ogilvie, A. E. J., Barlow, L. K., and Jennings, A. E.: 2000, 'North Atlantic climate c. A.D. 1000: Millennial reflections on the Viking discoveries of Iceland, Greenland and North America', *Weather* **55**, 34-45.

Ólafsson, B. (1967) 'Harðindaárin 1751-1757' (The dearth years 1751-1757), Unpublished dissertation, University of Iceland, Reykjavík.

Overpeck, J., Hughen, K., Hardy, D., Bradley, R., Case, R., Douglas, M., Finney, B., Gajewski, K., Jacoby, G., Jennings, A., Lamoureux, S., Lasca, A., MacDonald, G., Moore, J., Retelle, M., Smith, S., Wolfe, A., and Zielinski, G.: 1997, 'Arctic environmental change of the last four centuries', *Science* **278**, 1251-1256.

Pettersson, O.: 1913, 'Klimatförändringar i historisk och förhistorisk tid', *Kungl. Svenska Vetenskapsakademiens Handlingar* **51**, No. 2

Pettersson, O.: 1914, 'Climatic variations in historic and prehistoric time', *Svenska Hydrografisk-Biologiska Kommisionens Skrifter* **5**, 1-26.

Pfister, C. Luterbacher, J., Schwarz-Zanetti, G., and Wegmann, M.:1998, 'Winter air temperature variations in western Europe during the early and middle Ages (AD 750-1300)', *Holocene* **8**, 535-552.

Porter, S. C.: 1981, 'Glaciological evidence of Holocene climatic change' in Wigley, T. M. L., Ingram, M. J. and Farmer, G. (eds), *Climate and History. Studies in past climates and their impact on Man*, Cambridge University Press, Cambridge, pp. 82-110.

Porter, S. C.: 1986, 'Pattern and forcing of northern hemisphere glacier variations during the last millennium', *Quartenary Research* **26**, 27-48.

Sebagenzi, M. N., Vasseur, G., and Louis, P.; 1992, 'Recent warming in southeastern Zaire (Central Africa) inferred from disturbed geothermal gradients', *Global and Planetary Change* **6**, 209-217.

Selsing, L.: 1998, 'Interdisciplinary palaeoclimatic research at the Museum of Archaeology, Stavanger, Norway', in Frenzel, B. (ed.) *Documentary Climatic Evidence for 1750-1850 and the 14th century. Paläoklimaforschung* **23**, *Special Issue: ESF Project "European Palaeoclimate and Man"* **15**, European Science Foundation, Strasbourg, Akademie der Wissenschaften und der Literatur, Mainz, pp. 63-69.

Shen, P. Y., Pollack, H. N., and Huang, S.: 1996, 'Inference of ground surface temperature history from borehole temperature data: a comparison of two inverse methods', *Global and Planetary Change* **14**, 49-57.

Skinner, W. R. and Majorowicz, J. A.: 1999, 'Regional climatic warming and associated twentieth century land-cover changes in north-western North America', *Climate Research* **12**, 39-52.

Sigfúsdóttir, A. B.: 1969, 'Hitabreytingar á Íslandi 1846 - 1968' (Temperature variations in Iceland 1846-1968), in Einarsson, M. Á. (ed.), *Hafísinn* (The Sea Ice), Almenna bókafélagið, Reykjavík, pp. 70-79.

Speerschneider, C. I. H.: 1915, *Om Isforholdene i danske Farvande i ældre og nyere Tid. Aarene 690 – 1860.* Meddelelser Nr. 2, Publikationer fra det Danske Meteorol. Institut, Copenhagen.

Tarand, A. and Nordli, P. Ø.: 2001, 'The Tallinn Temperature Series Reconstructed back half a Millennium by use of Proxy Data', this volume.

Thórarinsson, S.: 1943, 'Oscillations of the Iceland glaciers in the last 250 Years', Chapter XI in Vatnajökull. Scientific results of the Swedish-Icelandic investigations 1936-37-38. *Geografiska Annaler* **1-2**, 1–54.

Thórarinsson, S.: 1956a, 'The thousand years struggle', in Thórarinsson, S. *The Thousand Years Struggle Against Ice and Fire*, Bókaútgáfa Menningarsjóðs, Reykjavík 5-33.

Thórarinsson, S.: 1956b, 'Ísland þjóðveldistímans og menning í ljósi landfræðilegra staðreynda' (The Icelandic commonwealth and culture in the light of geographical facts), *Skírnir* **130**, 236-248.

Thórarinsson, S.: 1974, 'Sambúð lands og lýðs í ellefu aldir' (The coexistence of country and people through eleven hundred years), in Líndal, S., (ed.), *Saga Íslands, 1. bindi*, Hið íslenska bókmenntafélag, Sögufélagið, Reykjavík, pp. 29-97.

Thoroddsen, Þ.: 1892 - 1904, *Landfræðissaga Íslands. Hugmyndir manna um Ísland, náttúruskoðun og rannsóknir fyrr og síðar I-IV* (The Historiography of the natural sciences in Iceland), Hið íslenska bókmenntafélag, Reykjavík 1892-1896, Copenhagen 1898-1904.

Thoroddsen, Þ.: 1908 - 1922, *Lýsing Íslands I-IV* (A description of Iceland), Hið íslenska bókmenntafélag, Copenhagen 1908-1922.

Thoroddsen, Þ.: 1914, 'Islands Klima i Oldtiden', *Geografisk Tidskrift* **22**, 204-216.

Thoroddsen, Þ.: 1916 - 1917, *Árferði á Íslandi í þúsund ár* (The climate of Iceland through one thousand years), Hið íslenska fræðafélag, Copenhagen.

Thoroddsen, Þ.: 1922 - 1923, *Minningabók* (Memoirs), Hið íslenska fræðafélag, Copenhagen.

Vasey, D. E.: 2001, 'A Quantitative Assessment of Buffers among Temperature Variations, Livestock and the Human Population of Iceland, 1784-1900', this volume.

Vilmundarson, Þ.: 1969, 'Heimildir um hafís á síðari öldum' (Historical sources on sea ice during the early modern period), in Einarsson, M. Á. (ed.), *Hafísinn*, Almenna bókafélagið, Reykjavík, pp. 313-332.

Vilmundarson, Þ.: 1972, 'Evaluation of historical sources on sea ice near Iceland' in Karlsson (ed.), *Sea Ice, Proceedings of an international conference*, National Research Council, Reykjavík, pp. 159-169.

Wallevik, J. E. and Sigurjónsson, H.: 1998, *The Koch index: Formulation, correction and extension VÍ-***G98035**-ÚR28, Icelandic Meteorological Office, Reykjavík.

White, J. W. C., Barlow, L. K., Fisher, D. Grootes, P., Jouzel, J., Johnsen, S. J., Stuiver, M., and Clausen, H.: 1997, 'The climate signal in the stable isotopes of snow from Summit, Greenland: results of comparisons with modern climate observations', *Journal of Geophysical Research* **102, C12**, 26425-26439.

Wigley, T. M. L. and Raper, S. C. B.: 1995, 'Modeling low-fequency climate variability: the greenhouse effect and the interpretation of paleoclimatic data', in Martinson, K., Bryan, M., Ghil, M., Karl, T. R., Sarachik, E. S., Sorooshian, S., Talley, L. D. (eds), *Natural ClimateVariability on Decade-to-CenturyTtime Scales*. National Research Council, Washington, D. C., 169-174.

Williams, L. and Wigley, T. M. L.:1983, 'A comparison of evidence for Late Holocene summer temperature variations in the northern hemisphere', *Quarternary Research* **20**, 286-307.

(Received 24 July 2000; in revised form 22 September 2000)

THE INITIATION OF THE "LITTLE ICE AGE" IN REGIONS ROUND THE NORTH ATLANTIC

J. M. GROVE

Girton College, Cambridge, Cambridge CB2 3EN, United Kingdom

Abstract. The "Little Ice Age" was the most recent period during which glaciers extended globally, their fronts oscillating about advanced positions. It is frequently taken as having started in the sixteenth or seventeenth century and ending somewhere between 1850 and 1890, but Porter (1981) pointed out that the "Little Ice Age" may 'have begun at least three centuries earlier in the North Atlantic region than is generally inferred'. The glacial fluctuations of the last millennium have been traced in the greatest detail in the Swiss Alps, where the "Little Ice Age" is now seen as starting with advances in the thirteenth century, and reaching an initial culmination in the fourteenth century. In the discussion here, evidence from Canada, Greenland, Iceland, Spitsbergen and Scandinavia is compared with that from Switzerland. Such comparisons have been facilitated by improved methods of calibrating radiocarbon dates to calendar dates and by increasing availability of evidence revealed during the current retreat phase. It is concluded that the "Little Ice Age" was initiated before the early fourteenth century in regions surrounding the North Atlantic.

Key Words: glacial initiation; "Little Ice Age"; North Atlantic

1. Introduction

In this paper, the extent to which the "Little Ice Age" began at the same, or similar times, in different regions around the North Atlantic is explored. For a discussion regarding other regions see Grove (in press, a,b). The term "Little Ice Age" does not refer directly to climate but to the most recent period during which glaciers extended globally and remained enlarged, while their fronts fluctuated about more forward positions. In this paper, the "Medieval Warm Period" is defined in terms of the interval between the "Little Ice Age" and earlier general expansion episodes (Grove and Switsur, 1994). It is hypothesized that the time when the "Little Ice Age" began should be related, at least approximately, to the ages of moraines signifying the end of the "Medieval Warm Period" (Grove and Switsur, 1994). The term "Medieval Warm Period" has been the subject of considerable controversy. Its nature and even its existence has been queried (Hughes and Diaz, 1994), as has that of the "Little Ice Age" (Bradley and Jones, 1992a,b). It should be emphasised that they were not periods of unbroken cold and warmth respectively. Climate varied on small scales both spatially and temporally, as it has in the twentieth century. Nevertheless, climatic conditions were such during the "Little Ice Age" that mass balances were sufficiently predominant for the glaciers to remain enlarged,

Climatic Change **48**: 53–82, 2001.

although their fronts oscillated. Similarly, during the "Medieval Warm Period" climatic conditions caused mass balances to be negative, and volumes of glaciers to be reduced, so that they retracted substantially, though their fronts no doubt oscillated, as they have been observed to do during the warming of the twentieth century.

The "Little Ice Age" has frequently been seen as occurring during the last three or four centuries, and ending somewhere between 1850 and 1890 (Bradley and Jones, 1992a,b), but Porter (1981, p. 99) pointed out that it may 'have begun at least three centuries earlier in the North Atlantic region than is generally inferred'. It is now timely to consider whether Porter was correct, and whether the time of initiation of the "Little Ice Age" coincided or differed within the regions surrounding the North Atlantic. The timing attributed to it has varied considerably from author to author, perhaps reflecting real regional differences. However, this lack of agreement could be due in part to the uneven distribution and character of the evidence available, to the dating techniques used, and their resolution, and possibly to differing degrees of effort devoted to unravelling glacial history.

Ice retreat during the twentieth century has exposed much previously concealed evidence. Conversion of radiocarbon to calendar dates (Stuiver and Pearson, 1993; Stuiver and Reimer, 1993) makes it possible to compare evidence from different regions from radiocarbon and absolutely dated sources more satisfactorily than was previously possible. Development of dendroglaciological methods and studies and construction of long tree-ring chronologies has significantly improved dating in the Alps, Canada and Alaska. A mass of reliable data shows that "Little Ice Age" expansion in Switzerland was already well under way in the fourteenth century (e.g. Holzhauser, 1983a,b; 1995; Pfister et al., 1994). The results of recent studies elsewhere around the North Atlantic will be considered here in light of the Swiss evidence. All the radiocarbon dates quoted have been calibrated according to Stuiver and Reimer (1993) at 1s, and therefore some differ in detail from those originally published using earlier calibration programmes.

2. The Swiss Alps

The "Little Ice Age" has been traced in the greatest detail in the Swiss Alps, where historical records are unusually rich and possibilities of moraine dating good. Many ice fronts extend below the treeline and have been in full view of settlements for hundreds of years, or have abutted onto farmland. Written records, paintings and drawings made by both local observers and visitors are plentiful (e.g. Zumbühl and Holzhauser, 1988; 1990). Identification of the calendar dates at which many *in situ* trees were killed by advancing ice, together with their ages at death, has been made possible by multiple dendrochronological analyses (Holzhauser, 1997). Accordingly, the precision of some reconstructions has been much greater than can be achieved by radiocarbon dating alone. Deliberate and systematic monitoring of the fluctuations of Swiss glaciers began in 1880, earlier than in any other country, and has been continued consistently since then, providing a uniquely long and complete record, which demonstrates the general coherence of their histories.

Documentary evidence thought to relate to the position of glacier fronts in the fourteenth century has attracted attention over the years. The uncovering of the remains of a wooden irrigation channel, the Oberriederin *Bisse*, by the withdrawal of the Grosser Aletsch Glacier, showed that the glacier must have been less extensive when the *Bisse* was constructed than it was to be again until the twentieth century (Lamb, 1985). It was recorded by Kinzl (1932) as abandoned in 1385, but this did not indicate when the early "Little Ice Age" advances of the Grosser Aletsch began. Evidence that the Allalin Glacier advanced across Saastal, a left-bank tributary valley of the Rhône, in the thirteenth century, holding back a lake, the Mattmarksee, was found in a document dated 13 April 1300, concerning the tenancy of pastures at Distelalp, upstream of the site of the lake (Grémaud, 1878). This was interpreted by Lütschg (1926), Ladurie (1971) and Delibrias et al. (1975) as showing that the Allalin was sufficiently enlarged in 1300 to block Saastal and separate the upper mountain pastures from the lower valley of the Visp, as was observed to occur during the nineteenth century. But this gave no information about the time when the Allalin began to advance, though it could be surmised that it must have started to increase in volume before 1300.

Enlargement of the Allalin and Grosser Aletsch glaciers in the thirteenth and fourteenth centuries could not have occurred without expansion of other Alpine glaciers. The chronology of over twenty moraine sequences in the Canton of Valais was investigated by research students from the Geographical Institute of the University of Zurich, and ages obtained by radiocarbon dating of organic samples from beneath many moraines, published by Röthlisberger et al. (1980). The calendar age ranges of some of these, resulting from calibration according to Stuiver and Reimer (1993), are shown on Table 1. All the data in this table were obtained from soil samples[1], except that from Glacier de Zinal, which came from larch wood. None of these were very precise.

---

[1]Dates on wood provide more reliable guidance than those from soil samples; formation of a soil layer can take a very considerable time. Matthews and Caseldine (1987) examined a present day soil in Norway, and its continuation under a "Little Ice Age" moraine. Using 14 radiocarbon dates from chemically-fractionated soil samples, they found that ages ranged from 495+/-55 in the top centimetre to > 4000 BP within 13cm of the buried soil surface.

Table I. Switzerland

Calendar Ages (AD) indicating the initiation of Little Ice Age.
The numbers in the margin give the laboratory identification number, the ^{14}C age and the calibrated age for 68% probability.
Horizontal lines represent the calibrated ages.

Dates from soil samples beneath moraines.
The Valaisian Alps (Data from Röthlisberger et al, 1980)

Glacier de Fenêtre
Hv-6805 990 +/- 95 980-1190

Schwarzberg
UZ-165 950 +/- 115 990-1220

Glacier du Mont Miné
Hv-6800 900 +/- 75 1030-1255

Findelen
Hv-6791 845 +/- 225 990-1390

Glacier de Corbassière
Hv-7226 815 +/- 40 1210-1280

Glacier du Mont Durand
Hv-6811 790 +/-65 1210-1290

Glacier de Moiry
UZ-153 700 +/- 85 1260-1390

Glacier du Brenay
Hv-6801 670 +/- 85 1280-1400

Glacier de Zinal (Röthlisberger et al 1980)
B-3200 920 +/- 50 1030-1210

Rhône (Holzhauser 1990)
UZ-829 670+/-65 1280-1390
UZ-830 665+/-80 1280-1400

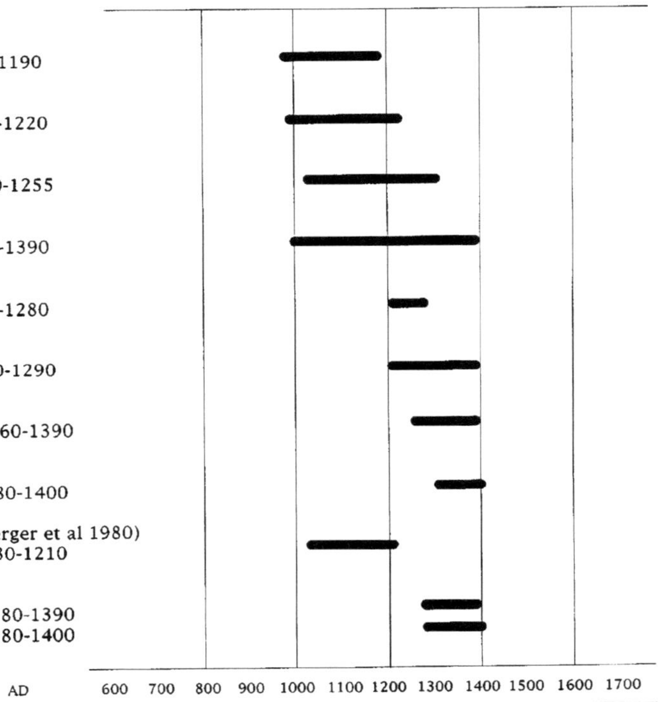

Since 1980 one of the members of the original Zurich group has concentrated on carrying out a remarkable series of intensive surveys of Swiss forefields, and has published outstandingly detailed and reliable data relating to the history of a number of glaciers (Holzhauser 1980, 1983a,b, 1984a,b, 1985, 1995, 1997). This has involved an unusual degree of persistence, as well as very detailed searches. Thus, unravelling of the history of the Lower Grindelwald Glacier over the last 2800 years has required the examination of nineteen profiles of the lateral Stieregg moraine and dating of no less than 46 samples extracted from them (Holzhauser and Zumbühl, 1996, p. 112). Useful guidance on the timing of advances can sometimes be recognised from the distribution of radiocarbon dated fragments of wood, even if they are not *in situ,* but precise information can be recovered if sufficient *in situ* tree stumps survive, especially if the bark is still present. Thus, the maximum extension of the Gorner glacier can be pinpointed exactly at 1385 with the aid of dendrochronological analysis of *in situ* larch stumps killed by the ice advance (Holzhauser, 1997, p. 47).

The most complete record of the Late Holocene fluctuation history of any glacier in the world comes from intensive investigations of the deposits of the Grosser Aletsch (Figure 1, Holzhauser, 1983a,b, 1984b, 1988, 1997). This reconstruction was based on a great variety of evidence, the heights above the ice at which tree stumps and samples were found forming an important element. Some samples were taken from the outer rings of trees overrun by the advancing ice, which could be absolutely dated by reference to dendrochronological series. This chronology is accordingly more accurate than one based only on radiocarbon samples. Retraction during the "Medieval Warm Period" was ended by rapid advance, starting after 1250 and culminating around 1350. Although the "Medieval Warm Period" here was broken by a marked advance between A.D. 1050 and 1150, this was not comparable in scale with those of later centuries and therefore has to be considered an interruption of medieval warmth rather than part of the "Little Ice Age". The Grosser Aletsch advanced about 40m per year after 1300, to reach its greatest extension around 1370-80 (Zumbühl and Holzhauser, 1988, p. 302). This fourteenth-century advance brought the front as far forward as did those in the seventeenth and nineteenth centuries. The major extension phases were separated by withdrawals which were insufficient to return the front to positions comparable with those of the warm period. It is evidently reasonable to view the whole period between the start of expansion in the thirteenth century and the great retreat of the late nineteenth and twentieth centuries as one complex "Little Ice Age", with each century scale fluctuation itself made up of smaller scale oscillations, such as those recorded during monitoring of the positions of glacier fronts. and more recently by mass balance measurements (Haeberli, 1995).

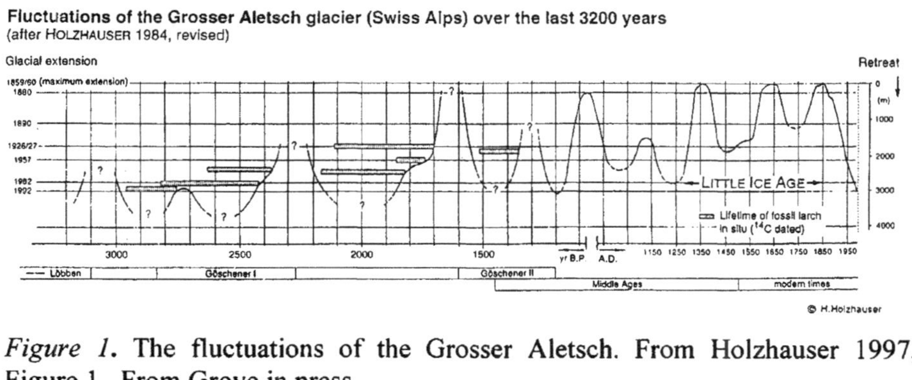

Figure 1. The fluctuations of the Grosser Aletsch. From Holzhauser 1997, Figure 1. From Grove in press.

The Gorner Glacier has advanced and retreated in harmony with the Grosser Aletsch, showing only very minor differences. Analysis of fossil larch trees has made possible an exact, calendar reconstruction of the fourteenth century advance for the first time in the Alps (Holzhauser, 1997, p. 47). The timing of later culminations of the Gorner differed only slightly from those of the Grosser Aletsch. Some minor differences have to be expected in view of the differences in topography (Kuhn, 1993; 1996) and inevitable contrasts in sensitivity and microclimate.

Other large Alpine glaciers have been shown to have been similarly affected by early "Little Ice Age" advances. The Grindelwald was advancing around A.D. 1338 (Holzhauser and Zumbühl, 1996, p. 112). The Rhône Glacier advanced to a maximum between 1350 and 1400 (Zumbühl and Holzhauser, 1988), which was slightly more extensive than any occurring later. Further east the Gurgl reached a maximum about 1300 (Patzelt, 1994, p. 298). In the Austrian Alps, the Gepatschferner, Gurglerferner and Simonykees are also known to have advanced during this period (Patzelt, 1974).

The 1980 team published radiocarbon dates from many forefields in the Swiss Alps (Röthlisberger et al., 1980), showing that enlargement of glaciers starting in the thirteenth century or earlier was not unusual (Table 1). Detailed investigations, comparable with those made on the forefields of the Grosser Aletsch and Gorner, continue as Holzhauser turns his attention to more of them (Holzhauser, in preparation). The detailed results are already too numerous to be presented here. It would be quite impractical to tabulate all the radiocarbon results, let alone the dendrochronological data that have been assembled.

The "Little Ice Age" in the Alps is now seen as starting in the second half of the thirteenth century (Holzhauser, 1997, p. 37). Its three culminations, including the first in the fourteenth century, were very similar in scale[2]. The "Little Ice Age" was not a unique event (O'Brien et al., 1995). Several phases of glacial expansion during the Holocene comparable in scale with those of the "Little Ice Age" have been traced in both the Swiss and Austrian Alps (e.g. Patzelt, 1974; Röthlisberger et al., 1980; Röthlisberger, 1986). It will now be considered whether the model of the "Little Ice Age" derived from Switzerland is applicable in other regions.

3. Canada

"Little Ice Age" advances in the Canadian Rockies began between the twelfth and thirteenth centuries; although evidence is fragmentary it is also definitive (Luckman, 1992; 1993; 1994; 1995; 1996a,b,c). Reconstruction of early "Little Ice Age" events depends on dating fossil wood. Historical accounts, dating back

---

[2]The climatic background to these events is discussed in papers by Pfister (1988, 1998).

no further than the nineteenth century, refer to only a small number of glaciers, most of which reached maximum "Little Ice Age" positions in the early eighteenth or mid-nineteenth centuries (Luckman, 1990). Only a few moraines predating 1700, consisting of small, locally preserved fragments, found at or near the margins of nineteenth century deposits, have been identified. Their dating is generally poorly constrained, but current retreat is uncovering progressively more fossil wood relating to earlier periods. It has therefore become possible to trace earlier glacial fluctuations, not only by radiocarbon dating of fragmentary material associated with moraines, but by making use of trees overwhelmed by the ice. Rooted stumps, showing evidence of having been killed by a glacier, such as shearing in the direction of ice movement, provide limiting ages for the glacial advances responsible for their deaths. Calendar dating of both *in situ*

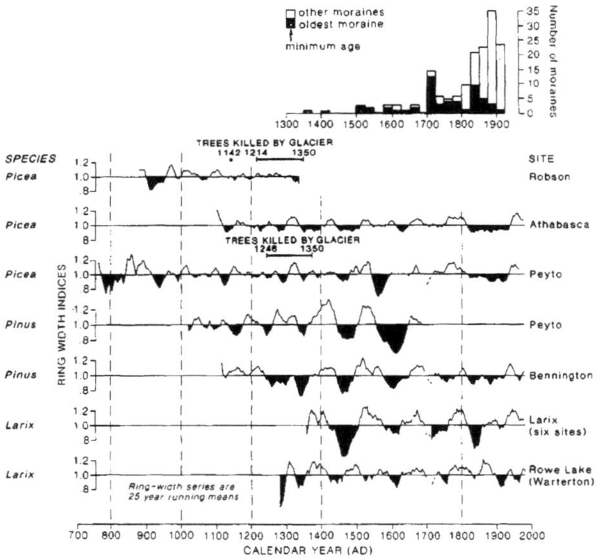

Figure 2. Glacial fluctuations and tree-ring chronologies over the last millennium in the Canadian Rockies. The seven indexed ring-width chronologies are plotted to the same vertical scale and are smoothed with a 25-year mean. Periods when tree-ring indices were below the mean for each record are shaded. The Larix chronology presented is the maen of six Larix chronologies from the Banff-Kananaskis region.The Rowe Lake chronology is from a site about 150 km further south. After Luckman 1996c, Figure 5. From Grove in press.

stumps and detrital wood has been made possible, as in Switzerland, by comparison of their ring sequences with long tree-ring chronologies, supplemented by radiocarbon dating (Figure 2) (Luckman, 1994; 1995; 1996a,b). Several long chronologies have been developed by Luckman and his associates, including one from *Pinus albicaulis* from a treeline site near the Bennington Glacier with an earliest date of A.D. 1112 (later revised to A.D. 1104), another from the Peyto Glacier (A.D. 948-1991), and, also at Peyto, a *Picea engelmannii* chronology covering the period A.D. 760-1992 (Figure 3). At Grey Pass, Purcell Mountains there is a *Larix lyallii* chronology from A.D. 801-1993 (Luckman, 1996a). The first calendrical dating of an early "Little Ice Age" advance in North America came from the Robson Glacier (Figure 3; Luckman, 1995). Heusser (1956) had obtained an early Libby radiocarbon date of 450+/- 150 BP on fossil wood over-ridden by the Robson and inferred that the "Little Ice Age" had begun between A.D. 1300 and 1600, a view revised after further dates were obtained by Luckman (1986), and calibrated (Table II). Note that the absolute age obtained by comparison with tree-ring sequences falls outside the range given by the radiocarbon date on B-3731. A tree overridden by the Kiwa Glacier (Figure 3) in the Premier Ranges of British Columbia, gave a similar age range (Luckman, 1986). Other types of evidence are in line with these indications of an early "Little Ice Age"; for example, variations in proglacial sedimentation rates were interpreted by Leonard (1997; 1985) as recording a glacial advance in the thirteenth century at Hector Lake.

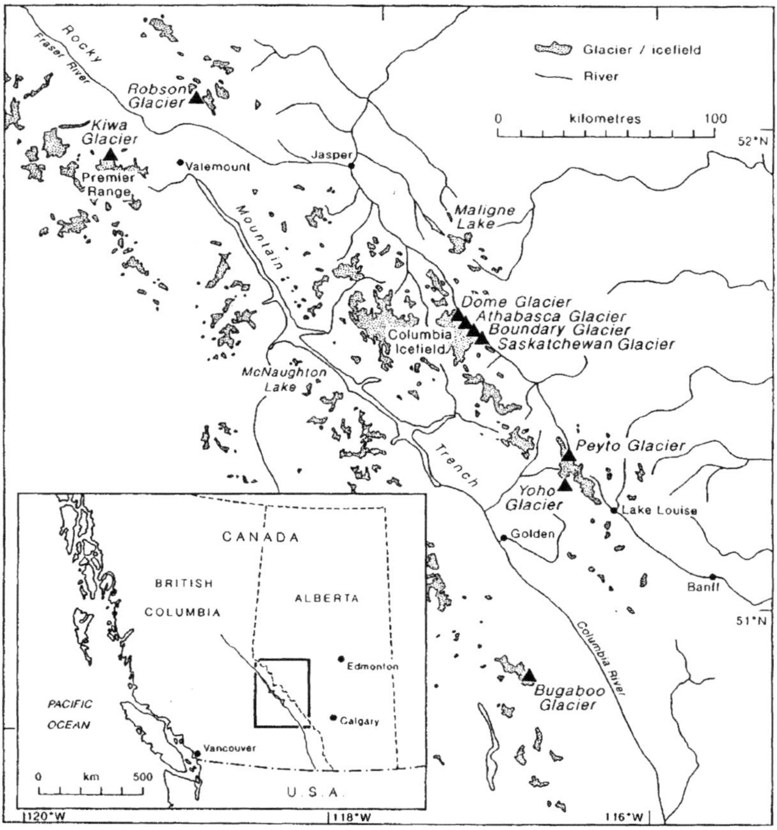

Figure 3. Map of the central Canadian Rockies showing the positions of many of the glaciers mentioned in the text. After Luckman et al 1993, Figure 1.

The most detailed evidence relating to the initiation of the "Little Ice Age" in Canada, giving precise calendar dates, comes from the forefield of the Robson Glacier, where *in situ* tree stumps, showing clear signs of having been sheared and killed by the ice, as well as detrital logs, have been exposed in abandoned stream channels inside the terminal moraine (Luckman, 1995). Ring width and maximum density chronologies, developed from *in situ* stumps, crossdated with long, living tree chronologies, show that the ice was advancing into forest between 1142 and 1150 and continued to do so until at least A.D. 1350 (see Table III). The front was only about 400m from its "Little Ice Age" maximum in the period 1150-1250. So many rooted stumps were located here that it proved possible to calculate rates of ice advance over time.

Table II. Canadian Rockies

Dates from trees killed by ice advances.
The numbers in the margin give the sample number, laboratory identification number, ^{14}C age and calibrated age (AD) for 68% probability.
Horizontal lines represent the calibrated ages.
X = year tree was killed given by crossdating from dendrological series.

Robson Glacier (Luckman 1995)				
R802S	B-2816	860+/-50	1175-1260	X1246
R816S	B-3731	900+/-60	1030-1220	X1253
Kiwa Glacier, Premier Ranges (Luckman 1986)				
K841	B-12416	840+/-80	1070-1280	
Peyto Glacier				
PS890STO	SRC-3111	920+/-60	1036-1210	
P893	B-33101	820+/-50	1160-1280	
PS8803	SRC-2991	505+/-70	1030-1210	

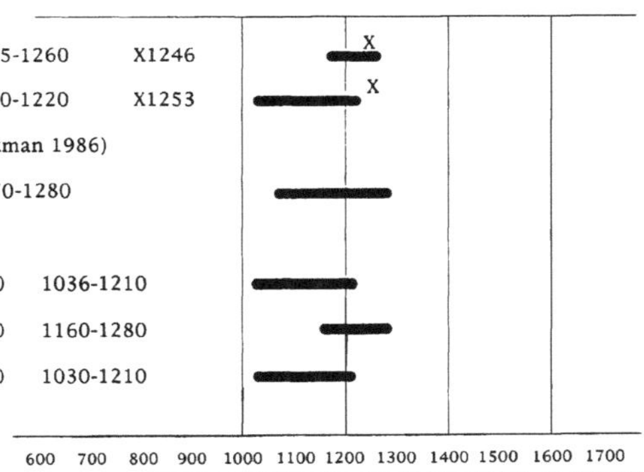

This early "Little Ice Age" expansion was slower than the advances of 6-20 m per year in the Cordillera region between the 1950s and 1980s (Harper, 1993). The Robson ice in the 'southern channel' advanced about 3.8 m per year between 1214 and 1261, the period within which over 80% of the kill dates lie. The oldest and youngest dates obtained were confirmed by replication in two trees. The terminal position of the advance is unknown but must have been within the outermost "Little Ice Age" moraine, about 500m downvalley of the fossil forest. No evidence was found for the ice position between ca. 1350 and ca. 1783. Luckman (1993) pointed out that once it is exposed, wood here may quickly decay, be buried or washed away, and underlined the importance of continued monitoring of glacier forefields for new material and exposures as glaciers recede.

Early "Little Ice Age" expansion of the Peyto Glacier between A.D. 1248 and about 1375 has been identified in a similar manner from detrital logs and *in situ* stumps, crossdated from the ringwidth record and killed by the advancing ice

Table III. Estimated rates of Glacial Advance across the Robson Glacier Forefield.

From	Date	To	Date	Distance (metres)	Years	Rate (m/yr^{-1})
a) Southern Channel						
R8713	1142	R8309	1214	175	72	2.4
R8309	1214	R8306	1246	113	32	3.5
R8306	1246	R8708	1261	65	15	4.3
R8708	1261	R9207	1300-1320	147	39-59	2.6-3.8
R8713	**1142**	**R9207**	**1300-1320**	**500**	**158-78**	**2.8-3.2**
R8309	1214	R8708	1261	178	47	3.8
b) Northern Channel						
R8713	1142	R8702	1214	220	72	3.1
R8702	1214	R8723	1350	322	136	2.4
R8713	**1142**	**R8723**	**1350**	**542**	**208**	**2.6**

Rates were estimated between selected *in situ* stumps, each recognised by a sample number.
Estimates in bold type are average values calculated using the maximum possible travel distance for each channel.
R8723 was not *in situ*, but jammed across the channel, but it provided a maximum rate for ice advance over the entire exposed area of the fossil forest and a maximum downvalley position for ice advance at the time.

after Luckman, 1995

between A.D. 1248 and about 1375 (Luckman, 1996b). Current evidence is insufficient to show whether this advance began later than that of the Robson. Radiocarbon dates from the forefield of the French Glacier in the Peter Lougheed Park, in the eastern Front Ranges of the Canadian Rockies, and the Stutfield Glacier (Luckman, 2000) indicate approximately contemporaneous advances (Smith et al., 1995).

The advances of the twelfth and thirteenth centuries, bringing ice fronts to positions 0.5-1.0 km upvalley of more recent moraines, are recognised as representing the start of the "Little Ice Age" (eg. Luckman, 1993; 1995). Ample evidence of more extensive glacial advances in the eighteenth and nineteenth centuries is provided by some sixty moraine chronologies detailed by Luckman (1996a). Advances before 1700 have been recognised in a few cases but the sample size is small and the dating poorly constrained. It is not possible on the basis of evidence so far available to compare adequately the importance and extent of glacial advances before 1700 in Canada with those in the Alps.

However, it is possible that more data will be forthcoming in the future in view of the small proportion of forefields in Canada that have been investigated in detail. Meanwhile, long tree-ring width records show that "there may have been two or three periods of glacial advance between 1300 and 1700 for which well-dated evidence has not yet been recovered" (Luckman 1994, p. 55).

While the "Little Ice Age" was underway in Canada in the thirteenth and fourteenth centuries, as it was in Switzerland, the early advances were less extensive than those which followed, rather than being on a similar scale. Another, and perhaps more important difference, is that while several Holocene events in the Alps are known to have been as extensive as those of the "Little Ice Age", this was not the case in Canada. Luckman (1993) has identified a regional advance period during which four glaciers overwhelmed trees dated between 2800 BP and 3200 BP at sites no farther downvalley than the ice front positions occupied in the 1990s, and a buried forest with trees dating from 3700-3500 BP is being released as the Robson tongue withdraws. Luckman (1994) concluded that during the last three thousand years multiple glacial advances became progressively more extensive, culminating in the eighteenth and nineteenth centuries. He attributed this to the influence of long-term decline in the receipt of summer solar radiation due to orbital changes superimposing an overall cooling trend on higher frequency climatic fluctuations controlling periods of glacial advance and retreat. Although buried forest sites are appearing progressively further upvalley, so far none of the terminal positions of earlier "Little Ice Age" or Holocene advances has been located.

4. Greenland

Greenland was settled in the course of Viking expansion during the "Medieval Warm Period" (Barlow et al., 1997). No details of the fluctuations of small glaciers during the last millennium are available from Greenland, but there is no doubt that serious climatic deterioration was associated with the waning and eventual disappearance of the Viking presence. Radiocarbon dates on moss from cairns in the Godthåbsfjord district, built by deer hunters, and recently melted out of semiperennial snow cover, indicate that they were built between A.D. 1290 and A.D. 1400 when snow cover in West Greenland was increasing (Weidick et al., 1992, p. 19). It is significant that burials made before the 1350s were penetrated by roots, while those made later were preserved by permafrost (Gad, 1970). Cooling between A.D. 1400 and A.D. 1700 is indicated by a decrease in oxygen isotope composition of enamel from Norse and Inuit teeth from archeological sites in Greenland (Fricke et al., 1995). But although the oxygen isotope composition of mammalian tooth enamel can be used as a proxy for local surface temperature (so long as the temperature conditions last long enough to be recorded in the teeth of men and animals living in an area), this method of oxygen isotope analysis of tooth enamel cannot be used to investigate seasonal, annual or very short-term temperature fluctuations. Such information has to come from elsewhere.

Analyses of the annual layering of the Greenland ice in the GISP2 core reveal many details of the course of events. Periods with low average accumulation, implying lower temperatures, occurred around A.D. 1200, 1500 and 1800, and periods with slightly higher accumulation, and higher temperatures, punctuated the record around A.D. 1400 and 1700 (Meese et al., 1994). Clusters of cold years occurring in 1308-18, 1324-29, 1343-62, and 1380-84 would have been extremely serious for a society living as precariously as did the Greenlanders, with their cultural inability to adapt to changing circumstances (McGovern, 1981; Buckland et al., 1996; Barlow et al., 1997; Pringle, 1997). It could have been the close spacing of abnormally cold summers in 1351 and 1352, and cold winters in 1351 and 1355[3], which finally triggered the abandon-

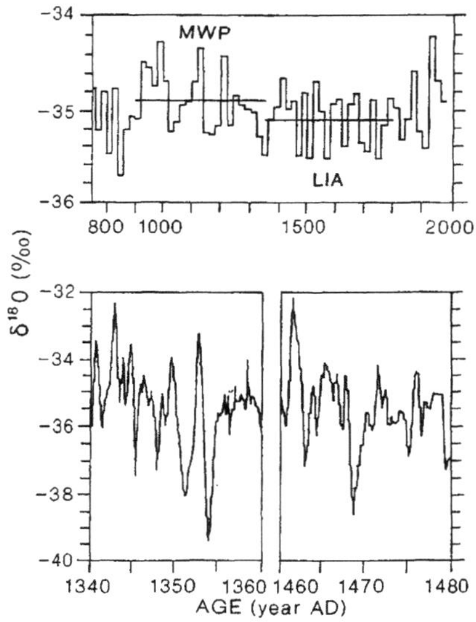

Figure 4A. Bidecadal oxygen isotope values from the GISP2 core record both the Medieval Warm Period and the "Little Ice Age". (The horizontal lines give average annual $\partial^{18}O$ values for the Medieval Warm Period and the "Little Ice Age")

Figure 4B. Interannual detail for the period 1340-1360 and 1460-1480. (From Stuiver et al 1995, Figure 4)

[3]According to Stuiver *et al*, "Samples were collected with sufficient density to obtain information on seasonal (monthly) $\partial\ ^{18}O$ change for most of the record". They admitted that "Although our $\partial\ ^{18}O$ fine structure is somewhat incomplete in that it probably does not register faithfully all winters, we note two very cold winters in A.D. 1352 and A.D.1355 in the GISP2 $\partial\ ^{18}O$ record. Development of the A.D. 1351 and A.D.1352 summers in the record also is very minimal and these summers were evidently very cold".

ment of the Western settlement, cutting off access to the hunting area of Nordsetur and thus the source of trade goods to the Eastern Settlement (see Figure 4, from Stuiver et al., 1995).

Early onset of cold conditions is also recorded in off-shore sediments. Dramatic changes have been found in both sediment type and faunal composition with depth in two cores from Nansen Fjord (Jennings and Weiner, 1996; see also Jennings et al., this volume). Its location, between Kangerdlugssuaq Fjord and Scoresby Sund, in southeast Greenland, makes it sensitive to changes in the relative strengths of water masses in the East Greenland current, which carries Polar Water and Atlantic Intermediate Water southward. The major faunal changes, which coincide with the main lithofacies changes, were interpreted by Jennings and Weiner (1996) as reflecting changes in character of the water masses carried in the Greenland current. Their chronology was based on Accelerator Mass Spectrometry (AMS) radiocarbon dates on calcareous benthic foraminifera (Table IV). No signs of bioturbation were observed. The dating control was somewhat sparse, but the samples were taken at critical points where major changes in the sediments and included fauna coincide. It emerged that, at least from the period A.D. 650-880 (AA-8332) until sometime between A.D. 1030 and 1220 (AA-10567), i.e., during the period which Lamb (1977) referred to as the "Medieval Warm Epoch", conditions were as warm as the warmest parts of the twentieth century and stable, with low sea ice in summer and Atlantic Intermediate Water present at the sea bed. The flux of Polar Water into the fjord during this period was diminished. A cold period beginning about 1270 and reckoned to have culminated about 1370, on the basis of the sedimentation rate, coincides with the failure of the Western Settlement. Conditions deteriorated further sometime between A.D. 1510 and 1680 (AA-12891) ushering in very variable conditions with frequent extreme cold events. Though this reconstruction is not closely dated, the implication of an early "Little Ice Age", with no evidence of a long period of sustained warmth separating it from even harsher conditions in later centuries, seems clear.

Table IV. Greenland

Dates relating to sediment changes in Nansen fjord, southeast Greenland.
The numbers in the margin give the laboratory identification number, the ^{14}C age and the calibrated age (AD) for 68% probability.
Horizontal lines represent the calibrated ages.
(Jennings & Weiner 1996, Kaufman & Williams 1992)

AA-12891	855+/-60	1510-1680
AA-8328	1125+/-50	1320-1430
AA-10565	1450 +/-60	1030-1200
AA-10567	1440 +/- 70	1030-1220

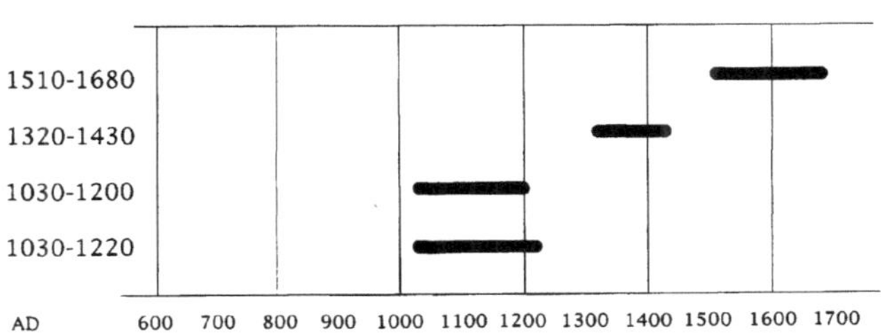

5. Iceland

Snowline lowering in Iceland of about 400 m during the last three centuries was greater than at any other time in the post-glacial period (Dugmore and Sugden, 1991), suggesting that, as in Canada, glacial extension would also have been maximal, and that evidence of thirteenth- or fourteenth-century glacial activity is likely to be fragmentary. The opportunities for dating moraines in Iceland are not as good as in Switzerland or Canada because wood is less abundant (though more abundant than has often been thought[4]), but numerous dated tephra layers help to counterbalance this (Thórarinsson, 1956; Dugmore, 1989). Use of dated tephra layers provides an important approach to reconstruction of glacial history. Studies on the forefields of five non-surging glaciers, all outlets of Eyjafjallajökull or Myrdalsjökull, using tephra layers of known date, support the conclusion that maximum glacial extension during the "Little Ice Age" was generally reached during the last three centuries (Dugmore and Sugden, 1991). Recent investigations around Myrdalsjökull have revealed that one of the units forming the compound moraine of Gigjökull was deposited around 700 BP, in the early "Little Ice Age" (Dugmore and Kirkbride, this volume). Detailed examination of the forefield of Svínafellsjökull, in Öræfi, has also produced evidence of thirteenth-century advances, the moraines formed lying less than 0.5 km outside those formed more recently (personal communication, H. Gudmudsson).

Surveying all the available documentary sources and selecting only those judged contemporary and reliable, Ogilvie (1991, 1998) concluded that there is circumstantial evidence that conditions were generally mild up to the late twelfth century, but that from then onwards until the sixteenth century relatively short periods of harsh climate occurred periodically (see Figure 5). She estimated that mean annual temperature during these phases may have fallen 1°C to 2°C below "typical twentieth-century temperatures". Low temperatures do not in themselves prove that glaciers advanced, but the coincidence between the incidence of cold seasons, and sea-ice occurrence is striking, and demands attention in view of plentiful evidence of close correlation between extensive sea ice around Iceland, low temperatures and glacial advances later in the "Little Ice Age". See also Ogilvie et al. (2000).

---

[4]In his *Land of Ice and Fire* (1938, p. 9), Ahlmann described large quantities of logs at Hoffell farm, near Hof, which he was told "had come from underneath Hoffellsjökull, carried out by the waters of the ice-river from the melting glacier ...This birch forest, judging from the logs, must have consisted of far mightier trees than those to be found in Iceland at the present time". In the 1930s birch logs were revealed in a stream cut through the outwash of Skaftafellsjökull, close to the front of the glacier (Ives, 1991, p.15). Ives was told that there "was a sufficient quantity to be 'mined' locally for fuelwood for some years". More recently pieces of wood have been found on the proximal side of the outermost moraine fronting Skaftafellsjökull, and on the forefield of Kviárjökull (personal communication, H.Gudmundsson, 8.12.1997).

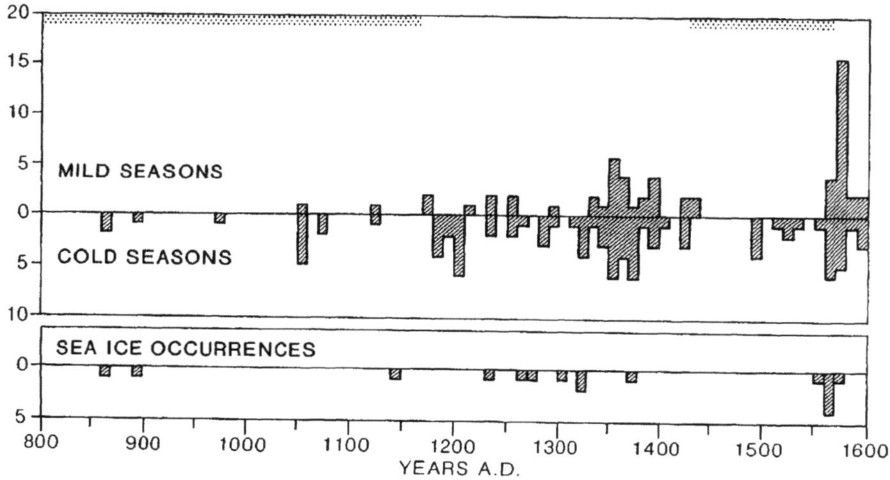

Figure 5. The number of mentions per decade of mild and cold seasons in Iceland, and of the presence of sea ice from A. D. 865-1598. Dotted black lines at the top of the diagram indicate periods of very poor data. From Ogilvie 1991, Figure 2.

6. Spitsbergen

Linnévatnet, near the mouth of Isafjorden, in western Spitsbergen (Svalbard), contains a sedimentary record of glacier fluctuations (Svendsen and Mangerud, 1997). The lake is fed by a meltwater stream from Linnébreen and several smaller glaciers inland. Though many glaciers in Spitsbergen are known to surge, there is no evidence that Linnébreen does; the ice front seems to have been localised within the "Little Ice Age" moraines throughout the Holocene. The uppermost part of the lacustrine sequence is characterised by continuous and well-developed laminations. Though the total organic carbon in the sediments is only 1-5%, a core taken close to the river mouth contained sufficient macro fossils of terrestial plants for AMS dating. It is argued that strongly laminated intervals with a high concentration of organic carbon reflect glacial maxima. When the glacial meltstream is dominant, the sediments entering the lake are enriched with particles from the coal seams within the Carboniferous sandstones underlying the ice. In periods of lesser glacial activity the sediments are enriched with detrital carbonate, because streams from the valley sides erode carbonate rock and carry no coal. The appearance of developed laminations, typical of proglacial lakes, was contemporary with increased coal content in the sediments.

'The sediment column covering the last 700-800 years, including the "Little Ice Age" stands out as a period of particularly high influx of glacially derived sediments' (Svendsen and Mangerud, 1997 p. 53). The glacial expansion that culminated in the nineteenth century is judged, on the basis of radiocarbon ages and calculated sedimentation rates, to have started in the thirteenth or fourteenth centuries (Table V). These findings are in line with the results of oxygen isotope analysis of an ice core from the Lomonosov ice plateau in east-central Spitsbergen. This indicated warm conditions in the twelfth and twentieth centuries, and lower temperatures, interrupted by brief warm intervals, in the interim. The coldest periods occurred between about A.D. 1180 and 1500, and between A.D. 1700 and 1900; the sixteenth century was generally warm. The dating error back to A.D. 1400 was estimated at no more than 10% (Gordiyenko et al., 1980).

Table V. Spitsbergen

Dates relating to sediment changes in Linnévatnet, Spitsbergen.
The numbers in the margin give the laboratory identification number, the ^{14}C age and the calibrated age (AD) for 68% probability.
Horizontal lines represent the calibrated ages.
(Data from Svendsen & Mangud 1997)

XXX	580+/- 70	1300-1430
XXX	730+/-45	1260-1300
XXX	745 +/-65	1220-1295
		AD

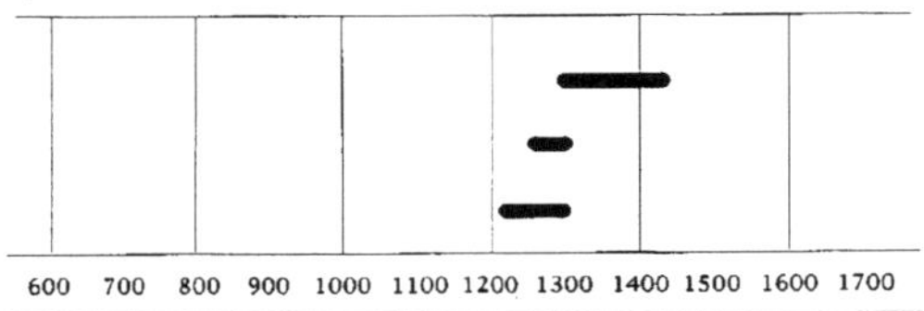

7. Scandinavia

In both northern and southern Scandinavia, the "Little Ice Age" culminated in the seventeenth and eighteenth centuries, but while in the south the glaciers then reached their greatest extent since the Holocene Climatic Optimum, in the north, several earlier Holocene expansion phases were comparable in scale or larger than those of the "Little Ice Age". During the twentieth century, and no doubt also in earlier centuries, the behaviour of glaciers in western maritime regions has differed somewhat from that in the more continental east.

7.1. NORTHERN SCANDINAVIA

In northern Scandinavia, initial climate deterioration seems to have occurred between A.D. 1350 and 1400. Evidence of a glacial advance around this time is inferred from the appearence of an increase in silt in lacustrine sediments of Vuolep Allakasjure (Karlén, 1984). The upper 2 cm of a peat deposit under the outermost moraine of Fingerbreen, an eastern tongue of the Svartisen icecap, was sampled by Karlén (1979). Seventy per cent of it consisted of wood fragments, giving a date of cal A.D. 1260-1390 (I-10364) as a maximum for the deposition of the moraine (Table VI). A peat sample from the upper 2 cm of peat under the foreset bed of a delta deposited in a lake dammed by expansion of one of the western tongues of Svartisen gave a maximum age of cal A.D. 1290-1430 (St-6757) for the extension of the ice. Karlén recognised that peats may be deposited gradually over a long period, and that the uppermost layer could be missing[5]. These dates, like St-7187 from Ritajekna in northern Sweden, are compatible with thirteenth- or fourteenth-century advances, but cannot be taken as conclusive. Lacustrine sediments deposited after the initiation of the "Little Ice Age" consistently show two well-separated periods during which large amounts of rock flour were deposited in downstream lakes (Karlén, 1976; 1991). The lack of suitable material for radiocarbon analysis has precluded moraine dating except by lichenometry.

Table VI. Scandinavia

Dates relating to Glacial Advances in Scandinavia.
The numbers in the margin give the laboratory identification number, the ^{14}C age and the calibrated age (AD) for 68% probability.
Horizontal lines represent the calibrated ages.
The symbol > indicates a maximum date.

Northern Scandinavia (Karlén 1979, 1984)		
Svartisen		
I-10364	695+/-75	1260-1390
St-6757	600+/-100	1290-1430
St-7187	640+/-90	1290-1410
Southern Scandinavia		
Jotenheimen (Harris et al 1987, Matthews 1991, Karlén & Matthews 1992.)		
Gjuvvatnet		
CAR-1131	750+/-60	1220-1295
Sagabreen		
CAR-750	910+/-55	1030-1210
Storbreen		
CAR-838	540+/-60	1330-1440
Bøverbreen		
CAR-782	305+/-60	1490-1660
Jostedalsbreen region.(Dahl & Nesje 1991, Nesje & Rye 1993, Dahl & Nesje 1994)		
Sandkarfonna		
T-10028	620+/-70	1290-1410
T-5813	890+/-60	1040-1210
T-8527	1110+/-60	890-1000
T-5814	1180+/-70	780-980
T-8528	1270+/-60	690-860
Hardangerjøkullen region (Nesje & Rye 1993, Dahl & Nesje 1994)		
Blåisen		
T-9440	575+/-75	1300-1430.
(Ovrefetene)		
T-8204	1040+/-60	900-1030
Omnesbreen (Elven 1978)		
T-1485	550+/-110	1300-1440
T-1478	440+/-120	1410-1640
T-1479	430+/-100	1410-1630

AD 600 700 800 900 1000 1100 1200 1300 1400 1500 1600 1700

[5]Other difficulties involved in radiocarbon dating of peat are discussed in Grove and Switsur (1994, p. 3).

7.2. SOUTHERN SCANDINAVIA

In southern Norway, particularly concentrated efforts to reconstruct climatic and glacial history have been made around Jostedalsbreen (Figure 6), in the Jotenheimen and around Hardangerjøkulen. Because glaciers in southern Norway were larger during the maximum of recent centuries than in the thirteenth and fourteenth centuries, only intensive investigations have revealed their earlier "Little Ice Age" history. Many forefields lie above the treeline, but proglacial terrestrial and lacustrine stratigraphic sections contain sequences reflecting glacier fluctuations (e.g. Matthews, 1991; Nesje et al., 1991; Nesje and Dahl, 1991a, b; Karlén and Matthews, 1992; Dahl and Nesje, 1994).

Clear evidence of the time when the "Little Ice Age" started comes from the lacustrine sediments of Gjuvvatnet, a high-altitude lake fronting a cirque glacier in mid-Jotenheimen. This catchment is particularly conducive to glacier formation and disappearance in response to relatively small climatic variations due to its high altitude, relief and aspect (Matthews and Karlén, 1992). Sediment changes consequently provide records of the switching on and off of glacier activity. An AMS date of 1220-1290 (CA-1131) marks the start of glacial activity in the thirteenth century, but changes in sediment type show that it was later interrupted briefly (Karlén and Matthews, 1992, p. 339). The characteristics of the sediments deposited during this interval suggest that the cirque was empty or the glacier extremely small, but as the ice has not completely melted away during the twentieth century, it is much more likely that it was very small. Identification of the initiation of corresponding advances of other small glaciers in south Norway is difficult as expansion was greatest in the seventeenth and eighteenth centuries, burying most of the evidence of earlier events, but Matthews (1991) succeeded in obtaining maximum dates for "Little Ice Age" moraine formation and some evidence of the earlier history of two cirque glaciers and the outlets of two small ice caps in Jotenheimen (see Table VI). Radiocarbon dates were obtained from samples of moss, plant remains and peat, as well as soil.

At Sagabreen, a small cirque glacier, the date of a sample of moss collected from the base of the outermost moraine, exposed in a stratigraphic section, (CAR-750) provides a minimum estimate for the onset of glaciofluvial sedimentation, suggesting that Sagabreen was expanding sometime between A.D. 1030 and 1210, probably from a position some distance away. A date of A.D. 1640-1950, also on moss, from higher within the moraine (CAR-751), demonstrates that the moraine itself was formed during the "Little Ice Age"[6]. Moss from a trench through the outermost lateral moraine of Storbreen, another small cirque glacier (CAR-838), shows that moraine formation was occurring

[6]The radiocarbon ages presented in Matthews (1991) were originally calibrated at 2 sigma using Stuiver and Reimer, 1986; revision 1,3, 1987. The values consequently differ from those given in this paper.

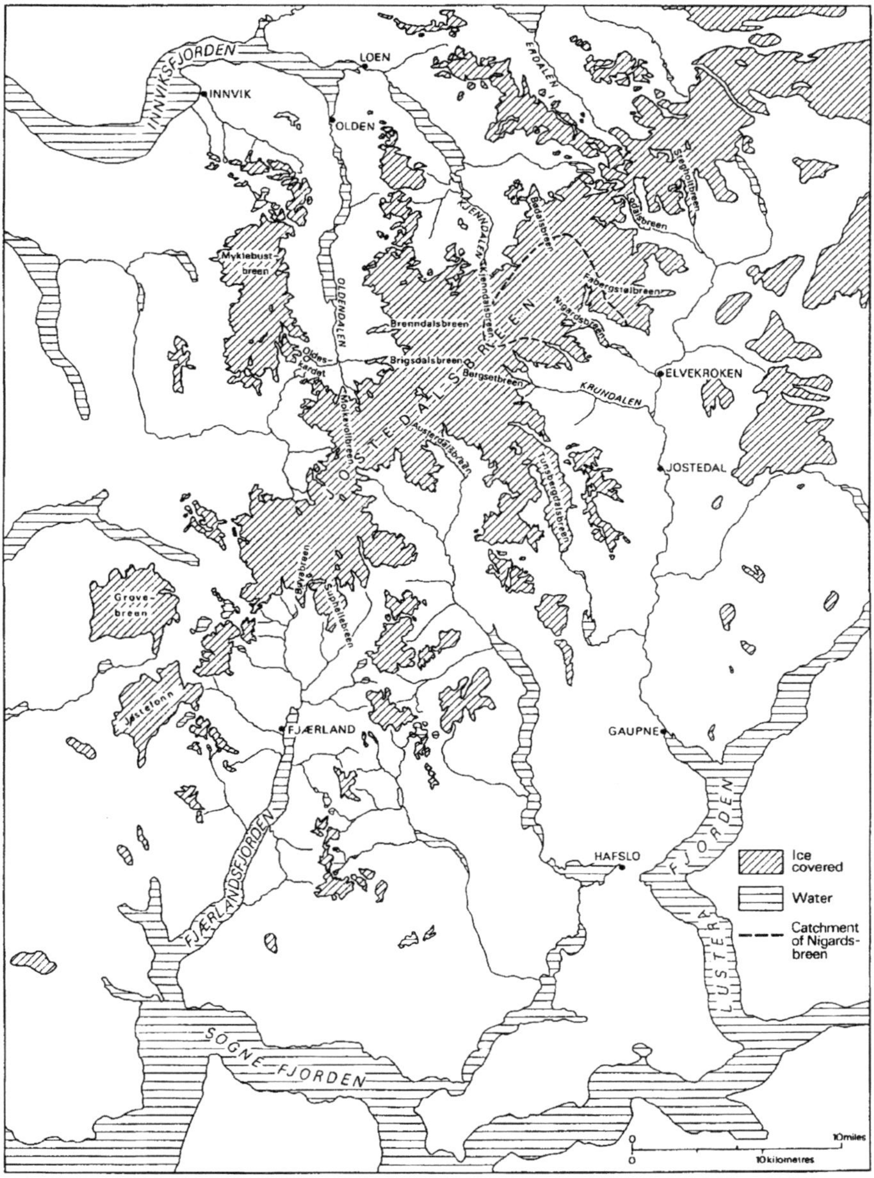

Figure 6. Jostedalsbreen and its surroundings. From Grove 1988, Figure 3.2.

during the period A.D. 1330-1440, although it may have begun somewhat earlier. Bøverbreen, an outlet from a small ice cap, is fronted by a moraine dammed lake. A date from moss immediately below lacustrine silts (CAR-782) gives an age range of 1490-1660, showing that the moraine damming the lake was formed sometime after 1490.

Nesje and Rye (1993) dated a tree stump from the uppermost of five palaeosols in a section through fan sediments deposited by a melt stream from the Sandskardfonna cirque glacier, in Sunndalen, northwest of Jostedalsbreen. This grew before the "Little Ice Age" expansion of the cirque glacier, which clearly occurred after A.D. 1040-1220 (T-5813).

In the Hardangerjøkull region, melting of the central part of Omnsbreen, a small glacier north of Hardangervidda, in the 1970s revealed an extensive area of fresh tills with humus and plant remains (Elven, 1978). The undisturbed nature of the plant remains suggests rapid climatic deterioration, causing the plants to be embedded in permanent snow. The species involved grow at somewhat lower altitudes today. Elven concluded that the climatic change was too fast for the vegetation to respond before being overwhelmed, and placed the event in the fourteenth century, having calibrated the ages according to Ralph et al. (1973). Calibrated according to CALIB 3 (Stuiver and Reimer, 1993), Elven's samples give calendar ranges between the fourteenth and seventeenth centuries (Table VI). In view of more recent and closer results from the same region, the fourteenth century is much more probable. Dahl and Nesje (1994) investigated the stratigraphic sequence in Øvrefetene, a small hollow that collects sediments from Blåisen, a northeastern outlet of Hardangerjøkulen, whenever the glacier enlarges and crosses a local water divide, but becomes a peat bog when the glacier is withdrawn behind the divide. They identified a brief Medieval advance sometime after cal A.D. 900-1030 (T-8204), then a period of glacier retreat during which peat accumulated in the basin 'until the initial "Little Ice Age" glacier advance occurred' after cal A.D. 1300-1430 (T-9440).

Nesje et al. (1991, p. 111) concluded that a minor climatic deterioration during the first part of the fourteenth century and a severe climatic deterioration from about A.D. 1650-1680 culminated in maximum glacier expansion during the mid-eighteenth century, noting that the shorter and steeper western outlets from Jostedalsbreen reached their most extended positions a few years before the more gently sloping eastern outlet valley glaciers. They recognised the "Little Ice Age", 'in many cases characterised by the most extensive Holocene glacier expansion, as occurring between the thirteenth and the early twentieth century' (Nesje and Dahl, 1991a, p .27).

Evidence of fourteenth- to early sixteenth-century expansion of small glaciers in southern Norway has been accumulating and, although for the most part not precise, it is now too substantial to be ignored. It has been argued that the outlet glaciers of Jostedalsbreen also advanced in the fourteenth century (e.g. Eide, 1955; Karlén, 1991). This view is based on a letter written in January 1340 revealing that some kind of disaster affected many farms in both Oldendal and Kjenndalen on the northwest side of Jostedalsbreen (*Diplomatarium Norwegicum* IX, 1340). This interpretation is not tenable in light of the very substantial documentary evidence to the contrary.

Nesdal and Bødal were amongst the farms in Kjenndalen specified as having had their taxes reduced in the 1340 document. They were undoubtedly the most vulnerable to damage from Kjenndalsbreen. But, unlike many others in the high valleys, they were occupied continuously after the plague of 1349-50, as were others further down Kjenndalen. Kjenndalsbreen advanced into Kjenndalen only in the late seventeenth century. Between 1667 and 1693 it carried away barns and enclosures belonging to the Indre Naesdal farms (Riksarkivet Oslo, Rentekammerarkivet, 1702). No evidence of glacial extension is to be found in fifteenth- or sixteenth-century documents. Evidently Kjenndalsbreen did not advance to any marked extent in the thirteenth or fourteenth centuries, though there may have been minor oscillations on too small a scale to be important to the farmers.

Oldendal, too, was never completely deserted after the Black Death; some farms remaining were occupied without interruption. Others were resettled especially between 1530 and 1620 (Aaland, 1973). By 1563, there were 23 farms in the valley and 16 more by 1667, some new ones having been established. Aabrekke and Tungøen were settled before 1562. They were built below a rock step at the junction of Oldendal and Brendalen, a right-hand tributary; both held grazing lands in Brendalen. Aabrekkebreen, at the head of Brendalen did not then reach the valley floor, but could just be seen on the skyline (Reksted, 1901). Documents written between 1563 and 1670 reflect no anxiety about the glacier and no damage to their land was recorded (Grove, 1985). However, by 1685 ice was spilling into Brendalen and accumulating, causing flood damage below. Its advance down valley can be traced in reports of the many courts of enquiry held in connection with applications for tax relief (Statsarkivet Oslo, Amtsregenskap; Bergens Stiftamt 1661; Rekstad, 1901; Aaland, 1932). In December 1723, the ice carried away the houses of Tungøen (Eide, 1955). Aabrekebreen advanced in the late seventeenth and early eighteenth centuries, but not in the fourteenth century.

In a similar way documentary evidence discloses that the Tverrabreen and Nigardsbreen outlets advanced onto farmlands in Jostedal and its tributary valleys east of Jostedalsbreen, in the late seventeenth and eighteenth centuries, submerging both grass and arable lands. Grasslands in use in the late-sixteenth century, and incurring taxes, must not only have been clear of ice at that time but for a long period before that as they were suitable for grazing (Grove, 1985; 1988). The outlet tongues of Jostedalsbreen have oscillated in general harmony with each other since measurement of their frontal positions began in the late nineteenth century (e.g. Faegri, 1948; Grove, 1988, Figure 3.9). Evidence of four of their fronts having advanced into surrounding valleys towards the end of the seventeenth century must therefore be taken as symptomatic of them all.

The desertions and tax reductions described in 1340 are most likely attributable, at least in large part, to increases in flooding, landslides and avalanches comparable with those that occurred in the seventeenth, eighteenth and nineteenth centuries (Grove and Battagel, 1983; Grove, 1972, 1988, Figure 12.2). Climatic conditions causing advances of small, sensitive, high-altitude glaciers and extension of permanent snow farther south on Hardangervidda probably also caused a succession of poor harvests and crop failures in the Nordfjord area, further contributing to social distress.

That the Jostedalsbreen outlets did not advance in the fourteenth century does not in any way disprove evidence of extension of small glaciers, which can react much more immediately to climatic fluctuations (Haeberli, 1995; Kuhn 1993; 1996). Matthews and Karlén (1992) showed that in Norway, during the last few thousand years, glaciers have formed at different times at different sites, depending on critical altitudinal thresholds in relation to the scale of atmospheric variation. High altitude cirque glaciers may be expected to react to even a minor climatic fluctuation; valley glacier outlets from an ice sheet are likely to react with substantial lag times, and may not register short-term climate fluctuations. Nesje and Kvamme (1991) and Nesje et al. (1991) have urged the view that Jostedalsbreen probably disappeared during the warmest part of the Holocene. There can be no doubt of major thinning. The late advances of the Jostedalbreen tongues may be attributed to the effects of the earlier reduction of the ice sheet, and the long time required for it to build up to the height required for outlet glaciers to form and move down their valleys. Their sensitivity thenceforth would have been less than that of small, high elevation cirque glaciers.

Matthews (1991) attributed the relatively large scale of "Little Ice Age" expansion of glaciers of widely different sizes in southern Scandinavia to the southerly position of the oceanic and atmospheric polar fronts and associated weather systems (Lamb, 1979, 1985). This would have caused southern Norway to be affected by increased frequency of depressions, lower summer temperatures and increased precipitation. The unexceptional extent of contemporary advances in northern Scandinavia can reasonably be attributed to shortage of winter precipitation resulting from the location of the polar front and associated weather patterns far away to the south.

8. Conclusion

The most detailed glacial histories of the last thousand years come from careful reconstructions based on dendrochronological and radiocarbon dating of *in situ* tree stumps. Radiocarbon dating of samples of wood, soil and vegetation retrieved from beneath or within moraines, especially from lateral positions, has provided much useful though less precise data. Such data must be treated critically, care being taken that the samples analysed relate closely to moraine formation, that is, that they give close limiting dates. Dates from immature soils are much more useful in this respect than those from mature soils.

Though no reconstructions of the glacial history of the last millennium are as complete as the most detailed from Switzerland, evidence that the "Little Ice Age" had begun in the thirteenth century, with the first culmination in the fourteenth century, is widespread. In the European Alps, the "Little Ice Age" is seen as starting with advances around 1280-1290, the subsequent fourteenth-century advances bringing tongues of large valley glaciers as far forward as in the seventeenth and nineteenth centuries. The first culmination was separated from later advance phases by withdrawal on a lesser scale than initial expansion.

In all the regions around the North Atlantic, physical evidence of pre-sixteenth-century glacial advances has been identified. In western Canada, precise data from fossil trees show that advances started in the twelfth century in some cases, earlier than elsewhere, and continued without interruption until at least 1350. There are no indications of interruptions during this expansion nor of the exact time when it culminated. Marine sediment records from southeast Greenland revealing a cold period beginning about 1270, and culminating around 1370, correlate with evidence of expanding snow cover, and permafrost formation. The failure of the Norse Settlement has to be attributed to the inability of the Norse, unlike the Inuit, to adapt their way of life to the stressful climatic conditions registered in the GISP2 core. Many Scandinavian glaciers in the north and some in the south, though not the major outlets of Jostedalsbreen, extended into the fourteenth century.

General glacial advances took place in the North Atlantic region in the thirteenth and fourteenth centuries. Swiss investigators have recognised several later "Little Ice Age" culminations, separated by minor recession phases. The picture from elsewhere is less precise but evidence of at least minor glacial withdrawal after the first culmination is forthcoming, except from Canada. In Greenland and Spitsbergen indications of a two-phase "Little Ice Age", with its first culmination in the fourteenth century, are clear but not precisely dated. Lack of close dating prevents satisfactory identification of the timing both of later culminations within the "Little Ice Age" period and of minor fluctuations superimposed on them. Ogilvie's Icelandic sea-ice reconstructions appear to imply several periods of glacial activity. In Scandinavia lacustrine sediments deposited after the initiation of the "Little Ice Age" consistently show two well separated periods during which large amounts of rock flour were deposited in downstream lakes (Karlén, 1976, 1991) That major outlets of Jostedalsbreen did not advance in the fourteenth century is probably attributable to its disappearance in the Holocene Optimum combined with its large size. Future field investigations may well permit more complete reconstructions of glacial history round the North Atlantic and reveal more satisfactory explanations for differences between them. The contrast between the situation in the Canadian Rockies, and Iceland, where the advances of recent centuries were on a larger scale than those which came earlier, and the European Alps where the various "Little Ice Age" culminations were on similar scales and the "Little Ice Age" itself comparable in magnitude with earlier Holocene fluctuations, has yet to receive much attention.

The data brought together by Bradley and Jones (1992a) clearly demonstrated that the "Little Ice Age" was not a long, synchronous, cold period, a circumstance which explains the complexity of the moraine sequences left by many glaciers. However, when Bradley and Jones (1993) came to combine their various series to give regional or hemispherical values, they produced a composite temperature anomaly series for the northern hemisphere, which was very much what might have been expected from the glacial evidence. The field evidence clearly shows that glaciers on all continents expanded and fluctuated about forward positions during recent centuries. (The term "Little Ice Age" relates to the behaviour of the glaciers, not directly to climate.[7]) In the North Atlantic region plentiful radiocarbon dated evidence reveals that the "Little Ice Age" was underway in the thirteenth and fourteenth centuries.

References

Aaland, J.: 1932, *Nordfjord fra gamle dagar til no. Dei einskilde bygder: Innvik-Stryn-Sandane*, Oslo, Ei Nemnd.

Aaland, J.: 1973, *Nordfjord fra gamle dagar til no. Dei einskilde bygder. Innvik-Stryn-Sandane,* Oslo, Ei Nemnd.

Ahlmann, H.W.: 1938, *Land of Ice and Fire*, Kegan Paul, Trench, Truber, London,

Barlow, L. K., Sadler, J. P., Ogilvie, A. E. J., Buckland, T. C., Amorosi, T., Ingimundarson, I., Skidmore, P., Dugmore, A. J. and McGovern, T. H.: 1997. 'Interdisciplinary investigations of the end of the Norse Western Settlement in Greenland', *The Holocene* **7**, 489-99.

Bradley, R. S. and Jones, P. D.: 1992a, *Climate since A.D.1500*, Routledge, London and New York.

Bradley, R. S. and Jones, P. D.: 1992b, 'When was the "Little Ice Age"?', in Mikami, T. (ed.), *Proceedings of the International Symposium on the "Little Ice Age" Climate*, Department of Geography, Tokyo Metropolitan University, 1-4.

Bradley, R. S. and Jones, P. D.: 1993, '"Little Ice Age" summer temperature variations: their nature and relevance to recent global warming trends', *The Holocene* **3**: 367-76.

Buckland, P. C., Amorosi, T., Barlow, L. K., Mayewski, P., McGovern, T. H., Ogilvie, A. E. J. and Skidmore, P.: 1996. 'Climate change, and the end of Norse Greenland', *Antiquity* **70**, 88-96.

Dahl, S.O. and Nesje, A.: 1994, 'Holocene glacier fluctuations at Hardangerjøkulen, central-southern Norway: a high-resolution composite chronology from lacustrine and terrestial deposits', *The Holocene* **4**, 269-77.

Delibrias, G., Ladurie, M. Le Roy: 1975, 'Le forêt fossile de Grindelwald: nouvelles datations', *Inter-Sciences* **1**,137-47.

---

[7] This definition has also been employed by Luckman in several publications, for instance in his 1994 paper (p. 43) he used the term "Little Ice Age" 'to describe glacier advances between the twelfth and nineteenth centuries'. See also Luckman, 1998.

Dugmore, A. J.: 1989, 'Tephrochronological studies of Holocene glacier fluctuations in south Iceland', in Oerlemans, J. (ed.), *Glacier Fluctuations and Climatic Change*, Kluwer Academic Publishers, Dordrecht-Boston-London, 37-55.

Dugmore, A. J. and Sugden, D. E.: 1991, 'Do anomalous fluctuations of Sólheimajökull reflect ice-divide migration?' *Boreas* **20**, 106-13.

Eide, T. O.: 1955, 'Breden og Bygda', *Norske Tidsskrift for Folkelivsgransking* **5**, 1-42.

Elven, R.: 1978, 'Subglacial plant remains from Omnsbreen glacier area, south Norway', *Boreas* **1**, 83-9.

Eythórsson, J.: 1931, 'On the present position of the glaciers in Iceland. Some preliminary studies and investigations in the summer of 1930', Reykjavík, *Vísindafélag Üslendinga* (Societas Scientiarum Islandica) **10**, 1-35.

Fægri, T. O.: 1948, 'On the variation of west Norwegian glaciers during the last 200 years', *Procès-verbaux de séances de l'Assemblée Générale d'Oslo de l'Union Géodésique et Géophysique Internationale*, 293-303.

Fricke, H. C., O'Neil, J. R., and Lynnerup, N.: 1995, 'Oxygen isotope composition of human tooth enamel from medieval Greenland: linking climate and society', *Geology* **23**, 869-72.

Gad, F.: 1970, *A History of Greenland*, vol. 1, London, Hurst and Co..

Gordiyenko, F. G., Kotlyakov, V. M., Punning, Ya,-K. M. and Vairmäe, R.: 1980, 'Study of a 200 m core from the Lomonosov Ice Plateau in Spitsbergen and the paleoclimatic implications', *Polar Geography and Geology* **5**, 242-51.

Grémaud, J.: 1878, *Documents relatifs à l'histoire du Valais 111*, Mémoires et documents publiés par la Société de la Suisse Romande **31**.

Grove, J. M.: 1972, 'The incidence of landslides, avalanches and floods in western Norway during the "Little Ice Age", *Arctic and Alpine Research* **4**, 131-8.

Grove, J. M.: 1985, 'The timing of the "Little Ice Age" in Scandinavia', in Tooley, M. J. and Sheail, G. M. (eds.), *The Climatic Scene*, George Allen and Unwin, London, 132-53.

Grove, J. M.: 1988. *The Little Ice Age*, Methuen, London and New York.

Grove, J. M.: In Press a. 'The initiation of the Little Ice Age', in *Climate and Climate Impacts through the last 1000 Years*, in Jones, P. D., Davies, T. D., Ogilvie, A. E. J. and Briffa, K. R. (eds.), Climatic Research Unit, University of East Anglia, U.K.

Grove, J. M.: In Press b. *The Little Ice Age*, second edition, London and New York, Routledge.

Grove, J. M. and Battagel, A.: 1983, 'Tax records from western Norway as an index of Little Ice Age environmental and economic deterioration', *Clim. Change* **5**, 265-82.

Grove, J. M. and Switsur, R.: 1994, 'Glacial geological evidence for the Medieval Warm Period', *Clim. Change* **30**, 1-27.

Gudmundsson, H. J.: 1997, 'A review of the Holocene environmental history of Iceland', *Quaternary Science Reviews* **16**, 81-92.

Haeberli, W.: 1995, 'Glacier fluctuations and climate change detection - operational elements of a worldwide monitoring strategy', *World Meteorological Organisation Bulletin* **44**, 23-31.

Harper, J. T.: 1993. 'Glacier terminus fluctuations on Mount Baker, Washington, U.S.A, 1940-1990', *Arctic and Alpine Research* **25**, 332-40,

Harris, C., Caseldine, C. J. and Chambers, W.: 1987. 'Radiocarbon dating of a paleosol buried by sediments of a former ice-dammed lake, Leirbreen, southern Norway', *Norse Geografiska Tidsskrift* **41**, 81-90.

Heusser, C. J.: 1956, 'Postglacial environments in the Canadian Rocky Mountains', *Ecological Monographs*, **26**, 263-302.

Holzhauser, H.: 1980, 'Beitrag zur Geschichte des Grossen Aletschgletschers', *Geographica Helvetica* **35**, 17-24.

Holzhauser, H.: 1982, 'Neuzeitliche Gletscherschwankungen', *Geographica Helvetica* **37**, 115-26.

Holzhauser, H.: 1983a, 'Beitrag zur Geschichte des Grossen Aletschgletscher', *Geographica Helvetica* **38**, 17-23.

Holzhauser, H.: 1983b, 'Die Geschichte des Grossen Aletschgletschers während der letzten 2500 Jahre', *Bulletin de la Murithienne* **101**, 113-34.

Holzhauser, H.: 1984a, 'Rekonstruktion von Gletscherschwankungen mit Hilfe Fossiler Holzer', *Geographica Helvetica* **39**, 3-15.

Holzhauser, H.: 1984b, 'Zur Geschichte der Aletschgletscher und Fieschgletscher', *Physische Geographie* **13** (Geographisches Institut der Universität Zürich) 1-448.

Holzhauser, H.: 1985, 'Gletscher-und Klimageschichte seit dem Hochmittalalter', *Geographica Helvetica* **40**, 168-85.

Holzhauser, H.: 1988, 'Methoden zur Rekonstruktion von Gletscherswankungen', *Die Alpen* **64**, 135-65.

Holzhauser, H.: 1995, 'Gletschschwankungen innerhalb der letzten 3200 Jahre am Beispiel des Grossen Aletsch-und Gornergletschers', Neue Ergebnisse, in: *Gletscher im ständigen Wandel,* Publikationen der Schweizerischen Akademie der Naturwissenschaften (SANW/ASSN), 6.vdf Hochschulverlag AG an der ETH Zürich, 101-22.

Holzhauser, H.: 1997. 'Fluctuations of the Grosser Aletsch Glacier and Gorner Glacier during the last 3200 years: new results', *Paläoklimaforschung/Palaeoclimate Research* **24** (special issue 16), 36-58.

Holzhauser, H.: (in preparation) 'Gletschgeschichte der letzen 3200 Jahre in en Walliser und Berner Alpen, Glazialmorphologische und dendrochronologische Untersuchungen'.

Holzhauser, H. and Zumbühl, H. J.: 1996, 'To the History of the Lower Grindelwald Glacier during the last 2800 years - paleosols, fossil wood and pictorial records - new results', *Zeitschrift für Geomorphologie* N.F. Suppl. Bd. **104**, 95-127.

Hovstad, H.: 1971, 'Pest og Krise', *Den Norske Turistforening Årbok* 33, 33-9 and 41-9.

Hughes, M. K. and Diaz, H. F.: 1994, 'Was there a "Medieval Warm Period", and if so, Where and When?', *Climatic Change* **26**, 109-142.

Ives, J. D.: 1991, 'Landscape change and human response during a thousand years of climatic fluctuations and volcanism: Skaftafell, southeast Iceland', *Pirineos* **137**, 5-50.

Jennings, A. E. and Weiner, N. J.: 1996, 'Environmental change in eastern Greenland during the last 1300 years: evidence from foraminifera and lithofacies in Nansen Fjord, 68 °N', *The Holocene*, **6**, 179-91.

Karlén, W.: 1976, 'Lacustrine sediments and tree-limit variations, Kebnekaise Mountains, Swedish Lapland', *Geografiska Annaler* **58A**, 1-34.

Karlén, W.: 1979, 'Deglaciation dates from northern Swedish Lapland', *Geografiska Annaler* **61A**, 203-10.

Karlén, W.: 1984, 'Dendrochronology, mass balance and glacier front fluctuations in northern Sweden', in: Mörner, N. A. and Karlén W. (eds.), *Climate changes on a yearly to millennial scale*, Dordrecht, Boston, Reidel, 263-71.

Karlén, W.: 1991, 'Glacier fluctuations in Scandinavia during the last 9,000 years', in Starkel, L., Gregory, L. and Thornes J. B. (eds.), *Temperate Paleohydrology. Fluvial Processes in the Temperate Zone during the last 15000 years,* Chichester, New York, Brisbane, Toronto, Singapore, John Wiley, 395-412.

Karlén, W. and Matthews, J. A.: 1992, 'Reconstructing Holocene glacier variations from glacial lake sediments: Studies from Nordvestlandet and Jostedalsbreen-Jotenheimen, southern Norway', *Geografiska Annaler* **74 A**, 327-48.

Kaufman, D. S. and Williams, K. M. (compilers): 1992, 'Radiocarbon date list VII:Baffin Island N.W.T.' *INSTAAR Occasional Paper* **48**, Institute of Arctic And Alpine Research, University of Colorado.

Kinzl, H.: 1932, 'Die grössten nacheiszeitlichen Gletschervorstösse in den Schweizen Alpen und in der Mont Blanc-Gruppe', *Zeitschrift für Gletscherkunde* **20**, 269-97.

Kuhn, M.: 1993, 'Die Mieminger Schneeferner, ein Beispiel eines Lawinenernährten Kargletschers', *Zeitschrift für Gletscherkunder und Glazialgeologie,* **29**, 153-71.

Kuhn, M.: 1996, 'The mass balances of very small glaciers', *Zeitscrift für Gletscherkunde und Glacial geologie* **31**, 171-79.

Ladurie, E. Le R.: 1971, *Times of Feast, Times of Famine*, New York, Doubleday.

Lamb, H. H.: 1977, *Climate: Present, Past and Future*, vol. 2: *Climatic History and the Future*, London, Methuen.

Lamb, H. H.: 1979, 'Climatic variation and changes in wind and ocean circulation in the "Little Ice Age" in the northeastern Atlantic', *Quaternary Research* **11**, 1-20.

Lamb, H. H.: 1985, 'The Little Ice Age period and the great storms within it', in Tooley, M. J. and Sheil, G. M. (eds.), *The Climatic Scene*, London, George Allen and Unwin, 104-31.

Leonard, E. M.: 1985, 'Glaciological and climatic controls on lake sedimentation', *Zeitscrift für Gletscherkunde und Glacial geologie* **21**, 35-42.

Leonard, E. M.: 1997, 'The relationship between glacial activity and sediment production: evidence from a 4450-year varve record of Neoglacial sedimentation in Hector Lake, Alberta', *J. Paleolimno* **17**, 319-330.

Leonard, E. M. and Reasoner, M.A.: 1999, 'A Continuous Holocene Glacial Record Inferred from Proglacial Lake Sediments in Banff National Park, Alberta, Canada', *Quat. Res.* **51**, 1-13.

Luckman, B. H.: 1986, 'Reconstruction of Little Ice Age events in the Canadian Rockies', *Géographie physique et Quaternaire* **40**, 17-28.

Luckman, B. H.: 1990, 'Mountain Areas and Global Change: A View from the Canadian Rockies', *Mountain Research and Development* **10**, 183-195.

Luckman, B. H.: 1992, 'Glacier and dendrochronological records for the Little Ice Age for the Canadian Rocky Mountains', in Mikami, T. (ed.), *Proceedings of the Symposium on "Little Ice Age" Climate*, Department of Geography, Tokyo Metropolitan University, 75-80.

Luckman, B. H.: 1993, 'Glacier fluctuations and tree-ring records for the last millennium in the Canadian Rockies', *Quaternary Science Reviews* **12**, 441-50.

Luckman, B. H.: 1994, 'Using multiple high-resolution proxy climate records to reconstruct natural climate variability: an example from the Canadian Rockies', in Beniston, M. (ed.), *Mountain environments and changing climates*, London and New York, Routledge, 42-59.

Luckman, B. H.: 1995, 'Calendar-dated, early "Little Ice Age" glacier advance at Robson Glacier, British Columbia, Canada', *The Holocene* **5**, 149-59.

Luckman, B. H.: 1996a, 'Reconciling the glacial and dendrochronological records for the last millennium in the Canadian Rockies', in Jones, P. D., Bradley, R. S. and Jouzel, J. (eds.), *Climate variations and forcing mechanisms of the last 2000 years*. NATO ASI Series, Vol. **14**, 85-108.

Luckman, B. H.: 1996b, 'Dendroglaciology at Peyto Glacier, British Columbia', in Dean, J. S., Meko, D. M. and Swetnam T. W. (eds.), *Tree Rings, Environment and Humanity, Radiocarbon* 1996, 679-88.

Luckman, B. H.: 1996c, 'Dendrochronology and global change, I', in Dean, J. S., Meko, D. M. and Swetman, T.W. (eds.), *Tree Rings, Environment and Humanity. Radiocarbon,* 1996, 1-24.

Luckman, B. H.: 1998, 'Landscape and climate change in the central Canadian Rockies during the 20th century', *The Canadian Geographer* **42**, 319-36.

Luckman, B. H.: 2000, 'The Little Ice Age in the Canadian Rockies, *Geomorphology* **32**, 357-384.

Luckman, B. H., Holdsworth, G. and Osborn, G. D.: 1993, 'Neoglacial glacier fluctuations in the Canadian Rockies', *Quaternary Research* **39**, 144-53.

Lütschg, O.: 1926, *Über Niederschlag und Abfluss im Hochgebirge, Sonderstellung des Mattmarkgebietes,* Schweizerische Wasserwirtschaftverband, Verbandschrift C No.14, Veröffentlichung der Schweizerischen meteorologischen Zentralanstalt in Zürich. Sekretariat des Schweizerischen Wasserwirtschaftverbandes.

Matthews, J. A.: 1984, 'Limitation of C^{14} dates from buried soils in reconstructing glacier variations and Holocene climate', in Mörner, N. A. and Karlén,W. (eds.), *Proceedings of the*

Second Nordic Symposium on Climatic Changes and Related Problems, Stockholm. Dordrecht, Reidel, 2991-90.

Matthews, J. A.: 1985, 'Radiocarbon dating of surface and buried soils; principles; problems and prospects', in Richards, K. S., Arnett, R. and Ellis, S. (eds.), *Geomorphology and Soils,* London, George Allen and Unwin, 269-88.

Matthews, J. A.: 1991, 'The late Neoglacial ("Little Ice Age") glacier maximum in southern Norway: new ^{14}C-dating evidence and climatic implications. *The Holocene* **1**, 219-33.

Matthews, J. A. and Caseldine, C. J.: 1987, 'Arctic-alpine brown soils as a source of paleoclimatic information: further ^{14}C dating and palynological evidence from Vestre Memurubreen, Jotenheimen, Norway', *Journal of Quaternary Science* **2**, 59-71.

Matthews, J. A. and Karlén, W.: 1992, 'Asynchronous Neoglaciation and Holocene climatic change from Norwegian glaciolacustrine sedimentary sequences', *Geology* **20**, 991-4.

McGovern, T. H.: 1981, 'The economics of extinction in Norse Greenland', in Wigley, T. M. L., Ingram, M. J. and Farmer, G. (eds.), *Climate and History. Studies in past climates and their impact on man,* Cambridge, Cambridge University Press, 404-33.

Meese, D. A., Gow, A. J., Grootes, P., Mayewski, P. A., Ram, M., Stuiver, M., Taylor, K. C., Waddington, E. D., Zielinski, G. A.: 1994, 'The accumulation record from the GISP2 core as an indicator of climate change throughout the Holocene', *Science* **266**, 1680-2.

Messerli, B., Messerli, P., Pfister, C. and Zumbühl, H. J.: 1978, 'Fluctuations of climate and glaciers in the Bernese Oberland, Switzerland, and their geoecological significance, 1600-1975', *Arctic and Alpine Research* **10**: 247-60.

Nesje, A. and Dahl, S. O.: 1991a, 'Holocene glacier variations of Blåisen, Hardangerjøkulen, central southern Norway', *Quaternary Research* **35**, 25-40.

Nesje, A. and Dahl, S. O.: 1991b, 'Late Holocene glacier fluctuations in Bevringsdalen, Jostedalsbreen region, western Norway (ca 3200-1400BP)', *The Holocene* **1**, 1-7.

Nesje, A. and Kvamme, M.: 1991, 'Holocene glacier and climatic variations in western Norway: evidence for early Holocene glacier demise and multiple Neoglacial events', *Geology* **19**, 610-12.

Nesje, A., Kvamme, M., Rye, R. and Løvlie, R.: 1991, 'Holocene glacial and climatic history of the Jostedalsbreen region, western Norway; evidence from lake sediments and terrestrial deposits', *Quaternary Science Reviews* **10**, 87-114.

Nesje, A. and Rye, R.: 1993, 'Late Holocene glacier activity at Sandeskardfonna, Jostedalsbreen area, western Norway', *Norges Geografiske Tidsskrift* **47**, 21-30.

Nörlund, P.:1924, 'Buried Norseman at Herjolfsnes. An archeological and historical study', *Meddelelser om Grønland* **55**,186-9.

O'Brien, S. R., Mayewski, P. A., Meeker, L. D., Meese, D. A., Twickler, M. S. and Whitlow, S. I.: 1995, 'Complexity of Holocene climate as reconstructed from a Greenland Ice Core', *Science* **270**,1962-4.

Ogilvie, A. E. J.: 1991, 'Climatic changes in Iceland A.D.c.865 to 1598', *Acta Archaeologica* **61**, 233-51.

Ogilvie, A. E. J.: 1998, 'Historical accounts of weather events, sea ice, and related matters in Greenland and Iceland, A.D. c.1250 to 1430', in: Proceedings of the 18th EPC Workshop on "Documentary climatic evidence for 1750-1850 and the 14th century", Tallinn, Estonia, 1994, *Paleoklimaforschung /Paleoclimatic Research* **23**, 25-45.

Ogilvie, A. E. J., Barlow, L. K. and Jennings, A. E.: 2000, 'North Atlantic climate c. A.D. 1000: Millennial reflections on the Viking discoveries of Iceland, Greenland and North America', *Weather*, 55, 34-45.

Oeschger, H. and Röthlisberger, H.: 1961, 'Datierung eines ehemaligen Standes des Aletschgletshers durch Radioaktivitätsmessung an Holzproben und Bemerkungen zu Holzfunden an weiteren Gletschern', *Zeitschrift für Gletscherkunde und Glazialgeologie* **4**, 191-204.

Patzelt, G.: 1974, 'Holocene variations of glaciers in the Alps', Colloques Internationaux du Centre National de la Recherche Scientifique No. 219, *Les méthodes quantitatives d'étude des variations du climat au cours du Pléistocene*, 61-9.

Patzelt, G.: 1994, 'Former lakes dammed by glaciers and resulting floods in the Øtztal, Tyrol', *Mountain Research and Development* **14**, 298-301.

Pfister, C.: 1988, 'Variations in the spring-summer climate of central Europe from the high Middle Ages to 1850. I. H. Wanner and U. Siegenthaler (eds.), *Long and Short Term Variability of Climate*, Berlin, Springer, 57-82.

Pfister, C., Holzhauser, H. and Zumbühl, H. J.: 1994, 'Neue Ergebnisse zur Vorstossdynamik der Grindelwaldgletscher vom 14. bis zum 16.Jahrhundert', *Mitteilungen der Naturforschenden Gesellschaft in Bern*, N.F.51, 55-79.

Pfister, C., Schwarz-Zanetti, G., and Wegman, M.: 1996, 'Winter severity in Europe: the Fourteenth Century', *Clim. Change* **34** , 91-108.

Pfister, C., Luterbacher, L., Schwartz-Zanetti, G. and Wegmann, M.: 1998, 'Winter air temperature variations in western Europe during the Early and High Middle Ages (AD 750-1300)', *The Holocene* **8**, 535-52.

Porter, S. C.: 1981, 'Glaciological evidence of Holocene climatic change', in: Wigley, T. M. L., Ingram M. J. and Farmer, G. (eds.), *Climate and History, Cambridge*, Cambridge University Press, 82-110.

Pringle, H.: 1997, 'Death in Norse Greenland', *Science* **275**, 924-6

Ralph, E. K., Michael, H. N. and Han, M. C.: 1973, 'Radiocarbon dates and reality', *Masca Newsletter* **9**, 1-20.

Rekstad, J.: 1901, 'Iagttagelse fra Bræer i Sogn og Nordfjord', *Norges Geologiske Undersøgelse*, **34** *Aabog 3*, . Reprinted 1902.

Röthlisberger, F., Hass, P., Holzhauser, H., Keller, W., Bircher, W. and Renner, F.: 1980, 'Holocene climatic fluctuations - radiocarbon dating of fossil soils (fAh) and woods from moraines and glaciers in the Alps', in Geography in Switzerland, *Geographica Helvetica* **35**, 21-52.

Röthlisberger, F. (ed.): 1986, *10000 Jahre Gletschergeschichte der Erde*. Aarau, Frankfurt am Maine, Salzburg. Saulander.

Sacco, F.: 1918: 'I ghiacciai italiani del Gruppo del Monte Bianco', *Bollettino del Comitato Glaciologico Italiano* ser.1, **3**, 21-102.

Smith, D. J., McCarthy, D. P. and Colenutt, M. E.: 1995. 'Little Ice Age glacial activity in Peter Lougheed and Elk Lakes provincial parks, Canadian Rocky Mountains', *Canadian Journal of Earth Sciences* **32**, 579-89.

Stuiver, M. and Reimer, P. J.: 1986, 'A computer program for radiocarbon age calibration', *Radiocarbon* **28**, 1022-30.

Stuiver, M. and Pearson, G.W.: 1993, 'High-precision bidecadal calibration of the radiocarbon timescale A.D.1950-500B.C.', *Radiocarbon* **35**,1-323.

Stuiver, M. and Reimer, P.J.: 1993, 'Extended ^{14}C data base and revised CALIB 3.0 ^{14}C age calibration program', *Radiocarbon* **35**, 215-30.

Stuiver, M. and Brazunias, T. F.: 1993, 'Modelling atmospheric ^{14}C influence and ^{14}C ages of marine samples to 10,000 BC', *Radiocarbon* **35**, 137-90.

Stuiver, M., Grootes, P. M. and Braziunas, T. F.: 1995, 'The GISP2 climate record of the past 16,500 years and the role of the sun, ocean and volcanoes'. *Quaternary Research* **44**, 341-54.

Svendsen, J. I. and Mangerud, J.: 1997, 'Holocene glacial and climatic variations on Spitsbergen and Svalbard', *The Holocene* 7, 45-57.

Thórarinsson, S.: 1956, The Thousand Years Struggle against Ice and Fire, Museum of Natural History, Dept. of Geology and Geography, miscellaneous papers, No. 14, Reykjavik.

Weidick, A., Bøggild, C. E., and Knudsen, N. T.: 1992, 'Glacier inventory and Atlas of West Greenland', *Grønlands Geologiske Undersøgelse*. Rapport 158.

Zumbühl, H. J. and Holzhauser, H.: 1988, 'Alpengletscher in der Kleinen Eiszeit-Sonderheft zum 125 jährigen', Jubiläum des SAC. Herausgegeben vom Schweizer Alpen-Club, Bern *Die Alpen.*
Zumbühl, H. J. and Holzhauser, H.: 1990, 'Alpengletscher in der Kleinen Eiszeit. Katalog und ^{14}C-Dokumentation', *Geographia Bernensia* **31**, Geographisches Institut der Universtät Bern. **36**, 129-322.

Archival References

Riksarkivet Oslo. Mattrikel over Nordfjord, 1667, 44.
Riksarkivet Oslo. Mattrikel over Nordfjord, 1723, 147, 89a-121a.
Riksarkivet Oslo. Rentekammerarkivet, Fogedregnskaper Sunnfjord-Nordfjord, 1702.
Statsarkivet Oslo, Amtsregenskap, Bergens Stiftamt 1661, Pakke 2, folder 2.
Statsarkivet Oslo, Ordnings Avdeling Affaednings-Forretninger Bergens Stiftamt 1702-84, Pakke 3.

(Received 15 June 1998; in revised form 25 September 2000)

OCEANOGRAPHIC CHANGE AND TERRESTRIAL HUMAN IMPACTS IN A POST A.D. 1400 SEDIMENT RECORD FROM THE SOUTHWEST ICELAND SHELF

A. E. JENNINGS[1], S. HAGEN[2], J. HARÐARDÓTTIR[3], R. STEIN[4], A. E. J. OGILVIE[1] and I. JÓNSDÓTTIR[5]

[1] *INSTAAR, University of Colorado, Campus Box 450, Boulder, Colorado, 80309-0450, USA*
[2] *Department of Geology, IBG, University of Tromsø, N-9037, Norway*
[3] *Science Institute, University of Iceland, Dunhagi 3, 107 Reykjavik, Iceland*
[4] *Alfred-Wegener-Institute for Polar and Marine Research, Columbustrasse, 27568 Bremerhaven, Germany*
[5] *Department of Geosciences, Faculty of Science, University of Iceland, Reykjavik, Iceland*

Abstract. Environmental proxies of soil erosion on Iceland, and oceanographic conditions on the adjacent shelf, were measured on a 50 cm box core taken from the southwest Iceland shelf in 1993 during cruise 93030 of the Canadian ship, *CSS Hudson*. These data, covering the last several centuries, are compared with the documentary record of sea-ice changes around Iceland since A.D. 1600. The site is under the influence of the Irminger Current, which carries warm, saline, Atlantic water northward along the shelf. Because of the relative warmth of this current, sea ice rarely occurs off southwest Iceland, even during the most severe sea-ice intervals of the historical record. In severe sea-ice years, however, the ice drifts clockwise around Iceland from the northeast and east and, in rare cases, reaches the southern coasts (Ogilvie, 1992). The chronology of the core was established by converting the basal radiocarbon date to calendar years and assuming a linear sedimentation rate from the base of the core to the year of collection, 1993. Organic carbon, stable C and O isotope ratios, planktonic foraminiferal assemblages, and sediment magnetic parameters were measured on samples from the core, plotted against calendar years and compared to the Icelandic sea-ice index. The environmental proxies suggest that increased soil erosion, reduced salinity, and, possibly, decreased marine productivity prevailed during the severe sea-ice interval lasting from the last quarter of the eighteenth century to around 1920. Such a situation could develop with climatic cooling, increased storminess, and loss of vegetation cover to stabilise the soil. Although the core site generally lies outside the sea-ice limits, the evidence clearly shows the influence of sea ice and fresh water, and is sensitive to the overall climatic deterioration manifested by the sea-ice record.

Key Words: "Little Ice Age"; Soil erosion; Foraminifera; Palaeoceanography; Sea Ice

1. Introduction

Iceland's location near the boundaries of atmospheric and oceanic polar fronts make it sensitive to small changes in the oceanic and atmospheric circulation that affect climate (Figure 1). Small movements in these features

Climatic Change **48**: 83–100, 2001.

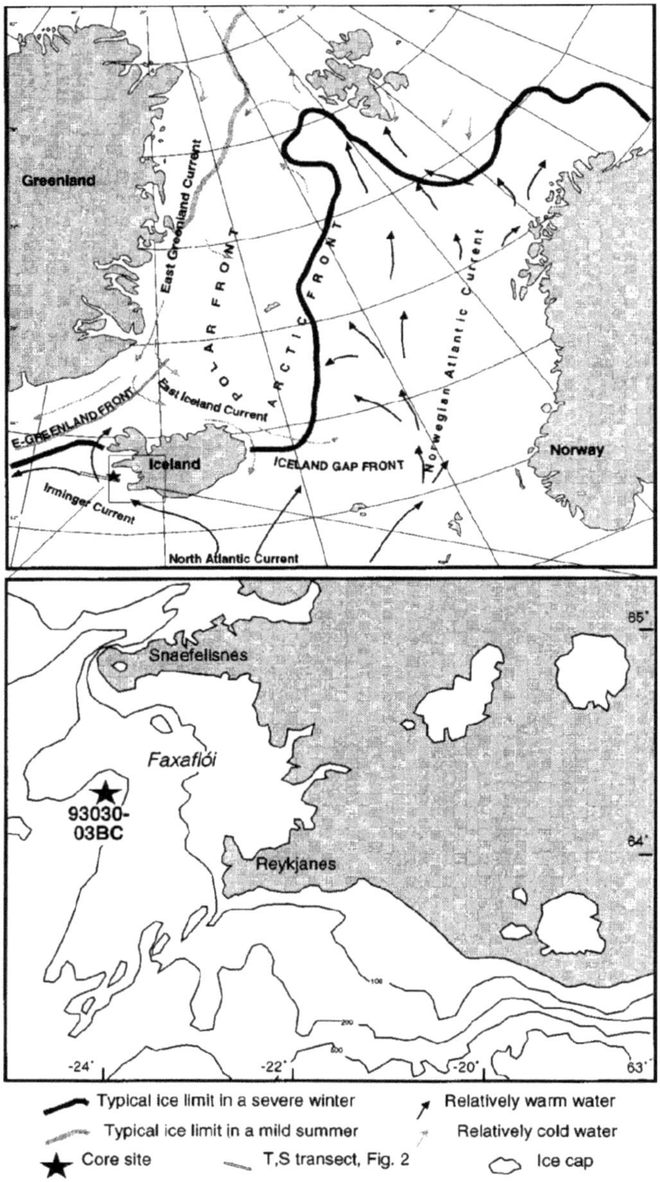

Figure 1. Location map showing bathymetry of western Iceland and location of 93030-03BC. Inset map showing surface currents in the Nordic Seas (modified from Mosby, 1968).

produce extreme and rapid environmental changes that impact the success of farming and fishing, historically the basis of Iceland's economy (Ogilvie, 1982, 1984a, 2000; Bergthórsson et al., 1988). Documentation of environmental and climatic changes in Iceland since Settlement (in the late ninth century A.D.) have been studied through research on historical documentary data (e.g. Ogilvie, 1984b; 1991; 1992), natural archives of past vegetation, such as peat bogs and lakes (e.g. Hallsdóttir, 1987), glacier fluctuations (e.g. Caseldine, 1987; Guðmundsson, 1997), and more recently, by high resolution sedimentological and magnetic analysis of lacustrine sediments (Harðardóttir et al., in press). Marine sediment records of the late Holocene environments, including post Settlement records, have not yet been studied.

As part of the investigation of the late Quaternary history of the southwestern Iceland shelf (Syvitski et al., 1999; Jennings et al., in revision) a box core of the uppermost sediments in Jökuldjúp, a bathymetric basin in Faxaflói, was recovered (Figure 1). The box core, 93030-03BC (latitude: 64° 18.00'; longitude: 24° 13.90'; water depth: 250 m) was taken in order to provide a detailed record of environmental conditions during the last several centuries, including the so-called "Little Ice Age", and to investigate the relations between modern proxies and environment in the "Atlantic" realm of the Iceland shelf, the area under the influence of the Irminger Current.

In this paper, results of measurements of carbon flux, stable C and O istotope ratios, planktonic foraminiferal assemblages, and sediment magnetic parameters from the core are presented. In order to document environmental changes on the southwest Iceland shelf from ca. A.D. 1450 to the year of collection, 1993, these data were then compared to the record of sea-ice variability. The results show increased delivery of soil materials to the shelf, indicating increased soil erosion on Iceland, and reduced salinities on the shelf associated with the advancing sea-ice edge during the period of severe sea ice between ca. 1780 and 1920, the "Little Ice Age" interval of this study.

2. Environmental Setting

The southwestern Iceland shelf is situated in the core of the Irminger Current, a branch of the North Atlantic Current, that carries warm, saline Atlantic water northwards along the western Iceland shelf (Figure 1). Temperature and salinity measurements of the water column in Faxaflói show little vertical variation, with salinities of 35 per mil and temperatures below 50 m of 7°C (Figures 1 and 2). The southern limit of sea ice during a severe winter is north of the box core site in Faxaflói. To the north and west, temperatures and salinities change in a steep gradient, coinciding with the marine Arctic and Polar fronts west and north of Faxaflói (Figure 1). The positions of these fronts vary within each year and between years. Severe sea-ice years occur when Polar surface water in the East Greenland Current is entrained in the East Iceland Current and impinges on the north coast of Iceland carrying sea ice and lower salinity water (e.g. Sigtryggsson; 1972; Malmberg, 1985). The severity of sea ice during the Great Salinity Anomaly years (1965-1972), when an increased flux of multi-year pack ice from the Arctic Ocean was carried out of its normal bounds in the East Greenland Current to reach the north coast of Iceland, provides a modern example of severe sea-ice conditions.

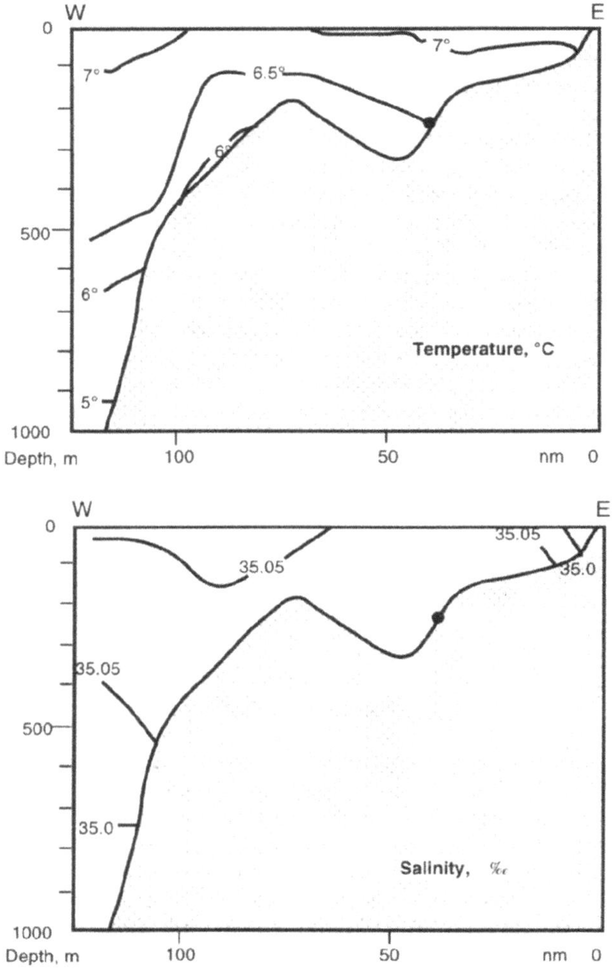

Figure 2. Temperature and salinity profile along transect in Faxaflói-location shown in Figure 1. The dot marks the approximate location of core 93030-03BC on the profile (modified from Helgadóttir, 1984).

Other intervals of severe sea ice around Iceland have been reconstructed from documentary data, including a significant sea-ice period between the late eighteenth century and the early twentieth century (Ogilvie and Jónsdóttir, 1996; Koch, 1945). Air temperatures recorded in Iceland are well correlated with sea-ice incidence (Bergthórsson, 1969; Ogilvie, 1984b; 1992).

The influence of the Irminger Current is so strong on the southwestern Iceland shelf, that even during the most severe sea-ice intervals, sea ice has rarely been observed south of Snaefellsnes (Figure 1). Historical documentary data on sea-ice distribution around Iceland covering the period from 1601 to 1850 indicate that Faxaflói was also generally out of the direct influence of sea ice during that time interval (Ogilvie, 1992; 1996). A rare exception may have been the unusually severe sea-ice year of 1695 (Vilmundarson, 1972). Furthermore, a series of sea-ice maps from 1940 through 1971, including the Great Salinity Anomaly years (Sigtryggsson, 1972) also shows that Faxaflói was consistently free of sea ice during this time. Consequently, because of its position relative to the surface current circulation and the reconstructed sea-ice history, box core 93030-03 appears to be in a position to record only the most severe sea-ice events and associated ocean-front driven environmental changes in the late Holocene. The sensitivity of the site is tested with the palaeoceanographic proxies measured in box core 93030-03.

3. Materials and Methods

During cruise 93030 of the Canadian research vessel, *CSS Hudson*, a benthos box core was raised from the western Iceland shelf in Jökuldjúp, a bathymetric basin in the outer part of Faxaflói (Figure 1; Asprey et al., 1994). Several subcores were taken from the boxcore by inserting 11-cm diameter PVC pipes into the box core. A vacuum-pump system was used to reduce compaction of the sediments. One of the subcores, 3I, was immediately sampled for foraminiferal analysis. Samples were taken every 1 cm from 0 to 10 cm, every 2 cm between 11 and 20 cm, and every 5 cm from 20 cm to the base of the core at 50 cm. These samples were treated with rose bengal stain and preserved in a 60% alcohol and water solution in order to allow determination of living foraminifera in the sediment. These samples were subsequently prepared within two months for planktonic foraminiferal analysis and stable O and C isotope measurements. Five bulk density measurements were made onboard on another subcore (Kate Moran, the Bedford Institute of Oceanography, written communication, 1996).

An additional subcore, 3H, was obtained in order to increase the sampling frequency from subcore 3I below 20 cm, and subsequently shipped unopened for additional analyses. In November 1995, this subcore was split in half longitudinally, visually described, x-rayed, and sampled at 2 cm intervals for bulk density, sediment magnetic parameters and carbon content, along the whole length of the 54 cm core. Samples for planktonic foraminiferal concentrations and stable C and O isotopes were taken at 2 cm intervals from 20 cm to the base of the core in order to increase the sampling density of the 20 to 50 cm interval in subcore 3I.

The number of planktonic foraminifera per ml wet sediment in each sample vial was used to estimate the foraminiferal flux through time. To determine the initial volumes of foraminiferal samples in subcore 3I, the level of sediment on the vial was marked, and after the sediment was removed from the vial for sieving, it was filled to the line with water and the volume of the water was measured in a graduated cylinder. The initial volume of subcore 3H samples was determined by volume displacement in a graduated cylinder. Foraminiferal samples were washed with distilled water on a 63 µm screen. The material larger than 63 µm was air dried and sieved on a 125 µm screen. The material > 125 µm was used for foraminiferal and isotope studies. In the samples from 3I, the species composition and number of planktonic foraminifera in each sample was determined by identifying to species level ca. 100 specimens in a representative split of the sample (Hagen, 1995). The sample was spread onto a picking tray and the foraminifera were identified in a randomly chosen sequence of squares. The equation used to determine the concentration of foraminifera was:

Equation 1:

Planktonic Foraminifera/ml wet sediment = ((#counted/split)/#squares counted/total squares))/ml sed sample.

For subcore 3H, the procedure differed in that the total number of planktonic foraminifera in a split was counted. Species data were not obtained in subcore 3H.

Using foraminiferal data from subcore 3I, summer paleotemperatures of the surface water in Faxaflói were estimated using the transfer function developed by Johannessen (1987) as detailed in Hagen (1995). Temperatures are estimated with an error of ± 1.13 degrees and a correlation coefficient, r, of 0.89. Stable isotope measurements were made on *Neogloboquadrina pachyderma* dextral picked from the sediment size fraction >125 µm to 1000 µm. The foraminifera were less than 250 µm (Hagen, 1995). The analyses were made on a Finnigan MAT 251 mass spectrometer at the Geological Mass Spectrometer Laboratory, University of Bergen, with a reproducibility of 0.07‰ for $\delta^{18}O$ and of 0.06‰ for $\delta^{13}C$.

Bulk density measurements were made using 5 cm^3 paleomag cubes sampled using a volumetric sampler that removes a cube of sediment without compaction. The cubes were weighed immediately after sampling, and then weighed again after air drying in a 30°C oven. The dry bulk density (g/cm^3) was used in the flux calculations. Comparison of the bulk density measurements made on the ship from another subcore with those made later on the subcore 3H averaged 5 percent larger. An error of ±5 % was assumed for the bulk density measurements.

Magnetic analyses were performed on the wet bulk density samples to evaluate changes in the concentration, grain size and type of magnetic minerals in the sediment. Mass magnetic susceptibility (X) was measured on a Bartington MS.2 susceptibility bridge and was used primarily to determine concentration of magnetite in the sediment, although magnetic grain size may affect the values. (In this paper the term "magnetite" is used for ferromagnetic material, and the term "hematite" for antiferromagnetic material). Anhysteric remanent magnetization (ARM) (here represented as mass susceptibility of ARM (Xarm)) was induced to the samples in a bias direct field of 0.1 mT and in an alternating field with a peak induction of 100 mT. The remanence was then measured on a Molspin spinner magnetometer. When normalized for magnetite content, the ratio Xarm/X can be used to evaluate grain size of magnetic minerals within the sediment. Finally, isothermal remanent magnetization (IRM) was induced to the samples at 1.2 T in forward direction, and at 0.3 T and 0.1 T in the backfield direction. Their remanent magnetization was also measured on a Molspin spinner magnetometer. Other magnetic variables, such as the S-parameter and HIRM (Hard IRM), were derived from these IRMs according to King and Channel (1991), and are used to evaluate the relative amounts of magnetite to hematite and the amount of hematite, respectively.

Carbon contents were measured on the air-dried sediments from the paleomag cubes. The sediments were disaggregated using a rubber mortar and a pestle. Measurement of total carbon (TC) and inorganic carbon (TIC) were made using a UIC Coulometrics carbon analyzer. Organic carbon (OC) content was calculated by the difference between the TC and TIC. The error on the carbon measurements was estimated by replicate runs; these were ± 1.9% for TC and ± 0.7% for TIC.

Rock-Eval pyrolysis was used to get information on the composition of the organic matter in the sediment. In particular it was used to determine the terrigenous vs. marine origins of the organic matter (Espitalié et al., 1977; Tissot and Welte, 1984; Liebezeit and Wiesner, 1990; Stein, 1991; Fahl and Stein, 1997). Using Rock-Eval pyrolysis on bulk sediment samples we determined the quantity of hydrocarbons initially present in the samples (S_1 peak in mg hydrocarbon per gram sediment), the quantity of hydrocarbons produced by pyrolytic degradation of the kerogen during heating to 550 °C (S_2 peak in mg hydrocarbon per gram sediment), and the amount of carbon dioxide generated during heating to 390 °C (S_3 peak in mg carbon dioxide per gram sediment), and the temperature of maximum of hydrocarbon generation during pyrolysis (Tmax value in °C) (Espitalié et al., 1977). These quantities were used to calculate the hydrogen index (HI) and the oxygen index (OI) values, which are used to classify, by empirical correlation, a terrigenous or marine origin for organic matter using a "van-Krevelen-type" diagram (Espitalié et al., 1977; Tissot and Welte, 1984; Stein, 1991). The HI value corresponds to the quantity of pyroloyzable hydrocarbons (S_2) per gram total organic carbon (mgHC/gC). The OI value corresponds to the quantity of carbon dioxide (S_3) per gram TOC (mgCO_2/gC). In immature, organic-carbon-rich (TOC> 0.5%) sediments, HI values of < 100 mg HC / g C are typical of terrigenous organic matter (kerogen type III), whereas HI values of 300 to 800 mg HC/gC are typical of marine organic matter (kerogen types I and II) (Figure 3).

4. Chronology and Flux Calculations

A radiocarbon date (AA-14213) on foraminifera at the base of subcore 3I of 880 ± 65 BP (uncorrected for marine reservoir age) was obtained. This date was calibrated to calendar years using the radiocarbon calibration program, Calib 3.0.3 using the marine calibration data set (Stuiver and Reimer, 1993). Delta R was set to 0 + 20 years because of the well mixed Atlantic water in the

Irminger Current (Stuiver and Braziunas, 1993). The program calculated a mean basal calibrated age of A.D. 1456, with one sigma error bars giving a range of A.D. 1421 to A.D. 1498 and two sigma error bars of A.D. 1368 to A.D. 1557. The top of the box core was assumed to be 1993, as the sediment/water interface was well preserved and contained living foraminiferal fauna (Jennings et al., 1994). Linear sedimentation was assumed from the top of the core to the base. This was supported by the single mostly homogeneous lithofacies present, although slight changes in sedimentation are probably smoothed out by bioturbation. The linear sedimentation rate was calculated at 10.74 yrs/cm from the data in subcore 3I. Additional radiocarbon dates were not submitted because of the difficulty in obtaining dates with small two sigma errors for samples dating over the last 500 years (Stuiver and Becker, 1993). The samples from subcore 3H were placed into the chronological framework provided by subcore 3I by assuming a one-to-one correspondence with depth.

The fluxes of OC, planktonic foraminifera, and diatoms used in this study are subject to errors in the measurement of each of these parameters, as well as in the calculation of the sedimentation rate. The root mean square error was used to estimate the errors on the fluxes, such that the total error was calculated as the square root of the sum of the squares of each error. This proportion was then multiplied by each measurement to provide error bars on each flux data point. Assuming a one sigma error of ± 50 years on the calibrated basal age, the flux data are subject to an error of ± 19% on the basis of the sedimentation rate term alone. The errors on the other parameters are in the range of ± 1% of each measurement.

5. Environmental Proxies: Depth and Time

The environmental proxies measured from 3I and 3H were plotted against depth and age in the box core and compared with variations in the sea-ice index in progress by Ogilvie and Jónsdóttir (1996), which extends back to A.D. 1600 (Figure 4). The sea-ice index shows an interval between A.D. 1780 and 1920 during which sea-ice conditions were probably more severe relative to earlier and later times (Figure 4). A short interval of lower sea-ice incidence occurs within the high sea-ice interval between A.D. 1840 and 1860 (Ogilvie and Jónsdóttir, 1996; Jónsdóttir and Ogilvie, 1997). The changes that occur in many of the environmental proxies at 20 cm corresponds with the beginning of the high sea-ice interval ca. A.D. 1780. Using the two sigma errors on the calibrated basal age, and extending this error envelope towards the core top age of 1993, the two sigma error envelope at 20 cm shows a mean age of A.D. 1780, with a two sigma error range of A.D. 1745 to 1820 (Figure 4).

5.1. CARBON AND SEDIMENT MAGNETIC DATA: PROXIES FOR SOIL EROSION

Total organic carbon percentages rise from values between 0.42 and 0.70 % in the lower part of the core to values as high as 0.97 % in the upper part of the core; the change in TOC percentages begins by 21 cm (Figure 4). Rock-Eval data clearly indicate that the organic matter has a terrigenous origin ("kerogen type III"; Figure 3), and that significant amounts of marine organic matter are not preserved in the sediments. The Hydrogen Index (mgHC/gC) values largely vary between 50 and 100 throughout the core (Table 1). The high Oxygen Index (mgCO2/gC) values (Figure 3; Table 1) are probably related to a high amount of labile polysaccharides typical for recent marine sediments (Liebezeit and Wiesner, 1990). The Tmax-values of < 400 °C (Table 1) clearly indicate that the organic matter is immature, bearing no reworked ancient material.

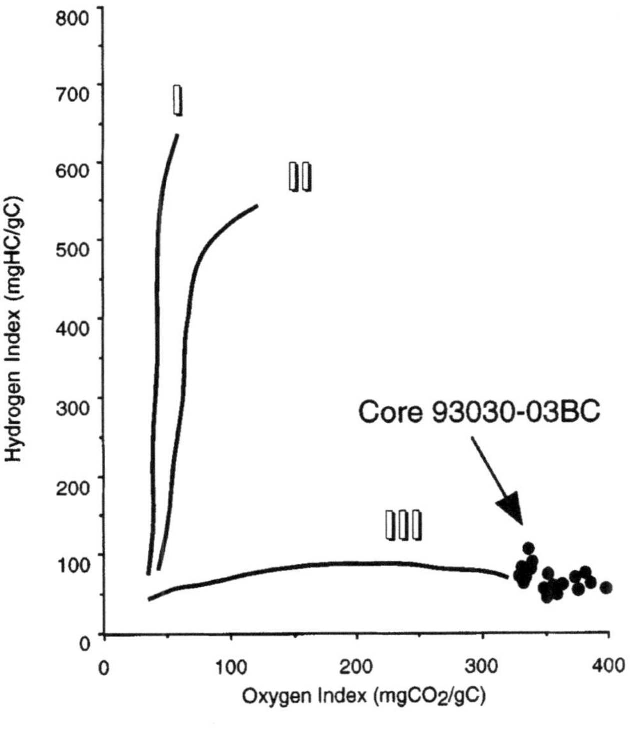

Figure 3. Hydrogen index versus oxygen index ("van Krevelen type") diagram of Core 93030-03BC. Roman numbers mark different kerogen types: I and II marine, III terrigenous organic matter (classification after Espitalié et al., 1977). The samples from box core 93030-03BC (black dots) clearly fall within the range of kerogen type III, terrigenous organic matter.

TABLE I

Rock-Eval Data from 93030-03H BC

Depth (cm)	Tmax (°C)	Hydrogen Index (mgHC/gC)	Oxygen Index (mgCO2/gC)
1.00	396	62	386
6.00	381	80	338
8.50	388	72	329
11.00	401	60	363
13.50	399	53	375
16.00	395	59	356
18.50	389	55	348
21.00	388	70	373
23.50	373	106	336
26.00	388	53	376
28.50	389	63	332
30.50	392	55	398
35.50	390	48	359
38.00	392	83	621
40.50	371	89	339
43.00	396	75	381
45.50	401	74	352
48.00	391	83	331
50.50	398	70	334
53.00	392	44	350

The terrestrial carbon flux (Figure 4) is subject to large one sigma error bars because of the one sigma error on the calibrated basal radiocarbon date. When plotted with the one sigma error bars, the carbon flux data points overlap one another, which, strictly speaking, suggests that there is no difference in the carbon flux through time. However, visually, there is a trend toward higher fluxes during the interval of severe sea ice because there are successive points with increased carbon flux relative to the pre 1780 data points.

The sediment magnetic variables that indicate the type of magnetic materials (S-parameter and HIRM) and the magnetic grain size (Xarm/X) in the sediment show a change starting at 24 to 26 cm (two sigma age range between A.D. 1670 to 1785) (Figure 4). The change from values of the S-parameter of > 0.91 to values < 0.91 indicates a small increase in hematite relative to magnetite upward in the core. A corresponding increase in HIRM values from

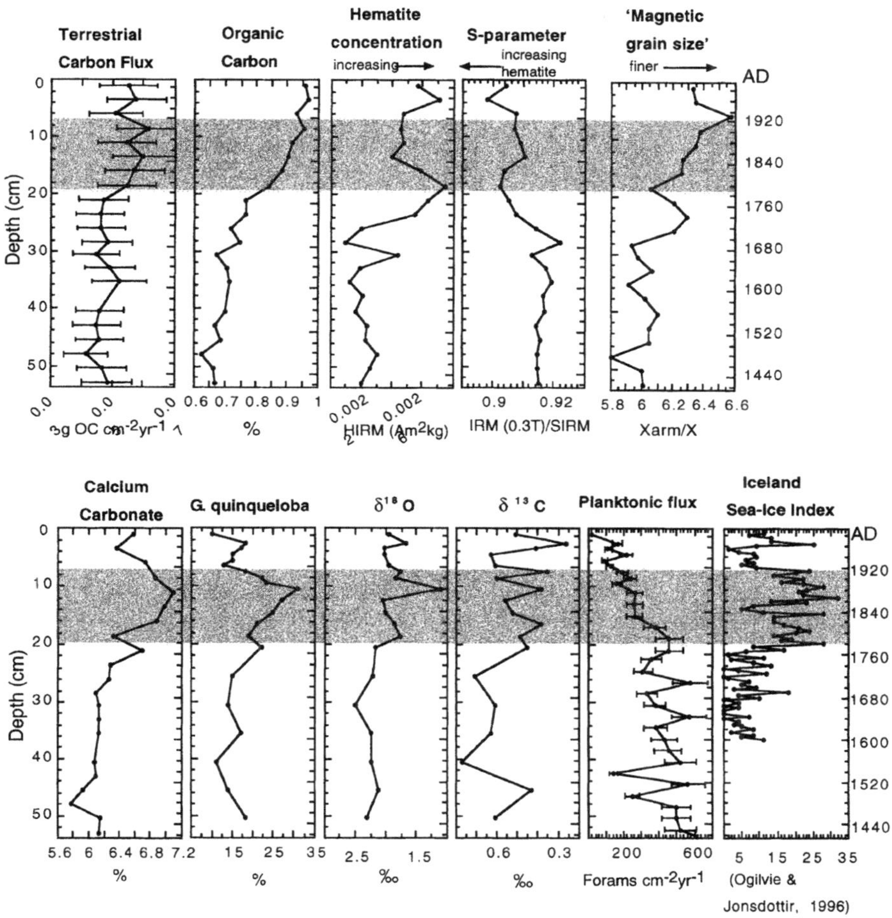

Figure 4. Soil erosion proxies (upper) and palaeoceanography proxies (lower) against depth in the core and mean calibrated age. The gray band represents the time period 1780 to 1920 when sea-ice conditions were severe as shown by the sea-ice index (Ogilvie and Jónsdottir, 1996). The two sigma error band of the age of the onset of changes in magnetic parameters at 24 cm is estimated to be A.D. 1692 to 1785, and the change in the other parameters at 20 cm is estimated to be between A.D. 1740 and 1820.

0.0023 to > 0.0026 suggests an increase in hematite concentration at the same level. Higher values of Xarm/X above 26 cm suggest a finer magnetic grain size in the top part of the core. The Xarm/X rises over this interval indicating finer magnetic grain size coinciding with the increase in hematite. This observation can be attributed to the tendency for hematite to occur as single domain grains or as coatings on sediment grains. Although these changes are exceedingly small, they are consistent among the variables and their timing agrees well with changes observed in the other chemical and biological proxies.

The sediment magnetic data indicate that, although magnetite is consistently the dominant magnetic mineral, more hematite was entering the marine environment beginning slightly before the increase in sea-ice incidence and continuing to the present. On Iceland, hematite occurs as a primary mineral in basalts, and as a detrital mineral from erosion of basaltic bedrock, but it also may be concentrated in soils by chemical weathering of magnetite to hematite. In combination with the increase in terrestrial carbon, the small increase in hematite in the core is consistent with erosion of soils on Iceland. The carbon and magnetic data taken together suggest an increase in the delivery of soil particles to the marine environment during the high sea-ice interval and continuing to the present.

5.2. STABLE ISOTOPE RATIOS AND FORAMINIFERAL DATA: PROXIES FOR OCEANOGRAPHIC CONDITIONS

Stable isotope analyses in subcore 3I have been presented in Hagen (1995) and are compared in this paper with additional environmental proxies (Figure 4). The new C and O stable isotope analyses on subcore 3H were offset systematically to lighter values relative to the original values in subcore 3I by ca. 0.3 per mil in $\delta^{13}C$ and by ca. 0.5 per mil in $\delta^{18}O$. Although the trends are similar, we do not use the stable isotope analyses from subcore 3H because it is unclear why, and by exactly how much, the values are offset.

The $\delta^{18}O$ values on *N. pachyderma* dextral in subcore 3I show a small variation around a mean value of 2.0 per mil, with a shift from slightly heavier values below 20 cm to ca. 0.3 per mil lighter values at about 20 cm (Hagen, 1995). The lightest isotope value of 1.1 per mil occurred at 10 cm. Values become heavier again above 7 cm (ca. A.D. 1920), corresponding with the end of the severe sea-ice interval, but do not return to the pre-sea-ice values. Although Hagen (1995) interpreted the shift to lighter values in terms of a 1 to 2 °C warming of surface waters, the shift can also be expained by a decrease in salinity. A light peak at 2.5 cm (A.D. 1965) corresponds in time with the Great Salinity Anomaly and with a high sea-ice index (Figure 4).

The $\delta^{13}C$ values on *N. pachyderma* dextral vary around a value of 0.5 per mil throughout subcore 3I, with a slight shift to lighter values at 20 cm (Figure 4). Hagen (1995) suggested that a decrease in productivity could generate this shift. A shift to the lightest $\delta^{13}C$ values in the core occurs at the 2.5 cm level (ca. A.D. 1965).

The planktonic foraminiferal assemblage data in subcore 3I were presented in Hagen (1995). The faunas change very little with depth, except for the significant rise in the percentages of *Globigerina quinqueloba* beginning at 20 cm and peaking at 10 cm, coinciding with the lightest $\delta^{18}O$ value (Figure

4). Percentages of *G. quinqueloba* decrease over the interval from 10 to 6 cm and rise to a low peak at 2 cm, corresponding with the peaks in $\delta^{18}O$ and $\delta^{13}C$. *G. quinqueloba* occurs commonly in large numbers at the Arctic and Polar fronts (Johannesen et al., 1994). The summer palaeotemperature reconstruction from these data shows little variation, with temperatures ranging between 11.5 and 12.0°C, and exhibiting no trend, suggesting that Arctic waters did not replace Atlantic waters at the site.

Planktonic foraminiferal flux declines from values as high as 700 $cm^{-2}yr^{-1}$ at the base of the core to a low value of 20 $cm^{-2}yr^{-1}$ at 1 cm depth in the core. The decline in flux is most obvious above 19 cm. The planktonic flux decreases throughout the interval corresponding to the high sea-ice period and continues to decline to the present, never recovering to earlier values (Figure 4). Benthic foraminiferal fluxes parallel those of the planktonic foraminifera, decreasing throughout the core, but more dramatically above 19 cm.

The calcium carbonate percentages are slightly elevated between 20 cm and 6 cm, coinciding fairly well with the interval of high sea ice between c.1780 and 1920. This situation was surprising considering the decrease in the planktonic and benthic foram fluxes over this same interval. The increase in calcium carbonate percentages must occur in the <125 μm size fraction, and could be accounted for by coccoliths.

6. Discussion and Conclusions

In order to reconstruct the environmental history on the southwest Iceland shelf during the last five centuries, two groups of environmental proxies have been analysed. One group represents changes in the oceanographic conditions on the shelf, whereas the second group provides information on erosion of soil particles from Iceland. The two groups of proxies show roughly synchronous changes that coincide with the interval of the increased sea-ice influx around Iceland between c. A.D. 1780 and 1920 determined from historical documentary data (Ogilvie and Jónsdóttir, 1996).

The shifts in the stable O and C isotope ratios and the rise in the percentages of *G. quinqueloba* all coincide with the onset of severe sea-ice conditions ca. 1780 (Figure 4). However, *G. quinqueloba* is the only parameter that shows a return to lower values after the end of the sea-ice interval. We interpret the lighter oxygen isotope ratios over the interval of more severe sea ice to freshening of the surface water rather than to sea surface warming. The rise in *G. quinqueloba* percentages is consistent with proximity of the Arctic front carrying sea ice and/or freshwater near to the site, either from the northwest, or around the island from the south. The faunas continue to reflect Atlantic water (Johannessen et al., 1994). The carbon isotope ratios becoming slightly lighter over the same interval may be attributed to decreased productivity, as suggested by Hagen (1995). Alternatively, or, in addition, increased sea ice in the region possibly would result in a reduced exchange between the surface water (where the planktonic foraminifers live) and the atmosphere, causing the lighter ^{13}C values (Johannessen et al., 1994). An overall decrease in foraminiferal production coincident with the severe sea-ice interval is indicated by the decline in foraminiferal flux, but the slight rise in calcium carbonate percentages may be accounted for in the finer sediment size fractions, perhaps by increased abundances of coccoliths, possibly concentrated by processes at the ocean front. On the southwest Iceland shelf, the core data and historical observations suggest that the ocean front and ice edge were nearby, but did not pass over the site.

Between 2 and 3 cm in the core (c. A.D. 1965), the trend in several of the palaeoceanography proxies is very similar to the trends that occur during the more severe sea-ice interval (Figure 4). The lightening in the $\delta^{18}O$, and the $\delta^{13}C$, and the rise in *G. quinqueloba* percentages appear to correspond closely in timing to the interval of high sea-ice incidence during the Great Salinity Anomaly. Although the record lacks the resolution to prove this association, it is likely that the proxies would respond similarly to the Great Salinity Anomaly as to times of more frequent sea-ice influxes.

The timing of the so-called “Little Ice Age”, represented in this paper by the period between A.D. 1780 and 1920, is often considered an important test for palaeoceanographic proxy data. This is because it is generally a small signal of cooling or decreased salinity relative to deglaciation. It appears from this data set that past changes in the sea-ice regime around Iceland of similar magnitude could be readily ascertained from analysis of Holocene sediments from this area.

The synchronicity of increased soil-particle delivery to the shelf and fresher marine conditions does not prove that increased soil erosion on Iceland is directly linked with climatic deterioration. However, it is possible that soil erosion underway through anthropogenic processes since the time of Settlement may have been exacerbated by the cooling linked with the increased sea-ice incidence.

Soil erosion on Iceland, mainly by wind, but also by rivers and slope wash, has been a serious problem ever since Settlement (e.g. Thórarinsson, 1961; Dugmore and Buckland, 1991; Dugmore and Erskine, 1994). The primary causes of soil erosion are generally agreed to be anthropogenic, with secondary climatic and volcanic forces also playing a role (Thórarinsson, 1961; Arnalds, 1987; Gerrard, 1991). Of the anthropogenic influences, livestock grazing is the most important in Iceland. The effects of grazing are negatively impacted further by the influences of the harsh climate and fragile volcanic soils. The climate of Iceland is highly marginal for agriculture. It has been suggested, for example, that it may only take a 1°C decrease in air temperature on Iceland, such as is likely to occur in a severe sea-ice year, to decrease the carrying capacity of the rangeland by 10 to 20%. This is because both the length of the grazing season and the vegetation yield is decreased (Bergthórsson et al., 1988). Overgrazing and grazing on marginal lands that are more prone to erosion occur during times of limited forage. Overgrazing results in increased soil erosion, especially in northern and highland ranges (Bergthórsson et al., 1988).

Dryland soils, the coarse-textured volcanic loess, are particulary prone to erosion by degradation of vegetation through the interaction of grazing with climate and volcanic activity. During cold spells when the mat of protective vegetation over the volcanic loess is disturbed, strong winds can disturb the soil, desiccating and freeze-drying it with cold winds. During this process, the finer organic particles are blown into the sea and carried out of the land system. Once started, the process of dryland soil erosion is difficult to reverse (Arnalds, 1987). The dryland soil erosion is particularly active in the south and northeast of Iceland.

Studies of changes in soil erosion with time indicate that the beginning of large-scale soil erosion on Iceland coincides with Settlement in the late ninth century (e.g. Thórarinsson, 1961; Dugmore and Buckland, 1991). A proposed decrease in vegetation cover from 65 to 25% is thought to have occurred with human impacts (Bjarnason, 1978; Arnalds, 1987). Since the first Settlement of Iceland, there is estimated to have been a 10-fold increase in the rate of aeolian sedimentation, a decrease in woody plant species, and an increase in grasses and sedges (Arnalds, 1987; Hallsdóttir, 1987). Furthermore, soil erosion increased during the eighteenth century, reaching a maximum in the nineteenth and early twentieth century when farms were abandoned because of drifting sand related to increased dryland soil erosion, especially in the south and northeast of Iceland (Runólfsson, 1987).

Dugmore and Erskine (1994) applied tephrochronology to soil profiles in order to study in detail the patterns of soil erosion in southern Iceland from the Settlement to 1947. These deposits provide a record of soil mobility, indicating times when wind erosion was active. Dugmore and Erskine's (1994) record shows that rates of soil erosion increase at all elevations at the time of the Settlement and, in general, continue to increase to the present. In the low country, less than 200m above sea level, the rates of erosion peak dramatically after 1823. Dugmore and Buckland (1991) interpret this pattern as the culmination of an erosion process that was initiated at Settlement and propagated downhill to lower elevation sites, possibly accelerated by a "Little Ice Age" climatic deterioration.

The carbon flux data in the boxcore show that soil particles were shed into the sea from Iceland land areas during the entire time period represented by the core, which, unfortunately does not extend back to the time of the Settlement. However, the data in the box core suggest that since ca. 1400, the period of greatest organic carbon flux occurred during the cold interval from 1780 to 1920. In the case of the sediment magnetic parameters, the change occurred slightly earlier, ca. 1740. The oceanographic proxies in the core suggest that between 1780 and 1920, Faxaflói was commonly influenced by the advancing sea-ice edge. During this time, the erosion of soil on Iceland may have been exacerbated by the climatic cooling. Several factors may have operated together to promote the soil erosion. Grazing on sensitive, erosion-prone areas may have become more intense as the vegetation yield was reduced and the biomass production was exceeded by the grazing animals. Trampling of the vegetation would also have been a contributing factor. Once the vegetation mat was disturbed, it would not easily have recovered, accelerating the erosion. The wind would have moved the coarser material in traction deposits, while it carried the soil particles out to sea as dust that was deposited on the continental shelf. Such processes would also have led to increased delivery of soil particles to the sea by rivers. After ca. 1920 the proxies for soil erosion do not return to previous values. Terrestrial carbon flux declines slightly, but the sediment magnetic parameters do not change, which is consistent with continuing soil erosion problems on Iceland today (Runólfsson, 1987).

Acknowledgements

We thank the Captain and the crew of the *CSS Hudson* for their cooperation during cruise 93030. This research was supported by grants from the US National Science Foundation (ATM-9531397 and OPP-9726510). The manuscript benefited from helpful reviews by Dr A. J. Dugmore and Dr C. J. Caseldine and by discussions with Dr C. Alexander. This is PALE contribution number 136.

References

Arnalds, A.: 1987, 'Ecosystem disturbance in Iceland', *Arctic and Alpine Research* **19**, 508-513.

Asprey, K.W., Syvitski, J.P.M., Andrews, J.T., and Dowdeswell, J.A.: 1994, 'CANAM-PONAM Cruise HU93030: West Iceland to East Greenland', *Geol. Surv. Can. Open File* 2824, 152 pp.

Bergthórsson, P.: 1969, 'An estimate of drift ice and temperature in Iceland in 1000 years'. *Jökull* **19**, 94-101.

Bergthórsson, P., Björnsson, H., Dýrmundsson, Ó., Gudmundsson, B., Helgadóttir, Á., and Jónmundsson, J.V.: 1988, 'The effects of climatic variations on agriculture in Iceland', in Parry, M.L., Carter, T.R., and Konijn N.T., (eds), *The Impact of Climatic Variations on Agriculture, Volume 1: Assessments in Cool Temperate and Cold Regions*. Kluwer Academic Publishers, Dordrecht, Holland, pp. 383-509.

Bjarnason, Á, H.: 1978, 'Erosion, tree growth and land regeneration in Iceland', in Holdgate, M. W. and Woodman, M. J. (eds), *The breakdown and restoration of ecosystems*, Plenum, London, pp. 241-248.

Caseldine, C.J.: 1987, 'Neoglacial glacier variations in northern Iceland: Examples from the Eyjafjördur area', *Arctic and Alpine Research* **19**, 296-304.

Dugmore, A.J. and Buckland, P.C.: 1991, 'Tephrochronology and late Holocene soil erosion', in Maizels, J.K. and Caseldine, C. (eds), *Environmental Change in Iceland: Past and Present*. Kluwer Academic Publishers, Dordrecht, pp. 37-55.

Dugmore, A.J. and Erskine, C.C.: 1994, 'Local and regional patterns of soil erosion in Southern Iceland', in Stötter, J. and Wilhelm, F. (eds), *Environmental Change in Iceland*, Münchener Geographische Abhandlungen, Reihe B., pp. 63-78.

Espitalié, J., Laporte, J.L., Madec, M., Marquis, F., Leplat, P., Paulet, J., and Boutefeu, A.: 1977, 'Méthode rapide de charactérisation des roches mәres, de leur potentiel pétrolier et de leur degré d'évolution'. *Rev. Inst. Franç. Pétrol* **32**, 23-42.

Fahl, K. and Stein, R.: 1997, 'Modern organic-carbon-deposition in the Laptev Sea and the adjacent continental slope: Surface-water productivity vs. terrigenous supply', *Org. Geochem*. **26**, 379-390.

Gerrard, J.: 1991, 'An assessment of some of the factors involved in recent landscape change in Iceland', in: Maizels, J.K. and Caseldine, C., (eds), *Environmental Change in Iceland: Past and Present*. Kluwer Academic Publishers, Dordrecht, pp. 237-253.

Guðmundsson, H.: 1997, 'A review of the Holocene environmental history of Iceland'. *Quaternary Science Reviews* **16**, 81-92.

Hagen, S.: 1995, *Watermass characteristics and climate in the Nordic Seas during the last 10,200 years: based on paleoceanographic proxies retrieved from sediment cores from the continental shelves off northern Norway and southwestern Iceland.* Candidatus Scientiarium Thesis, University of Tromsø, 114 pp.

Hallsdóttir, M.: 1987, *Pollen analytical studies of human influence on vegetation in relation to the Landnám tephra layer in southwest Iceland.* Lundqua thesis, volume 18, Lund University, Department of Quaternary Geology, 45 pp.

Harðardóttir, J., Geirsdóttir, Á., and Thórdarson, Th.: in press, 'Tephra layers in a sediment core from Lake Hestvatn, S-Iceland: implications for evaluating sedimentation processes and environmental impact of tephra fall deposits in the surrounding watershed', in White, J. and Riggs, N., (eds), *Lacustrine Volcaniclastic Sedimentation.* International Association of Sedimentologists (IAS) Special Publication. Blackwell, Oxford.

Helgadóttir, G.: 1984, *Senkvartaere Foraminifer og Sedimenter i Faxafloi-Jokuldjupomradet Vest for Island,* PhD thesis, University of Oslo, pp. 116 (plus appendices).

Jennings, A.E., Conklin, A.R., and Cooper, T.A.: 1994, 'Living and total benthic foraminifera from box cores across the East Greenland Front', *PaleoBios* **16** (supplement to No. 2), 38.

Jennings, A. E., Syvitski, J.P.M., Gerson, L., Grönvold, K., Geirsdóttir, Á., Harðardóttir, J., Andrews, J.T., and Hagen, S.: in revision, Chronology and paleoenvironments during the late Weichselian deglaciation of the SW Iceland shelf. *Boreas.*

Johannessen, T.: 1987, 'Resente planktoniske foraminiferer fra Norskehavet, Islandshavet og Nord-Atlanteren: Taksonomi, faunafordeling og stabil-isotopsammensetning'. Candidatus Scientiarium Thesis, University of Bergen.

Johannessen, T., Jansen, E., Flatøy, A., and Ravelo, A.C.: 1994, 'The relationship between surface water masses, oceaongraphic fronts and paleoclimatic proxies in surface sediments of the Greenland, Iceland, and Norwegian Seas', in Zahn, R. et al, (eds), NATO ASI Series, Vol I17, *Carbon Cycling in the Glacial Ocean: Constraints on the Ocean's Role in Global Change.* Springer Verlag, Berlin, Heidelberg.

Jónsdóttir, I. and Ogilvie, A. E. J.: 1997, 'Sea ice off the coasts of Iceland in historical times', *27th International Arctic Workshop, Program and Abstracts* 27 February-2 March, 1997. University of Ottawa, Canada, pp. 124-125

King, J.W. and Channell, J.E.T.: 1991, 'Sedimentary magnetism, environmental magnetism, and magnetostratigraphy', *Reviews of Geophysics Supplement,* U.S. National Report to International Union of Geodesy and Geophysics 1987-1990, American Geophysical Union, p. 358-370.

Koch, L.: 1945, 'The East Greenland Ice'. *Meddelelser om Grønland,* **130**. Copenhagen.

Liebezeit, G. and Wiesner, M.G.: 1990, 'Pyrolysis of recent marine sediments - I. Biopolymers', *Org. Geochemistry* **16**, 1179-1185.

Malmberg, S-A.: 1985, 'The water masses between Iceland and Greenland', *Journal of the Marine Research Institute* **9**, 127-140.

Mosby, H.: 1968, 'Surrounding Seas', in Sømme, A., (ed.), *A Geography of Norden,* pp. 18-26, map 7, J.W. Cappelen forlag, Oslo.

Ogilvie, A.E.J.: 1982, *Climate and Society in Iceland from the Medieval Period to the Late Eighteenth Century.* PhD thesis, School of Environmental Sciences, University of East Anglia, UK, 504 pp.

Ogilvie, A. E. J.: 1984a, 'The impact of climate on grass growth and hay yield in Iceland: A. D. 1601 to 1780', in Morner, N.-A. and Karlén, W., (eds), *Climate Changes on a Yearly to Millennial Basis,* Reidel, Dordrecht, pp. 343-352.

Ogilvie, A.E.J.: 1984b, 'The past climate and sea-ice record from Iceland, Part 1. Data to A.D. 1780'. *Climatic Change* **6**, 131-152.

Ogilvie, A. E. J.: 1991, 'Climatic changes in Iceland A. D. c. 1500 to 1598', in *The Norse of the North Atlantic* (Presented by G. F. Bigelow). *Acta Archaeologica* **61**-1990, Munksgaard, Copenhagen, pp. 233-251.

Ogilvie, A. E. J.: 1992, 'Documentary evidence for changes in the climate of Iceland, A. D. 1500 to 1800'. in Bradley, R.S. and Jones, P.D., (eds), *Climate Since A. D. 1500* Routledge, London and New York, pp. 92-117.

Ogilvie, A. E. J.: 1996, 'Sea-ice conditions off the coasts of Iceland A. D. 1601-1850 with special reference to part of the Maunder Minimum period (1675-1715)'. *AmS-Varia* **25**, Archaeological Museum of Stavanger, Norway, pp. 9-12.

Ogilvie, A. E. J.: 2000, 'Climate and Farming in Northern Iceland, ca. 1700-1850', in Sigurðsson, I. and Skaptason, J. (eds), *Aspects of the History of the Arctic and Sub-Arctic Region*. The University Press, Reykjavík, In press.

Ogilvie, A. E. J. and Jónsdóttir, I.: 1996, 'Sea-ice incidence off the coasts of Iceland: Evidence from historical data and early sea-ice maps', *26th International Arctic Workshop, Arctic and Alpine Environments, Past and Present Program with Abstracts* INSTAAR, Boulder, Colorado, 14-16 March, 1996, 109-110.

Runólfsson, S.: 1987, 'Land reclamation in Iceland', *Arctic and Alpine Research* **19**, 514-517.

Stein, R.: 1991, 'Accumulation of Organic Carbon in Marine Sediments', *Lecture Notes in Earth Sciences* **34**, Springer, Heidelberg, 217pp.

Stuiver, M. and Becker, B.: 1993, 'High-precision calibration of the radiocarbon time scale, A.D. 1950-6000 BC', *Radiocarbon* **35**, 35-65.

Stuiver, M. and Braziunas, T.F.: 1993, 'Modeling atmospheric 14C influences and 14C ages of marine samples to 10,000 BC', *Radiocarbon* **35**, 137-189.

Stuiver, M. and Reimer, P.J., 1993, 'Radiocarbon calibration program Rev 3.3.3A - Mac Test Version #6', *Radiocarbon* **35**, 215-230.

Sigtryggsson, H.: 1972, 'An outline of sea-ice conditions in the vicinity of Iceland'. *Jökull* **22**, 1-11.

Syvitski, J.P., Jennings, A.E. and Andrews, J.T.: 1999, 'High resolution seismic evidence for multiple glaciation across the southwest Iceland shelf'. *Arctic, Antarctic, and Alpine Research* **31**, 50-57.

Thórarinsson, S.: 1961, 'Uppblástur á Íslandi í ljósi öskulagarannsókna', *Ársrit Skógræktarfélags Íslands*, pp. 17-55.

Tissot, B.P. and Welte, D.H.: 1984, *Petroleum Formation and Occurrence*, 2nd ed., Springer Verlag Heidelberg, 699 pp.

Vilmundarson, Th.: 1972, 'Evaluation of historical sources on sea ice near Iceland', in Karlsson, T., (ed.), *Sea Ice, Proceedings of an International Conference*, National Research Council, Reykjavik, pp. 159-169.

(Received 22 September 1998; in revised form 28 February 2000)

THE TIME PERIOD A.D. 1400-1980 IN CENTRAL GREENLAND ICE CORES IN RELATION TO THE NORTH ATLANTIC SECTOR

L. K. BARLOW

INSTAAR, University of Colorado, Campus Box 450, Boulder, Colorado, 80309-0450, USA

Abstract. This paper presents a review of the time period A.D. 1400-1980 based on Greenland ice cores from the central west Greenland averaged record, and from winter and summer seasonal isotopic records from the Greenland Ice Sheet Project 2 (GISP2). This time period includes the so-called "Little Ice Age". The concept of the "Little Ice Age" has evolved from the idea of a simple, centuries-long period of lower temperatures to a more complex view of temporal and spatial climatic variability. In the central Greenland ice core isotopic signals, the fifteenth and early sixteenth centuries show multi-decadal excursions above and below the mean reference. The sixteenth and mid-eighteenth to mid-nineteenth centuries are notable for decade-to-decade swings (high-low) in the isotopic signal, while multi-decadal low excursions dominate the seventeenth century. The "subdued" nature of the "Little Ice Age" isotopic signal in central Greenland is probably influenced by the North Atlantic Oscillation (NAO), which presents opposing temperature excursions between west Greenland and northern Europe. Changes in the prevailing atmospheric circulation (Iceland Low) can explain some of the spatial and temporal variability between the central Greenland isotopic records and Iceland temperature.

Key Words: Greenland; ice cores; isotopes; "Little Ice Age"

1. Introduction

Although the original definition of the "Little Ice Age" was based on the advance of valley glaciers and referred to the last 4,000 years (see Ogilvie and Jónsson, this volume; J. M. Grove, this volume), it became popular to interpret century-scale proxy evidence of low temperature in the last 1000 years as a "Little Ice Age" signal. The search for a "Little Ice Age" in Greenland ice core isotopic signals ($\delta^{18}O$, δD) has generally used the approach of trying to fit the ice core record to time periods established from western European studies. Dansgaard et al. (1975), noted that 'the cold period from 1450 to 1700 (*sic*) often called the "Little Ice Age" was not particularly cold at Crête or ...(at) Milcent' (Figure 1). When interpreting the isotopic signal as a century-scale proxy temperature record, these cores suggest that Greenland experienced the lowest temperature of the last thousand years during the fourteenth century. Dansgaard et al. (1975) concluded that long-term (century-scale) changes in central Greenland are out of phase with changes in Europe and North America, possibly by a lead of 250 years. The more recent Greenland Ice Core Project (GRIP) and Greenland Ice Sheet Project 2 (GISP2; Figure 1) isotopic signals reconfirm that the fourteenth century is the lowest isotopically in the last 1000 years, and that a century-scale "Little Ice

Climatic Change **48**: 101–119, 2001.

Age" signal is weak in central Greenland (Johnsen et al., 1997). Stuiver et al. (1995) define a "Little Ice Age" period on the century time scale between

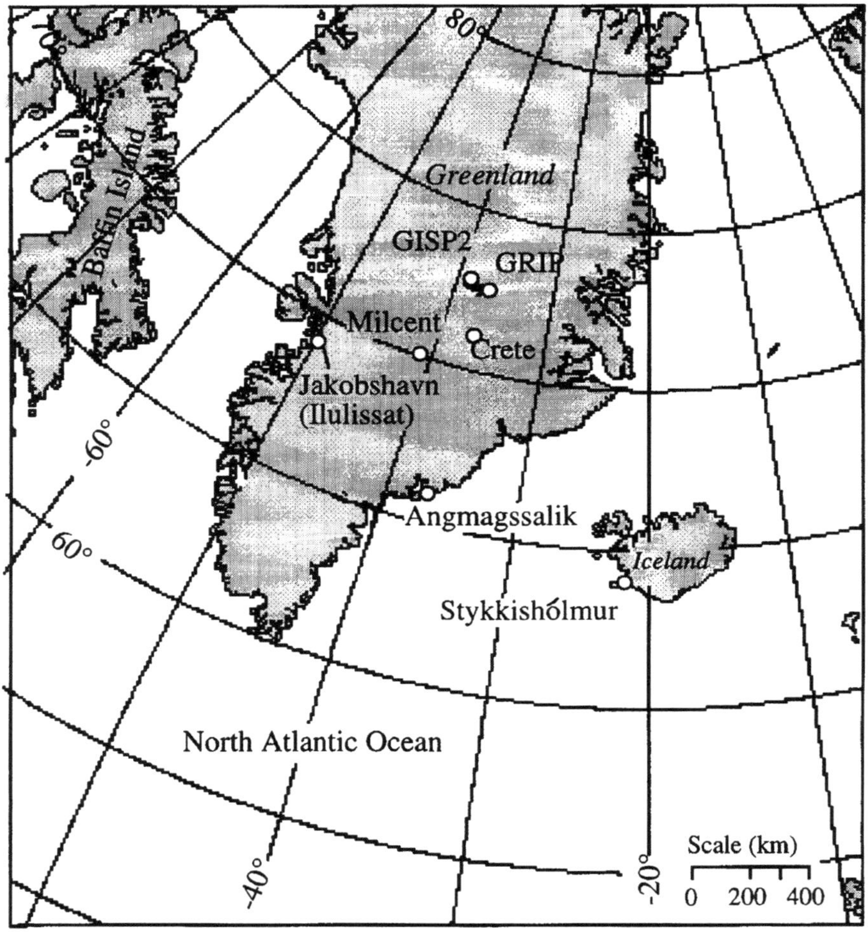

Figure 1. Location map of ice core and coastal temperature sites discussed in this paper. The central west Greenland isotopic record (CWG; Fisher et al., 1996) is an average of five records from the GRIP site, one from GISP2, one from Milcent, and five short and one long record from the Crête site. Coastal temperature records mentioned in the paper are Jakobshavn (Illulissat), Angmagssalik, and Stykkishólmur.

A.D. 1350 and 1800 in the GISP2 core, with the "Medieval Warm Period" ending in 1350. It is not clear why a distinct negative excursion in bidecadal isotopic values between 1320 and 1360 was not included in the "Little Ice Age" time period (see Stuiver et al., 1995). Fischer et al. (1998) define isotopically low excursions in North Greenland ice cores during the seventeenth and first half of the nineteenth centuries as a "clear" record of the "Little Ice Age". Fischer's North Greenland cores do not extend to the 1300s.

A bore-hole temperature profile from the GISP2 site showed lower temperatures between A.D. 1500 and 1900 (Alley and Koci, 1990), although the bore-hole was not long enough to incorporate all of the last 1000 years. Dahl-Jensen et al. (1998) defined the "Little Ice Age" from A.D. 1550 to 1850 based on combined bore-hole temperature profiles from the GRIP and Dye 3 sites. Bore-hole temperature represents a low resolution measurement, and temperature fluctuations are higher in frequency and probably larger than obtainable by this method (Alley and Koci, 1990). Two chemical parameters in the GISP2 ice core have been applied to the "Little Ice Age" question (Mayewski et al., 1993). An increase in calcium, which implies more intense atmospheric circulation over continental regions and/or the presence of more easily errodable terrain, is detected, beginning in the 1390s and ending in the 1840s. An increase in sea-salt sodium suggests a period of increased storminess in the marine environment from the 1580s to the 1910s. Through these parameters, Mayewski, and later work by Kreutz et al. (1997) define the "Little Ice Age" as a change in intensity of meridional atmospheric circulation that began around A.D. 1400 and continues to this day.

As early as the 1980s, researchers were beginning to realize that the concept of a North Atlantic-wide centuries-long cold temperature anomaly was not supported by the data (e.g. Williams and Wigley, 1983). Most Arctic paleoclimatologists are now in agreement with this conclusion; rather, the last 500-600 years are characterized by decadal-scale periods of relatively lower temperature interspersed with periods of relatively higher temperature, and that spatial and temporal differences exist in the data (e.g. Bradley and Jones, 1992; Ogilvie, 1992; Meese et al., 1994; Overpeck et al., 1997). This newer view comes partially out of practical necessity, as the uncertainty of impacts of future climatic change on humankind requires a focus on palaeoclimatic information at decadal- to subannual-scales. Further discussion in this paper, therefore, focuses on sub-century-scale anomalies over the last 500 years, possible reasons why the central Greenland isotopic signal is "subdued", and suggestions for appropriate application of the Greenland decadal- to subannual- scale ice core isotopic signals to areas outside central Greenland.

2. Central Greenland Isotopic Signals

Central Greenland provides an optimal site of high accumulation at seasonally definable resolution and extremely low incidence of snow melt. Recent investigations of subannual and annually resolved $\delta^{18}O$ and δD Greenland records confirm that there is a significant temperature signal in these records (Fischer et al., 1995; White et al., 1997; Jouzel et al., 1997; Grootes and

Stuiver, 1997). One area of active research involves the complex relationship between relative air temperature change and synoptic scale atmospheric circulation (see sections 3 and 4 below). Furthermore, given impeccable dating of individual cores, the noise associated with post-depositional processes can be reduced by averaging multiple records in close proximity (Fisher et al., 1996; Johnsen et al., 1997). The Central West Greenland (CWG) annual isotopic time series is the average of up to eleven cores from Central Greenland, including GISP2, with five of the cores located at the GRIP site on the ice divide (Fisher et al., 1996; Figure 1). Correlation between discrete decadal multi-core CWG and single core GISP2 for 1400-1900 is r= 0.64. The CWG annual signal is generally interpreted as a summer-weighted record (Fisher, 1997). However, annual central Greenland isotopic signals have also been shown to contain a statistically significant correlation with winter definition of the North Atlantic Oscillation (Barlow et al., 1993; White et al., 1997). Snow accumulation, which occurs in all seasons in the GISP2/GRIP region of central Greenland (Shuman et al., 1995), allows definition of low (i.e. lower temperature) and high (i.e. higher temperature) extremes for winter and summer isotopic signals. The CWG record provides the advantage of a higher signal/noise ratio, but lacks intra-annual timescale information, which is important for both atmospheric and historical applications. The GISP2 record provides seasonal resolution (average over three to four months), but with a lower signal/noise ratio. CWG annual and GSIP2 seasonal records for the fourteenth century are discussed in Barlow et al. (1997a).

The CWG is normalized to A.D. 553-1983. The decadal mean for the A.D. 1400-1980 interval is 0.02‰ above the reference. The decadal median for the A.D. 1400-1980 interval is 0.03‰ below the reference. For the CWG decadal time series, both mean and median values are shown in Figure 2. The median is a more conservative measure of central tendency as it is less influenced by anomalous years. The correlation between CWG decadal mean and median is r= 0.92. For the 58 decades of the CWG series, disagreement in excursion direction between mean and median values relative to the previous decade (i.e. relatively higher or lower compared to the previous decade) occurs on only 3 occasions, or 5% of the record (1430s, 1740s, 1770s). The greatest difference between decadal mean and median is 0.43‰ in the 1900s. Using a temperature conversion factor of 0.5 to 0.6‰ $\delta^{18}O/^{0}C$ (Cuffey et al., 1994; Johnsen et al., 1995), this discrepancy could mean as much as a $0.2^{0}C$ difference, with all other decadal discrepancies less than or near $0.1^{0}C$.

Fifty year running standard deviations of CWG decadal means and medians (Figure 3) show periods of greater and lesser interdecadal variability in the time series, for example, high variability in the middle sixteenth century and middle nineteenth century, and low variability surrounding the beginning of the sixteenth century and in the eighteenth century. 50 year running standard deviations of the single core GISP2 annual isotopic signal give the same time periods of higher and lower variability, with the exception of near-normal standard deviation during the mid-eighteenth century. The GISP2 winter and summer reference values are normalized over A.D. 1400-1900, to avoid seasonal bias in the near-surface record where the seasonal isotopic signal is undergoing significant vapour diffusion (Barlow, 1994). Correlation between the GISP2 summer and winter decadal records over this time period is r=0.68. Correlations between the GISP2 annual decadal record and summer and winter are r=0.91 and 0.89, respectively. Correlations between CWG and GISP2 summer and winter signals are r=0.51 and 0.58, respectively.

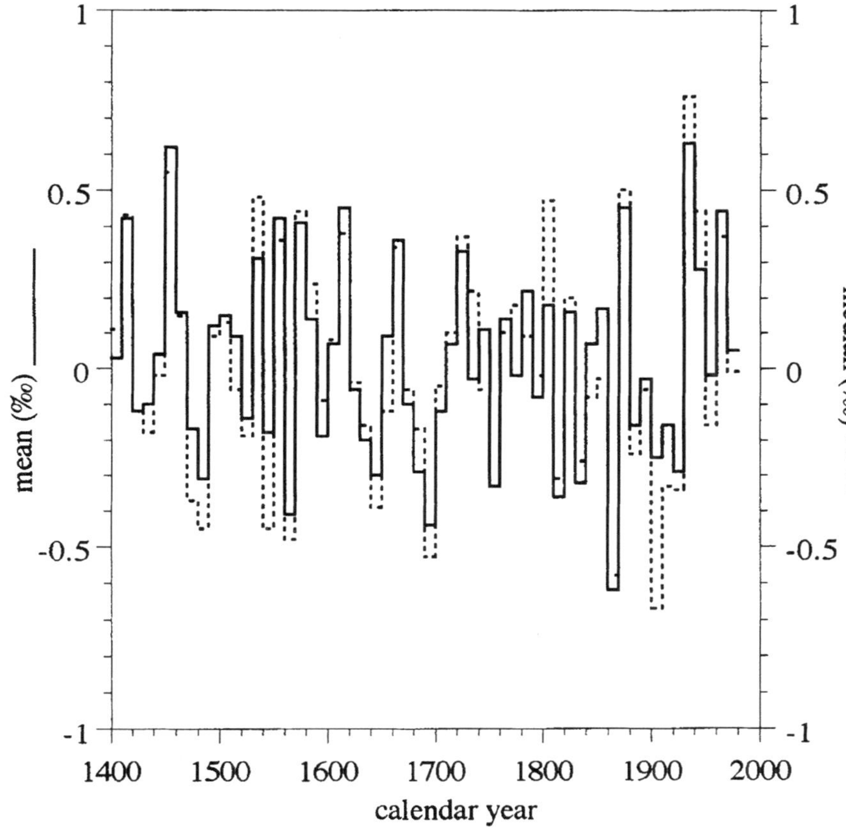

Figure 2. Normalized decadal means (solid line) and medians (dashed line) for central west Greenland isotopic signal (CWG), A.D. 1400-1900. See Figure 1 for location of individual ice core sites. Annual CWG data from D. Fisher (Fisher et al., 1996).

Overall, the decadal CWG signal shows clusters of isotopically higher or lower decades interspersed with periods of greater decade-to-decade variability (Figure 2). The 1410s appear to have experienced relatively higher than average temperature, followed by two decades of lower than average temperature and three decades of above average temperature, especially the 1450s. The 1470s and 1480s show a return to lower decadal averages, while the 1490s-1500s suggest warmer temperatures once more. Decade-to-decade variability is striking between the 1510s and the 1580s, with notably low values in the 1560s. The 1590s stand out as lower isotopically, and are flanked before and after by two decades of above normal values. The 1610s are notable for apparent warmth, followed by near-average values in the 1620s, low in the 1630s and 1640s, and a decade of above average temperature in the 1660s. Four consecutive isotopically low decades follow, from the 1670s to the 1700s, with the lowest in the 1690s. The 1720s appear to be anomalously warm in central Greenland, and the 1750s anomalously cold. The 1740s begin another period of high decade-to-decade variability that continues to the 1830s. (The 1770s CWG annual record shows six consecutive years above the reference for 1771-1776, then a negative excursion during the last three years of the decade.) The 1840s and 1850s appear as consecutive near-average decades, interpreted to be higher temperature than the surrounding 1830s and 1860s. The 1860s is the lowest decade isotopically in the whole 500 year record. The 1870s appear to be anomalously warm, followed by below average decades.

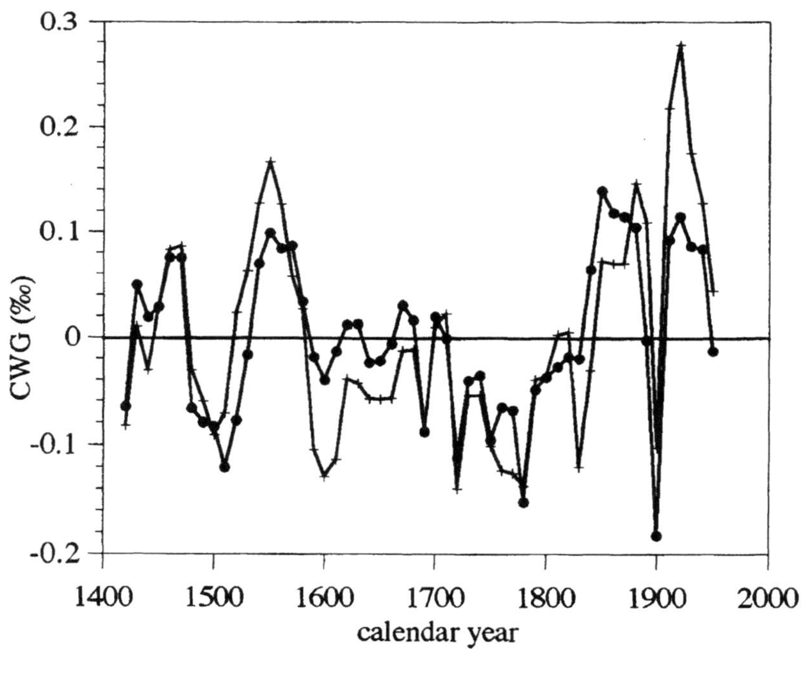

Figure 3. 50 year running standard deviations for CWG decadal means (.) and medians (+) A.D. 1400-1900.

The GISP2 decadal summer record (Figure 4) shows only three decades of above average summer warmth in the 1400s; the first decade, the 1450s, and the 1490s. Decade-to-decade swings in average isotopic value appear from the 1520s to the 1570s, corresponding with the CWG annual record. The 1570s to the 1670s, with the exception of the 1640s, are notable for their lack of low summer temperatures. Consecutive decades of lower than average values follow in the 1680s and 1690s, which are in turn followed by five decades of above average or average isotopic values. The 1750s appear to be anomalously lower in temperature, and the 1770s anomalously higher. The 1780s to 1810s show lower or average summer decades. The 1820s and 1830s are above average, the 1840s below average, and the 1870s begin three decades of above average isotopic values. One of the most striking features of the GISP2 winter isotopic record (Figure 4) is that there are no clusters of suggested lower temperature decades longer than thirty years, and that the 1650s to 1740s in particular show no extreme lower than average decades. The CWG annual, GISP2 summer and GISP2 winter signals show qualitative agreement in the 1400s, 1450s, 1490s (higher than average), and in the 1430s and 1480s (lower). Decade-to-decade swings between the 1530s and the 1570s agree in all three records. There is a striking contrast between the high signal of the 1550s and the low signal of the next decade. The 1560s to 1580s are lower than average, followed by three decades of above average warmth. The 1640s are isotopically low and the 1660s high in all three signals. The 1690s are strikingly low in the CWG and GISP2 summer series. The 1750s show agreement (low) in all three records. After this, there are no clear agreements between the three records until the 1860s, when CWG and GISP2 winters appear anomalously low, and the 1870s, when all three records suggest an unusually warm decade.

3. The Subdued "Little Ice Age" Signal in Central Greenland

The subdued nature of negative isotopic excursions in central Greenland ice cores during the last 500 years is disappointing to some researchers. Atmospheric circulation patterns suggest, at least on the inter-annual to decadal time scale, that direct correlation between central Greenland and Europe proxy temperature anomalies is not expected to be high. This lack of correlation may be linked to an important mode of natural climate variability, the North Atlantic Oscillation (NAO).

The NAO is defined as an oscillation in the mean sea-level pressure difference between the Iceland Low and the Azores High (Walker, 1924; van Loon and Rogers, 1978; Hurrell, 1995). It is linked with, among other factors, the strength of westerly winds, paths of winter storms through the North Atlantic, moisture and heat transport between the North Atlantic Ocean and adjacent continents, variability in North Atlantic sea-surface temperatures (SSTs) and sea ice extents, and the latitude of the Gulf Stream/North Atlantic Drift (see e.g., Rogers and van Loon 1979; Moses et al., 1987; Deser and Blackmon 1993; Rogers, 1997a; Hurrell and Van Loon, 1997; Serreze et al., 1997; Taylor and Stephens, 1998). Extreme modes of the NAO are mainly responsible for a "seesaw" in winter temperature between West Greenland (data from Jakobshavn; Figure 1) and northern and western Europe; lower winter temperatures in Europe correspond with higher winter temperatures in West Greenland, and vice versa (van Loon and Rogers, 1978). Although the sea-level pressure pattern of the NAO is most well defined in winter, teleconnections with the NAO are found in spring, summer, and autumn (Rogers and van Loon, 1979).

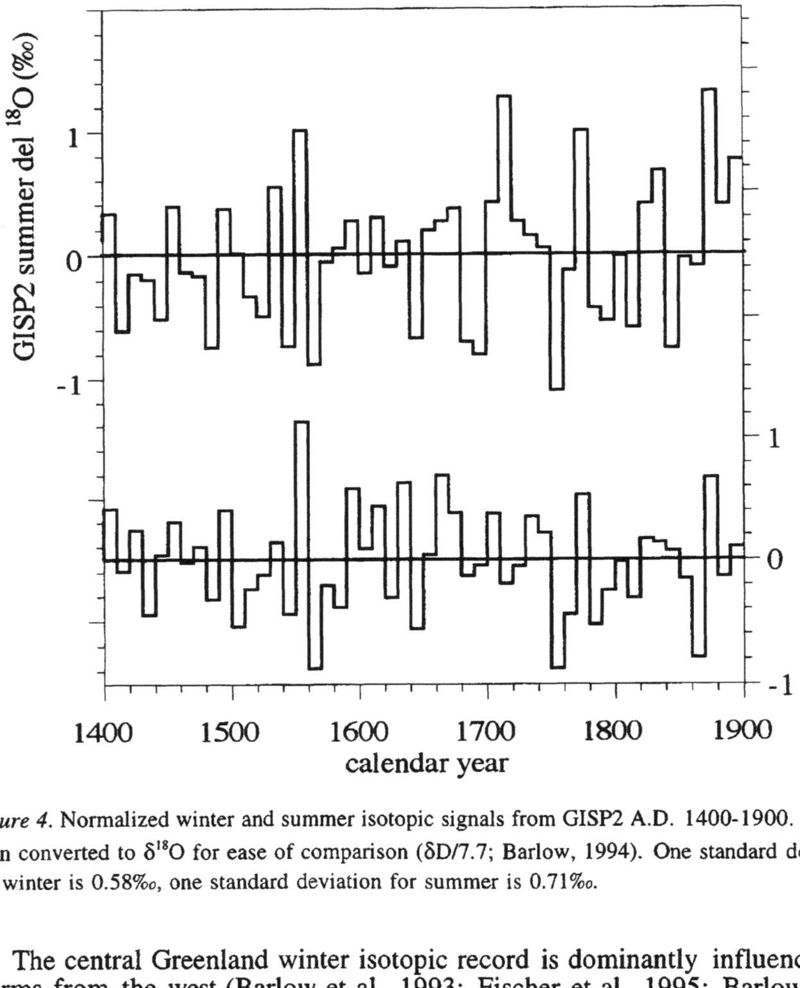

Figure 4. Normalized winter and summer isotopic signals from GISP2 A.D. 1400-1900. δD has been converted to $\delta^{18}O$ for ease of comparison (δD/7.7; Barlow, 1994). One standard deviation for winter is 0.58‰, one standard deviation for summer is 0.71‰.

The central Greenland winter isotopic record is dominantly influenced by storms from the west (Barlow et al., 1993; Fischer et al., 1995; Barlow et al., 1997b), and so the isotopic signal records an opposing temperature excursion to that of northern and western Europe during extreme NAO years. Isotopic excursions in extreme NAO years are statistically significant in winter, summer, and annual records (Barlow et al., 1993). Rogers (1997b) shows that two different atmospheric circulation patterns result in mild winter temperatures in Northern Europe. One pattern is associated with the NAO, which occurs about 60% of the time. The second is caused by westward displacement of the Siberian anticyclone, and is not directly NAO related. Western Greenland high and low temperature anomalies are both strongly linked with the NAO, as will be discussed later.

Multi-decadal dominance of one oscillation mode over the other is common (Hurrell, 1995; Hurrell and van Loon, 1997; Jones et al., 1997; Appenzaller et al., 1998; Kushnir, 1999) and can explain, for instance, lower than normal winter temperatures in the 1960s and early 1970s, and higher than normal temperatures in the 1980s. Recent modelling studies demonstrate that multi-annual to multi-decadal anomalies in North Atlantic SSTs reinforce, and indeed, force the thermal and geopotential structure of the NAO (Rodwell et al., 1999). Given the dominant association of the central Greenland winter isotopic signal with western Greenland, and multiple lines of evidence for low-frequency (decadal or longer) variability in the phase of the NAO, it is possible that decadal-scale temperature anomalies in northern and central Europe over the last 500 years have been negatively correlated with temperature anomalies in central Greenland.

A comparison of the Central West Greenland isotopic signal and Central England winter temperature in this regard is inconclusive. For decadal temperature anomalies relative to the mean of the full record, there is no clear anti-phase relationship between Central West Greenland and Central England (Figure 5). However, using the NAO-related temperature seesaw as a template, the sign of decadal isotopic anomalies concur with the occurrence of decadally persistent NAO modes, such as the 1980s, 1960s, 1940s to 1950s, and the 1900s-1930s (Hurrell and van Loon, 1997). In addition, nineteenth century NAO reconstructions (Jones et al., 1997), indicate clusters of positive NAO years in the 1880s, 1860s, and 1830s, which correspond with negative excursions in the CWG time series. The role of the NAO in influencing opposing temperature excursions in west/central Greenland and north/western Europe is complicated by other climatic influences such as western displacement of the Siberian anticyclone, the temporary cooling effect of volcanic emissions (Stuiver et al., 1995), solar variability (Fischer et al., 1998), as well as oceanic influences from SSTs and near-surface ocean temperatures (Rodwell et al., 1999). Possibly the best approach for understanding the influence of these factors on Greenlandic and European decadal-scale temperature would be modelling studies. The onset of the "Little Ice Age" detected in valley glacier advances (Grove, 1988), century-scale increases in dust, marine storminess, or ice core bore-hole temperature cannot be explained by the NAO. However, the NAO's multi-annual to multi-decadal influence on climate and climate variability in the North Atlantic and adjacent land areas is surely imprinted on the last 500 years.

4. Application of the Central Greenland Isotopic Record to Iceland

Century-scale features in a climate proxy record do not give the detail necessary for socially and economically relevant sub-annual to decadal-scale investigations. Neither is a climate proxy record with continual sub-annual resolution optimal if its applicability is limited to a local area that has never been, or ever will be, populated. As previously discussed, direct transfer of the central Greenland high-resolution isotopic signal for interpretation of temperature history in northern or western Europe is not recommended. Correlation with Iceland holds more promise.

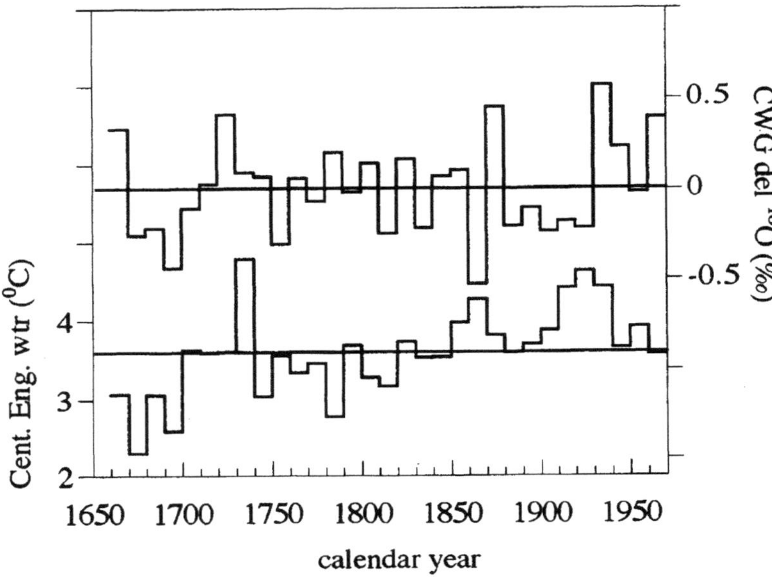

Figure 5. Comparison of CWG and central England winter temperature. Central England temperature is from Manley (1974).

During the past 100 years, Greenland and Iceland have shown time periods of agreement and non-agreement in instrumental temperature trends (e.g. Rogers, 1985; Briffa and Jones, 1993). Similar variability in strength of agreement exists between central Greenland isotopic records and Iceland (Barlow et al., 1997b). An understanding of the atmospheric and oceanic circulation patterns which underlie the variability in such correlations is necessary for confident continuous interpretation of Iceland climate from Greenland ice cores. Two approaches are explored here; agreement or non-agreement in the sign of excursion direction from the previous year or decade, and establishment of atmospheric circulation patterns which include both central Greenland and Iceland. For Stykkishólmur annual T/CWG $\delta^{18}O$ (1823-1982), r= 0.57. For Stykkishólmur annual T/GISP2 annual δD (1823-1982), r= 0.44.

Agreement in year-to-year excursion direction implies a qualitatively identical shift in temperature without quantifying the extent of temperature change. Decadal central west Greenland (CWG) and GISP2 isotopic data are shown in Figure 6 for the period 1830-1980, compared with temperature data from Stykkishólmur. CWG shows 9/14 decadal shifts in agreement with Stykkishólmur annual temperature, and 9/14 agreement with summer (JJA) temperature. 9/14 agreement also is seen for GISP2 summer/Stykkishólmur summer, and 8/14 agreement for GSIP2 winter/Stykkishólmur winter (DJF). Agricultural winter (September - May) shows 11/14 agreement with CWG, while agreement is less between GISP2 winter and Stykkishólmur agricultural winter (5/14, not shown). Figure 7 is an example of a time period of good year-to-year correlation from A.D. 1835 to 1845. Figure 7 shows 7/10 years of agreement in excursion direction between CWG and Stykkishólmur annual temperature, and 6/10 agreement between CWG and summer temperature. GISP2 summer/Stykkishólmur summer shows 6/10 agreement, and there is 7/10 agreement between GISP2 winter and Stykkishólmur winter. GISP2 annual shows 9/10 agreement with Stykkishólmur annual temperature (not shown). CWG winter/Stykkishólmur agricultural winter has 3/10 agreement (not shown).

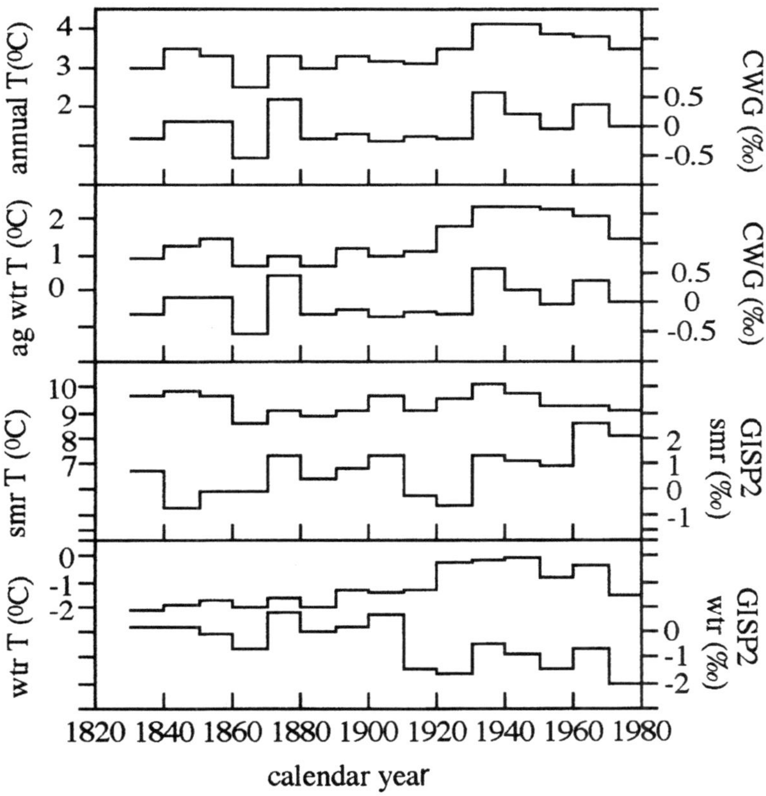

Figure 6. Comparison of Stykkishólmur, Iceland temperature and central Greenland isotopic records, A.D. 1830-1980. top: mean annual temperature and CWG; upper middle: mean agricultural winter (Sept.-May) and CWG; lower middle: mean summer temperature and GISP2 summer; bottom: mean winter temperature and GISP2 winter. Not shown are agricultural winter temperature and GISP2 winter, and summer temperature and CWG. Stykkishólmur temperature data are supplied by the Icelandic Meteorological Office.

Agreement in winter or summer seasonal excursion direction from one year to the next between the GISP2 isotopic signal, east Greenland temperature and west Iceland temperature is linked with different patterns in atmospheric circulation (Barlow et al., 1997b). For winters when GSIP2, Angmagssalik and Stykkishólmur (Figure 1) were uniformly warmer/higher isotopically than the previous winter, the seasonal mean Iceland Low is positioned west of Iceland nearer southern Greenland, the mean Baffin Trough (500mb) is weak and situated further west, and anomalous east-southeasterly flow brings warmer air to central Greenland, east Greenland, and Iceland. For winters when the three locations were uniformly colder/lower isotopically than the previous winter, the mean Iceland Low is positioned to

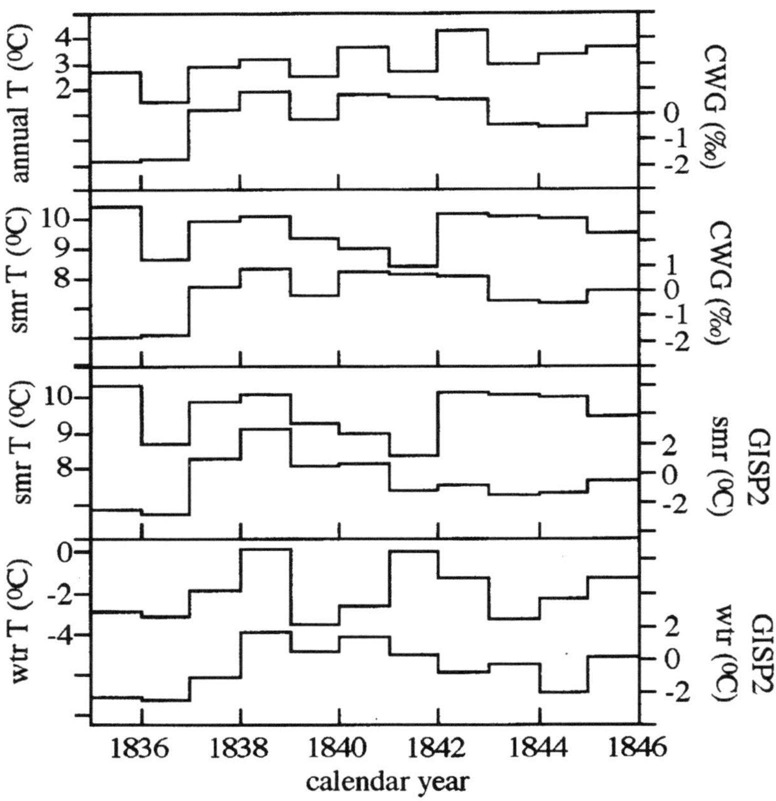

Figure 7. Comparison of Stykkishólmur, Iceland temperature and central Greenland isotopic records, A.D. 1835-45. top: annual temperature and CWG; upper middle: summer temperature and CWG; lower middle: summer temperature and GSIP2 summer; bottom: winter temperature and GSIP2 winter.

the north and east, further away from Greenland, the Baffin Trough shows stronger definition and is situated further east, and anomalous northeasterly geostrophic flow brings cooler air to the area. Similar atmospheric circulation variations are seen for uniformly warmer and uniformly cooler summer year-to-year excursions.

Rogers et al. (1998) find significant correlation between central Greenland isotopically low or high winter extremes and the position of the seasonal mean Iceland Low. Winter extreme single isotopic values correspond to between one-and-a-half and two months of time. Low isotopic winter extremes in central Greenland are associated with an anomalously deep Iceland Low to the west or east of Iceland. Either of these positions bring cold northerly geostrophic flow over Greenland. High isotopic winter values are associated with a weaker mean Iceland Low which is positioned more to the south and across, or even west of, the southern tip of Greenland. This brings southerly warm geostrophic flow over Greenland. The differing sea-level pressures and positions of the Iceland Low for high and low isotopic winters are the same as those noted for the temperature seesaw of the NAO, and partly explain the strong influence of the NAO on winter isotopic values in central Greenland. Summer high isotopic extremes are associated with ridging over east Greenland and the Arctic Ocean, bringing warm southerly air flow to central Greenland. Troughing over the same area brings cold northerly air flow to central Greenland during anomalously low isotopic summers.

Icelandic temperatures, however, are not strongly influenced by the temperature seesaw (Jones et al., 1997). Colder weather in Iceland is linked with displacement of the Iceland Low to the east or northwest of the island (Trausti Jónsson, personal communication, 1999). One might therefore expect a winter that was both isotopically low in central Greenland and cold in Iceland to be associated with a deep sea-level presure Low frequently positioned to the northwest or east of Iceland, while a winter that showed a relatively low isotopic signal in central Greenland and was mild in Iceland would be associated with a deep mean Iceland Low to the west of Iceland. When central Greenland and Iceland are under the influence of an atmospheric circulation that brings a uniform temperature response, a report of a cold or severe winter in Icelandic historical records could appear in the central Greenland record as either a negative excursion in the isotopic signal from the previous year, or as a negative isotopic value with respect to the mean reference. Two time intervals when Icelandic historical climate data are particularly rich are compared with the CWG annual and GISP2 winter records in Figures 8 and 9.

Nine-month Icelandic winters are defined by Ogilvie (1981, 1991, 1992) as severe/cold, or mild. Information on spring weather is included where it exists. Historical comments on summer weather focus on precipitation, with only the summer of 1695 reported in temperature (cold; Figure 8). Severe winters occurred in western Iceland in the years 1686, 1688, 1690, 1691, 1692, 1694, 1695, 1696, 1697, 1699, 1700, 1701, and 1703. Sea ice is noted 1685. 1685, 1687, and 1698 are noted as having cold springs. The central Greenland isotopic signal shows consecutive years below the mean reference level, suggesting persistence in the prevailing circulation pattern. A.D. 1694-1703 was mainly cold in Iceland with some heavy ice years, and was persistently below the mean for the Greenland isotopic data. A.D. 1694-1703 could represent a time when the mean Iceland Low was persistently deep and northwest of Iceland, resulting in northeasterly geostrophic flow to central Greenland and northwesterly flow to Iceland, thus bringing colder temperatures to both locations as well as sea ice to Iceland.

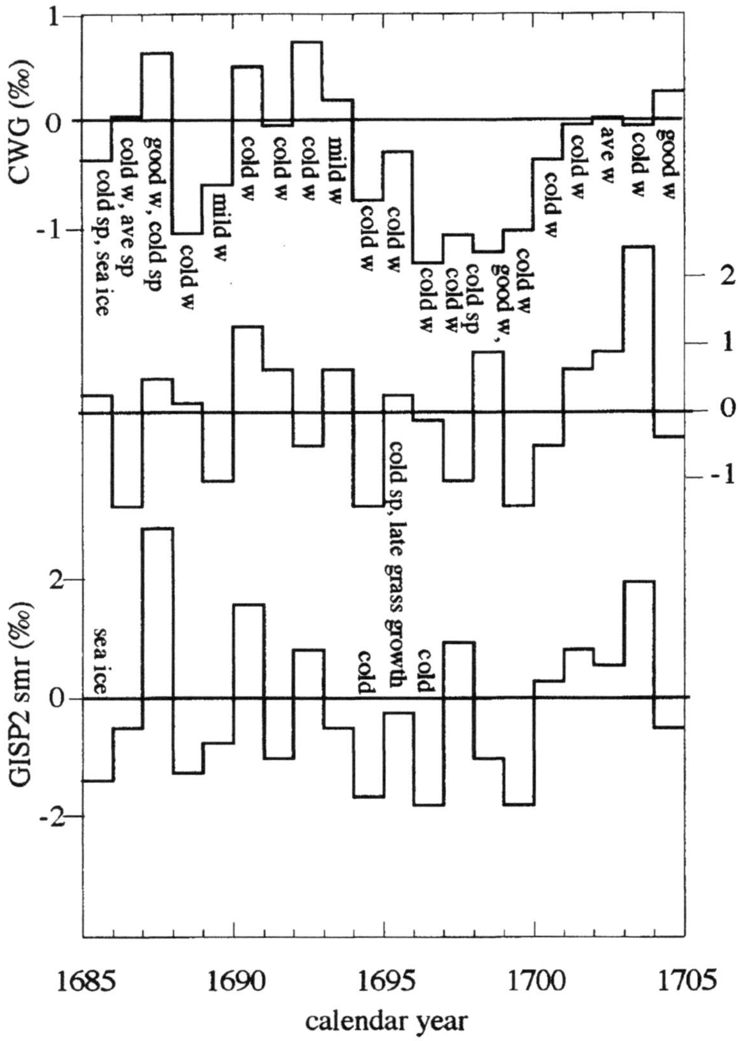

Figure 8. Comparison of Icelandic historical records with central Greenland isotopic signals, A.D. 1685-1704. Upper and middle: Iceland winter (w) and spring (sp), with CWG (annual) and GISP2 winter isotopic signals. Bottom: Iceland summer (s) with GISP2 summer. Blank years indicate no Iceland data. Iceland historical data are from Ogilvie (1981, 1991, 1992).

Figure 9 shows the time period A.D. 1560-1579. While there are no historical data for winters in 1560, 1562, 1563, or 1565, severe sea ice was reported in 1564, and severe winters with sea ice in 1566, 1567, and 1568. A cold spring is noted in 1565. 1561 and 1569 are the only winters mentioned as normal or good in this decade. Summer temperature data are too sparse to include. Ogilvie (1991) contrasts the generally cold 1560s with the 1570s, which are generally mild. While severe winters are mentioned in 1571 and 1577, good (mild) winters are mentioned in 1570, 1572, 1573, 1574, 1578, and 1579. A good autumn is mentioned in 1575, and a good spring in 1576. The central Greenland isotopic signal appears to agree in most years with Icelandic historical records. The 1570s could represent a time when the mean Iceland Low was generally weak and positioned to the southwest of Iceland, toward southern Greenland. A.D. 1576 shows opposing relative temperature anomalies in central Greenland and Iceland.

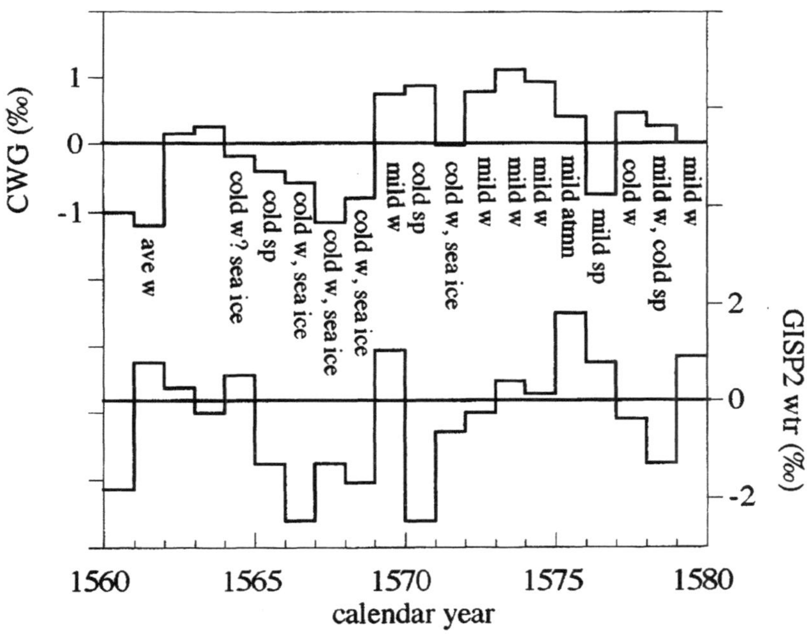

Figure 9. Comparison of Icelandic historical records for winter (w) and spring(sp) with CWG (annual) and GISP2 winter isotopic signals, A.D. 1560-1580. Blank years indicate no Icelandic data. Iceland historicalic data are from Ogilvie (1981, 1991, 1992).

5. Conclusions

Isotopic signals from Greenland ice cores are frequently interpreted as a proxy for relative temperature change, and have often been looked upon for a temperature record of a "Little Ice Age". On the decadal scale through the last 600 years, isotopically low time periods are intermittent. This intermittence agrees with other proxy climate indicators from the Arctic and beyond. In general, decadal-scale proxy climate records of the last 600 years tell a tale of temporal and spatial variability. It has been shown here how changes in the prevailing atmospheric circulation can explain some of these differences: we see patterns that resolve confusion and add confidence in extending the interpretation of areas rich in proxy climate data to those with gaps.

For investigations of the human dimensions of climate variability and climate change, the value of the central Greenland ice core isotopic signal is enhanced when it can be applied to more populated areas. Barlow et al. (1997b) and Rogers et al. (1998) demonstrate significant and specific linkage between central Greenland seasonal isotopic signals and atmospheric circulation patterns. Modern and historical applications to Iceland have been shown. These studies, however, are limited by the short time span of upper air and sea-level pressure data, and data on sea-surface conditions. Investigations using coupled ocean-atmosphere climate models may improve understanding of the different atmospheric circulation patterns that influence seasonal-to decadal- scale ice core isotopic signals and air temperatures. Incorporation of ice core sea-salt sodium, an indicator of marine storminess, may also improve our ability to interpret the past record in terms of synoptic-scale circulation patterns. Development of more continuous climate proxy records from Iceland and Greenland with seasonal or annual resolution would also help greatly in development of palaeosynoptic climatology of the Greenland/Iceland sector of the North Atlantic. As North Atlantic sea level pressure patterns show linkages with land and sea areas from the eastern United States to North Africa and north to the Arctic Ocean, further development of synoptic scale palaeoclimatology in the Greenland/Iceland area should help to improve our understanding of decadal- and annual-scale palaeoclimatology of this entire region.

Acknowledgements

This research is supported by the National Science Foundation, USA, Office of Polar Programs, ARCSS Program Grant no. OPP-9418845. The manuscript benefited from the thoughtful reviews of Drs Tom Wigley, Trausti Jónsson, and Ingibjörg Jónsdóttir, as well as the editorial comments of Astrid Ogilvie. The author also wishes to thank Drs Astrid Ogilvie, Jeff Rogers, David Fisher, Hubertus Fischer, John Andrews, Mark Serreze, Roger Barry, and Jim White for many stimulating discussions. Winston Seiler provided graphic support for some of the figures.

References

Alley, R. B. and Koci, B.: 1990, 'Recent warming in central Greenland?', *Ann. of Glaciol.* **14**, 6-8.

Appenzeller, C., Stocker, T. F., and Anklin, M.: 1998, 'North Atlantic Oscillation dynamics recorded in Greenland ice cores', *Science* **282**, 446-449.

Barlow, L. K.: 1994, '*Evaluation of seasonal to decadal scale deuterium and deuterium excess signals, GISP2 ice core, Summit, Greenland, A.D. 1270-1985*', Unpub. Ph.D. dissertation, INSTAAR, University of Colorado, Boulder, Colorado, USA

Barlow, L. K., White, J. W. C., Barry, R. G., Rogers, J. C., and Grootes, P. M.: 1993, "The North Atlantic Oscillation signature in deuterium and deuterium excess signals in the Greenland Ice Sheet Project 2 core, A.D. 1840-1970', *Geophys. Res. Lett.* **20**, 2901-2904.

Barlow, L. K., Sadler, J. P., Ogilvie, A. E. J., Buckland, P.C., Amorosi, T., Ingimundarson, J. H., Skidmore, P., Dugmore, A. J., and McGovern, T. H.: 1997a, 'Interdisciplinary investigations of the end of the Norse Western Settlement in Greenland', *Holocene* **7**, 489-499.

Barlow, L. K., Rogers, J. C., Serreze, M. C., and Barry, R. G.: 1997b, 'Aspects of climate variability in the North Atlantic sector: discussion and relation to the Greenland Ice Sheet project 2 high-resolution isotopic signal', *Journ. Geophys. Res.* **102** C12, 26,333-26,344.

Bradley, R. S., and Jones, P. D.: 1992, 'Climatic conditions over the last 500 years.' in Bradley, R. S. and Jones, P. D. (eds), *Climate Since A .D. 1500*, Routledge, New York, USA, 649-666.

Briffa, K. R., and Jones, P. D.: 1993, 'Global surface air temperature variations during the twentieth century: Part 2, implications for large-scale high-frequency paeleoclimatic studies', *Holocene* **3**, 77-88.

Cuffey, K. M., Alley, R. B., Grootes, P. M., Bolzan, J. M., and Anandakrishnan, S.: 1994 'Calibration of the $\delta^{18}O$ isotopic paleothermometer for central Greenland using borehole temperatures', *Journ. Glaciology* **40**, 341-349.

Dahl-Jensen, D., Mosegaard, K., Gundstrup, N., Clow, G. C., Johnsen, S. J., Hansen, A. W., and Balling, N.: 1998, 'Past temperatures directly from the Greenland Ice Sheet', *Science* **282**, 268-271.

Dansgaard, W., Johnsen, S. J., Reeh, N., Gundestrup, N., Clausen, H. B., and Hammer, C. U.: 1975, 'Climatic changes, Norsemen, and modern man', *Nature* **225**, 24-28.

Desser, C. and Blackmon, M. L.: 1993, 'Surface climate variations over the North Atlantic ocean during winter: 1900-1989', *Journ. Climate* **6**, 1743-1753.

Fischer, H., Wagenbach, D., Laternser, M., and Haeberli, W.: 1995, 'Glacio-meteorological and isotopic studies along the EGIG line, central Greenland', *Journ. Glaciology* **41**, 515-527.

Fischer, H., Werner, M., and Wagenbach, D.: 1998, 'Little ice age clearly recorded in northern Greenland ice cores', *Geophys. Res. Lett.* **25**, 1749-1752.

Fisher, D. A.: 1997, 'High resolution reconstructed Northern Hemisphere temperatures for the last few centuries using regional average tree ring, ice core and historical annual time series', *EOS Trans. Amer. Geophys. Union* fall mtg supp **78**(46), 32.

Fisher, D. A., Koerner, R. M., Kuivinen, K., Clausen, H. B., Johnsen, S. J., Steffensen, J. P. and Gundestrup, N.: 1996, 'Inter-comparison of ice core $\delta^{18}O$ and precipitation records from sites in Canada and Greenland over the last 3,500 years and over the last few centuries in detail using EOF techniques', in Bradley, R. S., Jouzel, J., and Jones, P. D. (eds) *Climate variations and forcing mechanisms of the last 2000 years*, Springer-Verlag, Berlin.

Grootes, P. M. and Stuiver, M.: 1997, Oxygen 16/18 variability in Greenland snow and ice with 10^{-3}-to 10^{5} -year time resolution', *Journ. Geophys. Res.* **102** C12, 26,455-26,470.

Grove, J. M.: 1988, *'The Little Ice Age'*, Methuen, New York.

Grove, J. M.: 'The Initiation of the "Little Ice Age" in Regions Round the North Atlantic', this volume

Hurrell, J. W.: 1995, 'Decadal trends in the North Atlantic Oscillation: regional temperatures and precipitation', *Science* **269**, 676-679.

Hurrell, J. W., and Van Loon, H.: 1997, 'Decadal Variations in Climate Associated with the North Atlantic Oscillation', *Climatic Change* **36**, 301-326.

Johnsen, S. J., Dahl-Jesen, D., Dansgaard, W., and Gundestrup, N.: 1995, 'Greenland paleotemperature derived from GRIP borehole temperature and ice core isotope profiles', *Tellus* **47B**, 624-629.

Johnsen, S. J., Clausen, H. B., Dansgaard, W., Gundestrup, N. S., Hammer, C. U., Anderson, U., Anderson, K. K., Hvidberg, C. S., Dahl-Jensen, D., Steffensen, J. P., Shoji, H., Sveinbjörnsdóttir, A. E., White, J., Jouzel, J., and Fisher, D.: 1997, 'The $\delta^{18}O$ record along the Greenland ice core project deep ice core and the problem of possible Eemian climatic instability', *Journ. Geophys. Res.* **102** C12, 26,397-26,410.

Jones, P. D., Jónsson, T., and Wheeler, D.: 1997, 'Extension of the North Atlantic Oscillation using early instrumental pressure observations from Gibraltar and south-west Iceland', *Int. Journ. Climatol.* **17**, 1433-1450.

Jouzel, J., Alley, R. B., Cuffey, K. M., Dansgaard, W., Grootes, P., Hoffman, G., Johnsen, S. J., Koster, R. D., Peel, D., Shuman, C. A., Stievenard, M., Stuiver, M., and White, J.: 1997, 'Validity of the temperature reconstruction from water isotopes in ice cores', *Journ. Geophys. Res.* **102** C12, 26,471-26,488.

Kreutz, K. J., Mayewski, P. A., Meeker, L. D., Twickler, M. S., Whitlow, S. I., and Pittalwala, I. I.: 1997, 'Bipolar changes in atmospheric circulation during the Little Ice Age', *Science* **277**, 1294-1296.

Kushnir, Y.: 1999, 'Europe's winter prospects', *Nature* **398**, 289-291.

Manley, G.: 1974, 'Central England temperatures: monthly means 1659-1972', *Quart. Journ. Royal Met. Soc.* **100**, 389-405.

Mayewski, P. A., Meeker, L. D., Morrison, M. C., Twickler, M. S., Whitlow, S. I., Ferland, K. K., Meese, D. A., Legrand, M. R., and Steffensen, J. P.: 1993, 'Greenland ice core "signal" characteristics: an expanded view of climate change', *Journ. Geophys. Res.* **98**, 12,839-12,847.

Meese, D. A., Gow, A. J., Grootes, P., Mayewski, P. A., Ram, M., Stuiver, M., Taylor, K. C., Waddington, E. D., Zielinski, G. A.: 1994, 'The accumulation record from the GISP2 core as an indicator of climate change throughout the Holocene', *Science* **266**, 1680-1682.

Moses, T., Kiladis, G. N., Diaz, H. F., and Barry, R. G.: 1987, 'Characteristics and frequency of reversals in mean sea level pressure in the North Atlantic sector and their relationship to long-term temperature trends', *Journ. Climatol.* **7**, 13-30.

Ogilvie, A. E. J.: 1981, *'Climate and society in Iceland from the medieval period to the late eighteenth century'*, Unpub. Ph.D. dissertation, University of East Anglia, Norwich, U.K.

Ogilvie, A. E. J.: 1991, 'Climatic changes in Iceland A.D. c. 865 - 1598', in Bigelow, G. F. (presentor), *The Norse of the North Atlantic, Acta Archaeologica* **61-1990**, Munksgaard, Copenhagen, 233-251.

Ogilvie, A. E. J.: 1992, 'Documentary evidence for changes in the climate of Iceland, A.D. 1500-1800', in Bradley, R. S. and Jones, P. D. (eds), *Climate since A.D. 1500*, Routledge, New York, 92-117.

Ogilvie, A. E. J., and Jónsson, T.: ' "Little Ice Age" Research: a Perspective from Iceland', this volume.

Overpeck, J., Hughen, K., Hardy, D., Bradley, R., Case, R., Douglas, M., Finney, B., Gajewski, K., Jacoby, G., Jennings, A., Lamoreux, S., Lasca, A., MacDonald, G., Moore, J., Retelle, M., Smith, S., Wolfe, A., and Zielinski, G.: 1997, 'Arctic environmental change of the last four centuries', *Science* **278**, 1251-1256.

Rodwell, M. J., Rowell, D. P., and Folland, K. C.: 1999, 'Oceanic forcing of the wintertime North Atlantic Oscillation and European climate', *Nature* **398**, 320-323.

Rogers, J. C.: 1985, 'Atmospheric circulation changes associated with the warming over the northern North Atlantic in the 1920s', *Journ. of Climate and Applied Met.* **24**, 1303-1310.

Rogers, J. C.: 1997a, 'North Atlantic storm track variability and its association to the North Atlantic Oscillation and climate variability in Northern Europe', *Journ. Climate* **10**, 1635-1647.

Rogers, J. C.: 1997b, 'On the causes of mild winters in northern Europe', in Madden, R. E. et al. (eds), *Harry van Loon Symposium - Studies in Climate II*, NCAR Tech Note 433, National Center for Atmospheric Research, Boulder, Colorado, 51-66.

Rogers, J. C. and van Loon, H.: 1979, 'The seesaw in winter temperatures between Greenland and Northern Europe, part II: some oceanic and atmospheric effects in middle and high latitudes'. *Month. Weather Rev.* **112**, 1999-2015.

Rogers, J. C., Bolzan, J. F., and Pohjola, V. A.: 1998, 'Atmospheric circulation variability associated with shallow-core seasonal isotopic extremes near Summit, Greenland', *Journ. Geophys. Res.* **103** D10, 11,205-11,219.

Serreze, M. C., Carse, F., Barry, R. G. and Rogers, J. C.: 1997, 'Icelandic Low cyclone activity: climatological features, linkages with the NAO and relationships with recent changes elsewhere in the Northern Hemisphere circulation', *Journ. of Climate* **10**, 453-464.

Shuman, C. A., Alley, R. B., Anandakrishnan, S., White, J. W. C., Grootes, P. M. and Stearns, C.: 1995, 'Temperature and accumulation at the Greenland summit: comparison of high resolution isotope profiles and satellite passive microwave brightness temperature trends', *Journ. Geophys. Res.* **100** D5, 9165-9177.

Stuiver, M., Grootes, P. M., and Braziunas, T. F.: 1995, 'The GISP2 $\delta^{18}O$ climate record of the past 16,500 years, and the role of the sun, ocean, and volcanoes', *Quat. Res.* **44**, 341-354.

Taylor, A. H. and Stephens, J. A.: 1998, 'The North Atlantic Oscillation and the latitude of the Gulf Stream', *Tellus* **50A**, 134-142.

Van Loon, H., and Rogers, J. C.: 1978, 'The seesaw in winter temperatures between Greenland and Northern Europe, part I: general discussion', *Monthly Weather Review* **106**, 296-310.

Walker, G. T.: 1924, 'Correlations in seasonal variations of weather IX', *Mem. Ind. Meteorol. Dept.* **24**, 275-332.

White, J. W. C., Barlow, L. K., Fisher, D., Grootes, P., Jouzel, J., Johnsen, S. J., Stuiver, M., and Clausen, H.: 1997, 'The climate signal in the stable isotopes of snow from Summit, Greenland: results of comparisons with modern climate observations', *Journ. Geophys. Res.* **102** C12, 26,425-26,440.

Williams, L. D., and Wigley, T. M. L.: 1983, 'A comparison of evidence for late Holocene summer temperature variations in the Northern Hemisphere', *Quat. Res.* **20**, 286-307.

(Received 2 July 1999; in revised form 25 September 2000)

THE "LITTLE ICE AGE" AND ITS GEOMORPHOLOGICAL CONSEQUENCES IN MEDITERRANEAN EUROPE

A.T. GROVE

Department of Geography, Downing Place, Cambridge CB2 3EN, United Kingdom

Abstract. Alpine glacier advances in the "Little Ice Age" took place in the decades around 1320, 1600, 1700 and 1810. They were the outcome of snowier winters and cooler summers than those of the twentieth century. Documentary records from Crete in particular, and also from Italy, southern France and southeast Spain point to a greater frequency in Mediterranean Europe's mountainous regions of severe floods, droughts and frosts at times of "Little Ice Age" Alpine glacier advances. Deluges, when more than 200 mm of rain fall within 24 hours, are most frequent on mountainous areas near the coast. An instance is given of the geomorphological consequences of a great deluge which struck the Tech valley in the eastern Pyrenees on 17 October 1940. An increased frequency of deluges, probably at times when Alpine glaciers were advancing in the "Little Ice Age" and earlier in the Holocene, in areas known to be tectonically unstable and underlain by soft sediments, could better explain the occurrence of fluvial terraces in Mediterranean Europe sometimes known as the "younger fill", than soil erosion resulting from deforestation.

Key Words: Mediterranean Europe; "Little Ice Age"; floods; soil erosion

1. Introduction

Historical and geomorphological studies of Mediterranean landscapes often point to rapid sedimentation in river valleys and deltas in medieval and early modern times and attribute it to soil erosion. The soil erosion is claimed to be the result of population growth, agricultural expansion and deforestation. The evidence for the timing of clearance, and its effectiveness in promoting erosion is rarely demonstrated. Rural population growth was generally fastest from the mid eighteenth to the end of the nineteenth century, later than the sedimentation. Human activity, it is argued here, is only part of the explanation for the erosion. Climatic extremes, especially a greater frequency of deluges, possibly associated with "Little Ice Age" ice advances, were more important than is usually recognized.

2. The "Little Ice Age" and the "Younger Fill"

The "Little Ice Age" is best defined as being the interval within the second millennium A.D. during which glaciers extended further down their valleys than in the twentieth century. Glacier advances were especially marked in Alpine Europe in the decades around 1320, 1600, 1700 (in some cases), and 1810, and minor advances interrupted a general recession about 1850, 1885, 1915 and 1965

Climatic Change **48**: 121–136, 2001.

(J.M. Grove, 1988, 1996, and this volume; Holzhauser, 1984a, 1984b, and 1997; Holzhauser and Zumbühl, 1996; Patzelt, 1996). The climatic conditions in the Alps associated with the advances were a greater frequency of cool summers and snowy winters. Certain individual years were especially severe, with great cold and very heavy snowfall. Others, especially between 1675 and 1715, are known for their severe dry winters (Pfister, 1994). Even within this period the 1690s were snowy. In particular, the winter of 1694-95 saw deep snow in Piedmont, as well as in Europe further north, and also flooding at places as far apart as Setubal, Bordeaux and Rome (Lindgrén and Neumann, 1981). During the severe winter of 1697-8 in western and central Europe, cold weather in Spain, brought by cold air masses pushing south-westwards, coincided with long rainy periods in the eastern Mediterranean (Pfister, 1994; Kington, 1999).

Rumsby and Macklin (1996) have presented evidence from historical sources which suggests that 'river basins in north, west and central Europe experienced enhanced fluvial activity between 1250 and 1550 and again between 1750 and 1900', What happened elsewhere in Europe during the decades of ice advance in Alpine Europe? Very few routine meteorological records go back beyond the beginning of the nineteenth century and few river flow records before the beginning of the twentieth century. In the records that do exist some regularities can be recognized. For example, Probst (1989), using 5-year moving averages, shows that fluctuations in the flow of the major French rivers this century, before 1980, were broadly synchronous. Furthermore, his diagrams for the period 1912-75 show a marked resemblance between the discharge curves of the Guadalquivir, Ebro and Po.

One might not expect the timing of climatic and river discharge extremes to correspond throughout Mediterranean Europe. The region's topography is complex. Globally it lies between the Azores High and the southwest Asian monsoon, with the west more under the influence of Atlantic weather systems than the east. There is a strong correlation of -0.6 to -0.7 between the North Atlantic Oscillation (NAO) index and December-March rainfall totals in Lisbon and Madrid, though the correlation weakens to -0.3 to -0.4 in the central and eastern Mediterranean (Hurrell, 1995).

Throughout Mediterranean Europe most of the rain is brought by eastward-moving cyclonic depressions. A minority are Atlantic depressions following southerly tracks, their number increasing when the NAO index is negative and winter storms move more freely into the Mediterranean (Kushnir, 1999). By far the majority of Mediterranean depressions are generated over the Sea itself, usually in certain well-recognized areas (Flocas, 1988; Wigley, 1992). Conceived by invading northerly air streams, some coming from the Atlantic, some originating in Scandinavia and Siberia, they are born in the lee of the mountains, especially in the Gulf of Genoa, the Gulf of Lions and west of Crete. The strength and frequency of the cold air outbreaks varies through time, increasing with a more meridional circulation, decreasing with a more zonal circulation.

Documentary records from Crete in particular, and also from Sicily, southern France and southern Spain point to a greater frequency of severe floods, droughts and frosts in the decades around 1600 and 1700, than have been experienced this century. This has stimulated a re-examination of the causes of rapid erosion and sedimentation in Mediterranean Europe.

A possible starting point for discussion is Vita-Finzi's (1969) *The Mediterranean Valleys* in which he pointed to the widespread occurrence of a medieval fluvial terrace. He called it "the younger fill", to distinguish it from older terraces of glacial times - "the older fill". He pointed out that fluvial terraces of Roman and modern age are generally lacking, and that in modern times downcutting rather than sedimentation has been the rule. He concluded that it was therefore unlikely that the "younger fill" was the result of erosion accelerated by human activity and was inclined to attribute its formation to a climatic fluctuation, either the "Medieval Warm Period" or the "Little Ice Age", both of which had recently been brought to attention by Lamb (1963) and Le Roy Ladurie (1965).

Neither Vita-Finzi nor others made a serious attempt to explain what features of the climate might have been responsible for the greater geomorphological activity. In general there was a tendency to think only in terms of changes in mean temperature, though some consideration was also given to the possibility of Atlantic depressions taking more southerly tracks and the circulation over Europe displaying more meridional features. An exception was Bintliff (1982), an archaeologist, who hypothesised that 'in the not too distant past the local climate was highly variable, with severe droughts inhibiting scrub colonisation and loosening the topsoil, alternating with dramatic rainfall, producing these great colluvial sediment flows; such a climate may have characterised the latter part of the "Little Ice Age" in the Mediterranean up till the middle of the last century;....' (p. 157).

Vita-Finzi's conclusions attracted much attention from archaeologists and geographers, the majority of whom were inclined to disagree with him (Grove, 1997). Most are still persuaded that the main cause of sedimentation in Mediterranean valleys is soil erosion, which they usually attribute to deforestation (e.g. Innocenti and Pranzini, 1993; Billi and Rinaldi, 1997; Guillén and Palanques, 1997). However, it is rare for evidence to be presented showing that the time of clearing corresponded with an acceleration of erosion, or to show that clearing has resulted in particular erosion episodes; such a rare example is the study of the Biferno valley in Molise, Italy, by Barker and Hunt (1995). In general, there has been a tendency not to distinguish geological erosion from soil erosion, to assume that alluvium consists of soil, and to take it for granted that trees are not capable of growing again when they have been felled. It is, admittedly, a difficult area in which to provide precise data but significant evidence is accumulating.

3. Deluges, Erosion and Deposition: A Particular Example

We know that in recent years much of the erosion that has occurred has been the result of unusually intense rainfall, deluges when more than 200 mm have fallen within 24 hours (Reichenbach et al., 1998). The disastrous mudflows in Campania, southern Italy, in April 1998 are a recent instance. An earlier, well-studied example is the very violent storm on 17 October 1940, which delivered over 500 mm of rain to 500 km^2 of the eastern Pyrenees, and as much as 1930 mm in 5 days to Saint-Laurent de Cerdans (Météo-France, 1995). It was calculated by Pardé (1941) that this "200-year storm" (the most violent in the eastern Pyrenees since that of 16 October 1763) removed the equivalent of 7.65 mm of sediment from the whole of the upper catchment of the river Tech, about as much as might be expected to be eroded over a hundred years without such an extreme event (Figure 1).

This should not automatically be attributed to soil erosion. Most of the sediment was derived, not from the wide-spreading slopes of the catchment, but from gullies, mass movements, and from Pliocene and Pleistocene valley fill, exposed by past faulting and uplift and now subject to mass movements and badland erosion (Calvet, 1993 and 1994). Pardé reckoned that about a half of the total product of erosion was deposited on the lower valley floor and a half was carried out to sea.

The river Tech escapes from its valley to reach the Mediterranean across a gently sloping alluvial fan. A recent study of documents applying for compensation payments by people in Elne has thrown light on the details of deposition on this fan area in October 1940. (Jacob, 1997). The floodwaters, spilling along former courses of the river and diverted by a railway embankment, scooped out deep hollows in some places and deposited spreads of sediment much more widely in others. Jacob calculated that 1.3 million m^3 were spread over the 700 ha of the Elne commune, and 290,000 m^3 were excavated and carried away from the flood plain by the floodwaters, giving a net accumulation 14 cm thick. The losses included a good deal of valuable soil; the gains consisted of pale sandy silt in some sectors and darker more organic alluvium in others, with greater thicknesses of coarse sand and gravel near channels and depressions where the floodwater moved fastest. At Bompas, in the same region but 15 km to the north between the Agly and Tet, about 3 m thickness of sediment somewhat similar to that at Elne covers a Roman road and the remains of medieval buildings (Calvet, 1993 and 1994). Presumably it represents several events like that of 1940 or, more likely, numerous lesser events.

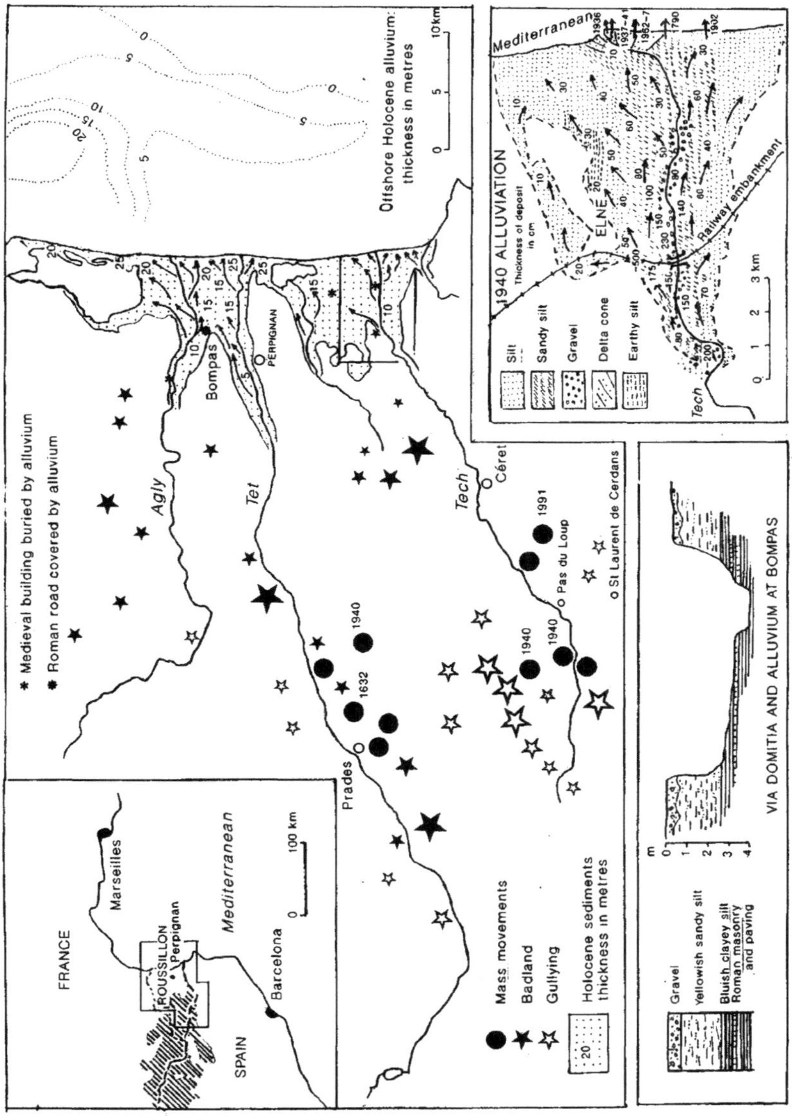

Figure 1. Alluviation in Roussillon. Sediment brought down from the eastern Pyrenees by the Agly, Tet and Tech in the course of the Holocene has accumulated offshore and behind the coast. At Bompas, 3m have accumulated on top of Roman remains (based on M. Calvet 1994, Figs. 219 and 222). Elsewhere, medieval buildings have been buried or partly buried. A deluge on 17 October 1940 resulted in the Tech depositing a net thickness of 14 cm of alluvium over 700 ha in the commune of Elne. (Derived from N. Jacob 1997, Figs. 1-4).

4. Location and Timing of Deluges and Erosion

Around the Mediterranean, very intense storms are generally confined to certain fairly well-defined mountainous areas: within 100 km of the coast of south-eastern Spain, in the eastern Pyrenees, at the southeast corner of the Massif Central, in Liguria and Corsica around the Gulf of Genoa, in the Appenines especially in Calabria, in Montenegro, the Pindos mountains and highland Crete (Llasat and Rodriguez, 1997). In all of these areas, deluges of between 500 and 1000 mm of rain within 48 hours have been recorded. In one of them, near Mont-Agoual overlooking the lower Rhone valley and near the source of several important rivers, the mean return period of 200 mm of precipitation falling in 24 hours (sometimes spread over two days) in the time span 1958 - 1994, was only about 2 years (Figure 2; see also Figure 1 in Obled and Tourasse, 1992).

Many of the elevated, deluge-prone areas are also susceptible to erosion because they are tectonically unstable. They are in situations where uplift in Late Cenozoic times has been accompanied by incision and the development of steep slopes, and by the accumulation of great thicknesses of sediments which remain unconsolidated. These are liable to be incised by gullies and to collapse as a result of slumping and landslides, all of which yield large volumes of sediment (e.g. Calzolari and Ungaro, 1998). The evidence of former erosion is in the form of depleted soil profiles, gullies, landslips and badlands. Such features are usually difficult to date. More readily dateable, from their content of archaeological material and organic material, are the colluvial deposits, alluvial fans, fluvial terraces and river deltas composed of the products of erosion.

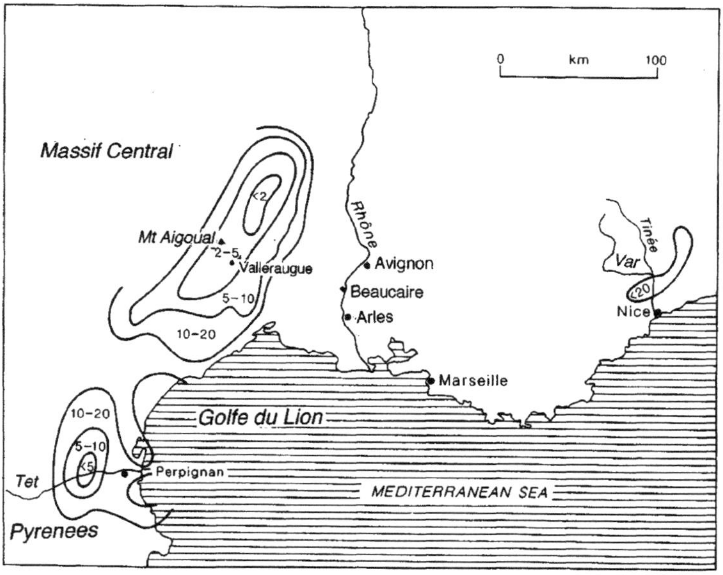

Figure 2. Isolines of the return period in years of falls of 200 mm of precipitation in 24 hours, southern France, 1958-1994. Based on data in Météo-France (1995).

The "younger fill" composing the fluvial terraces studied by Vita-Finzi, is for the most part broadly medieval, but there are indications in Etruria (Potter, 1976), southern Spain (Butzer, 1980), and Provence (Ballais and Crambes, 1992), that there were two main periods of deposition. The first followed the end of the Western Roman Empire, from about A.D. 500-800, when Alpine glaciers advanced during the Göschener Cold Phase II (Gamper and Suter, 1982, Holzhauser, 1997); the second lasted from the thirteenth to the seventeenth century A.D.

5. The Growth of Mediterranean Deltas

Mediterranean deltas are beginning to yield evidence of their rates of formation having varied through time. They began to accumulate about 7000 years ago when sea-level stabilized (Stanley and Warne, 1994). At first, estuaries and embayments were filled and then the deltas, notably the Ebro and Po but also a large number of others, began to prograde seawards (Trousset, 1987; Jeftic et al., 1992). Tectonic subsidence has affected the situation, especially in the case of the Rhone. In the western and central Mediterranean, delta growth was especially evident in the sixteenth and seventeenth centuries, presumably as a result of the greater supplies of sediment reaching the mouths of rivers at that time. In the eastern Mediterranean, around the Aegean, it was more apparent in Antiquity (500 BC - A.D. 200). Nowadays delta fronts are everywhere being eroded by the sea on account of dams holding back sediment, and as the result of the extraction of sand and gravel from river beds.

6. Weather Conditions Associated with Mediterranean Deluges

From instrumental and from documentary records, it is evident that the heaviest deluges in the western and central Mediterranean, although they can occur in almost any month, mostly occur in October and November, a little earlier in the year than this in southeast Spain and the south of France, later in southernmost Spain, northern Italy and around the Aegean (Lopez Gomez, 1983; Russo and Sacchini, 1994; Llasat and Rodriguez, 1997). They are associated with cyclonic depressions which involve both cold air from the north, and warm air moving a long distance over a Mediterranean Sea where surface temperatures in the autumn may still be over 22°C and even as high as 26°C (Alpert et al., 1990). The warm, moist air, in contact with cold air and lifted over mountain ranges 2000 m high, is extremely unstable and, especially if it is converging horizontally and the freezing level is high, can deliver an enormous mass of water within a few hours.

For intense rain in eastern Spain the warm winds need to be easterly; in most other deluge-prone areas, westerly winds are required (Millan et al., 1995).

For most deluges it would seem that the initial stimulus is an incursion of cold northerly air, generating a deep depression, which accentuates the uplift caused by convection. Some of the smaller storms are essentially convective, the larger ones are frontal; most are a combination of the two.

7. Relevant Features of the Climatic Record of Mediterranean Europe

Documentary records, most of which only begin to be of value for tracing climatic change about the thirteenth century, indicate that times of variable and extreme weather have been concentrated in certain decades when Alpine glaciers were advancing: 1. in the fourteenth century; 2. the decades around 1600; 3. the decades around 1700 and; 4. in certain decades of the nineteenth century.

7.1. THE FOURTEENTH CENTURY

The first major phase of extremes began in the 1310s and continued, with intervals, until the 1380s. It is recorded chiefly in northern Italy and the south of France. It was expressed in deluges and cold winters, but also in summer rain and occasional drought (Alexandre, 1987; Pfister et al., 1998).

Camuffo and Enzi (1996) point to a peak in Po flood frequency at the beginning of the fourteenth century and it may be significant that the rerouting of the Brenta to the south, the first diversion of rivers to prevent silting in the Venetian Lagoon, began in 1324 (Camuffo, 1987). There were floods all over middle Italy on 3 November 1333 (the present Ponte Vecchio in Florence replaces a bridge then swept away) and in early autumn 1345 (Porta, 1990-91). The Arno and the Po froze in the winters of 1354/5 (Pfister et al. 1998) and 1367/8. In Siena, processions and prayers for rain early in 1368 were followed by floods the following season. In 1371/2 there were processions for rain at Florence in December, followed by processions praying for the rain to stop in May.

7.2. THE DECADES AROUND 1600

There was frequent flooding of the Tiber at Rome in the last third of the fifteenth century and of the Po in the decades around 1500 (Camuffo and Enzi 1996). However, the greatest cluster of weather disasters in Mediterranean Europe began in the 1570s, reached a peak in the 1590s, and tailed off in the 1640s. Droughts, floods, cold winters, and other exceptional events appear wherever in the Mediterranean we have records. Eight winters between 1595 and 1614 are rated by Camuffo and Enzi (1992) as great or severe. Anomalous climatic events recorded in the chronicles of the Alessandria area of Piedmont were especially numerous between 1590 and 1620. (Pavese et al., 1992). In the lower valley of the Po, cereal yields were seriously reduced in the period 1590-1630, especially in the years between 1589 and 1600, in large part as a result of excessive rainfall and flooding (Guidoboni, 1998). The Venetians, afraid that increased silt deposition might close the mouth of their lagoon, diverted the mouth of the Po to the southeast between 1604 and 1607 (Cencini, 1998). Tiber flood frequency at Rome reached a maximum between 1589 and 1660 (Camuffo and Enzi, 1996). The flood of December 1598 carried away the ancient Æmilian Bridge and drowned 1400 people (Davidson, 1985). In Tuscany and the south of France the 1590s were marked by crop failures, attributed variously to drought, cold, and out-of-season rain; these led to wheat being imported, unusually, from northern Europe, which was itself in trouble (Braudel, 1966). Font Tullot (1988) records many unusual snowfalls in Spain. Rodrigo et al. (1994, 1996) find 1575-1650 to have been a generally wet period in southeast Spain, punctuated by occasional droughts such as that of 1604-5. Catastrophic floods were unusually frequent between 1571 and 1630, especially in Catalonia (Barriendos Vallvé and Martín-Vide, 1998). 1617 was a 'deluge year' throughout eastern Spain, with intense rain in October; this was repeated throughout Spain early in 1626.

Weather is well documented for Crete for the period 1548—1648, possibly better than for any other part of the Mediterranean. Venetian administrators sent home a great mass of reports and letters mentioning, among other things, droughts, cold winters, great heat, and out-of-season rainfall. These are precise, first-hand descriptions made by observers present at the time. The records demonstrate a high frequency of intense rain (and droughts) especially in the last decades of the sixteenth century (J. M. Grove and Conterio, 1995).

Some of the storms had a more severe and more widespread effect than any in recent years. In 1576 the Governor-General reported that

> the very heavy and unusual rains we have had this year in June, July and August have greatly ruined the vintage . . . the said rain has also destroyed the salt-works of Suda and Spinalonga.
>
> (Foscarini, 14.9.1576, *Archivio di Stato, Venice, Provveditori da Terra e da Mar, Filza 738*)

These places are 150 km apart.

On 14 October 1590 a deluge hit the city of Réthymnon, which being on a promontory is not ordinarily susceptible to flooding:

> Most of the inhabitants of Rethimno fled out of their houses and sheltered themselves as well as they could, especially in some of the churches. All the rivers flooded the town of Rethimno, and the countryside, destroying a lot of buildings and causing high water. Eighty old buildings have fallen down.. . . A lot of lands have been completely ruined.
>
> (Nicol Priuli, Rector of Rettimo, 22.11.1590, *Archivio di Stato, Venice, Provveditori da Terra e da Mar, Filza 755*)

7.3. THE DECADES AROUND 1700

After a period of quiescence in the mid-seventeenth century, the extreme events in and around 1600 were repeated a century later in Crete and elsewhere. On the smallish Ægean island of Tinos the Venetian authorities reported that the winter of 1682/3 was one of strong winds, continuous rain, heavy snow and 'very intense cold'; 480 cattle and 3427 sheep and goats perished. The Turkish authorities in Crete reported that torrential rain had caused serious damage to aqueducts. Matters got worse in the 1690s. In July 1692 the Governor-General of the remaining Venetian islands reported that:

> the scantiness of a very thin crop of wheat having increased the worry that this current year of famine is almost universal, so that there is very little help from the mainland and from the kingdom of Morea [Peloponnese].
>
> (Vendramin, 28.7.1692, *Archivio di Stato, Venice, Provveditori da Terra e da Mar, Filza 1175*)

There was similar trouble in 1694.

Camuffo and Enzi (1994) mention numerous floodings in northern Italy between 1682 and 1707, very cold winters between 1684 and 1694 and the particularly cold winter of 1709. In Sicily there was a run of disastrous harvests in the 1690s (J.M.Grove and Conterio, 1994). In 1691, 1692, and 1694 the wheat stored in a depot at Termini Imerese was spoilt in summer, either because of rain during the harvest or because out-of-season rain got into the warehouses. In 1693 and 1694 wheat and barley crops were scanty. In 1700 the Duke of Gampilerio noted that even less wheat had been brought down from the mountains to the various stores than in the past few years 'even though the crop then was poor'; he speculated that this was due to the climate having become colder and wetter. In 1692 heavy snowfall was reported from Castroreale.

Mediterranean Spain experienced intense rainfall in 1684 (Rodrigo et al., 1996). According to Barriendos Vallvé and Martín-Vide (1998), much of this region was spared catastrophic floods between 1691 and 1720, though Molina Sempere et al. (1992) refer to Segura floods being numerous at this time.

7.4. THE NINETEENTH CENTURY

The records of flooding and freezing of the Rhone at Arles brought together by Pichard (1995), show a good correspondence between freezing of the river and the frequency of floods reaching over 4 m. Flood frequency on the lower river during the later stages of the "Little Ice Age" was much higher than in this century, especially in the late sixteenth century, 1701-1710 and 1811-1820. The frequency of autumn and winter floods over 6 m at Beaucaire, upstream of the Rhone delta, mainly generated by deluges in the Massif Central and the Durance, is associated with freezing of the river and glacier advances. Of the 29 floods reaching to over 6 m between 1800 and 1920, 19 were in three decades; four between 1836 and 1846, six between 1882 and 1891, and nine between 1907 and 1917 (Pardé, 1924-5).

In southeast Spain, the Segura flooded on numerous occasions in the decades either side of 1900 (Molina Sempere et al., 1992). Between 1888 and 1895 there were five occasions when peak discharges of the Ebro at Zaragoza exceeded 3000 m^3sec^{-1}, a value which was not to be reached again until 1926 (Gellatly et al., 1995). The clusterings of high floods in the middle and again towards the end of the nineteenth century coincided with glacier advances in the Alps and Pyrenees.

8. Holocene Glacier Advances and River Alluviation

Starkel (1991) has referred to what he calls "second order" fluctuations in the regimes of central European rivers in the Holocene. He argues that rivers draining mountainous areas (as do practically all the rivers of Mediterranean Europe) are under the "upstream control" of floods and that, as a result, their channel forms and fills have responded to Holocene climatic fluctuations similar to that of the "Little Ice Age".

A number of other geomorphologists have begun to attribute more active episodes of erosion and sedimentation in southern Europe and the southwest USA to a greater frequency of deluges associated in some cases with glacier advances. A study in the Cantabrian region of northern Spain indicates the possibility that mass movements there were more frequent at certain periods in the Holocene; in the Pre-Boreal (8500-7500 BP), in the Sub-Boreal (3000-2500 BP), and in the last few centuries (500-300 BP) when rainfall totals are believed to have been higher than they are now (Díez et al., 1996). Enzel and Wells (1997) provide evidence from southern California that 'the most extreme modern storms and floods were more frequent by at least an order of magnitude during specific time periods in the course of the Holocene'. Similarly, Ely (1997) finds that the largest floods in Arizona and southern Utah over the last 6000 years cluster into distinct time periods: 5000-3600 14C yr BP (i.e. 3800-2200 BC), 1100-900 BP (A.D. 900-1100) and after 500 BP (A.D. 1400). Over the last 1000 years, she notes, there is a positive relationship between the paleofloods and long-term variations in the frequency of El Niño events.

In Mediterranean Europe, archaeological and geomorphological studies have provided dates for fluvial terraces, constituting the "younger fill", which accumulated at various periods in the Holocene. Some of the terraces, judging by their dates, were the results of slope erosion and accompanying valley floor alluviation during periods of Holocene ice advance before the "Little Ice Age". For instance, alluviation on riverine plains in northern and central Italy was widespread during the glacier advances of Göschener Cold Phase II, between A.D. 500 and 800, burying Roman towns, bridges, roads and centuriation networks (Veggiani 1983, Cremaschi et al., 1994). In the Peloponnese (southern

Greece), both the Pikródaphne alluvium in the Argolid, which has been dated to 2300-1600 BC (Jameson et al., 1994), and a 10 m gravel terrace of the Alpheiós, overlying wood dated, (Q-2942), to 3660±30 BP, about 2000 BC (Grove and Rackham, in press) were deposited, at least in part, about the time of the Löbben ice advance of 1930-1250 BC (J. M. Grove, 1997; Holzhauser, 1997). More dates from these and other prehistoric alluviations are needed to confirm a correspondence between them and Alpine ice advances earlier in the Holocene, similar to those of the "Little Ice Age".

9. Conclusion

There appears to be a link between glacier advances and some types of extreme weather close by in the Mediterranean. Glaciers advance when a succession of snowy winters is reinforced by reduced ablation losses in spring and summer. The same weather systems that gave heavy autumn and winter snowfall in the Alps in the "Little Ice Age" and contributed to glacier advances of a similar kind earlier in the Holocene could also account for much of the flooding and alluviation responsible for the "younger fill" of river valleys in southern Europe. Such extreme weather events also occurred in regions like Crete, which, though remote from the meteorological systems of the western Alps, are also liable to be affected by outbursts of cold air into the Mediterranean generated by high pressure systems over central Europe (Alpert and Reisin, 1986). Rarely does it seem necessary to invoke forest clearance, the spread of agriculture, or the abandonment of land to explain the sedimentary features.

References

Alexandre, P.: 1987, *Le Climat en Europe au Moyen Age,* École des Hautes Études en Sciences Sociales, Paris.

Alpert, P. and Reisin, T.: 1986, 'An early winter polar airmass penetration to the eastern Mediterranean', *Monthly Weather Rev.* **114,** 1411-1418.

Alpert, P., Neeman, B.U. and Shay-el, Y.: 1990, 'Climatological analysis of Mediterranean cyclones using ECMWF data', *Tellus* **42,** 65-77.

Ballais, J-L. and Crambes, A.: 1992, 'Morphogenèse holocène, géosystèmes et anthropisation sur la montagne Sainte-Victoire', *Méditerranée* **75**.1.2, 29-41.

Barker, G.W. and Hunt, C.O.: 1995, 'Quaternary valley floor erosion and alluviation in the Biferno Valley, Molise, Italy: the role of tectonics, climate, sea level change, and human activity', in Lewin, J., Macklin, M.G. and Woodward, J.C. (eds.), 1995 *Mediterranean Quaternary River Environments*, Balkema, Rotterdam, 145-57.

Barriendos Vallvé, M. and Martín-Vide, J.: 1998, 'Secular climatic oscillations as indicated by catastrophic floods in the Spanish Mediterranean coastal area (14th-19th centuries)', *Clim.Change* **38,** 473-91.

Billi, P. and Rinaldi, M.: 1997, 'Human impact on sediment yield and channel dynamics in the Arno basin (central Italy)'in *Human Impact on Erosion and Sedimentation,* IAHS Publ. **245**, 301-311.

Bintliff, J.L.: 1982, 'Climatic change, archaeology and Quaternary science in the eastern Mediterranean region', in Harding, A. (ed.), *Climatic Change in Later Pre-History*, Edinburgh University Press, 143-161.

Braudel, F.: 1966, *The Mediterranean and the Mediterranean World in the Age of Philip II*, 2 vols, Second revised edition, Collins Sons, London.

Butzer, K.W.: 1980, 'Holocene alluvial sequences: problems of dating and correlation', in Cullingford, R.A., Davidson, D.A. and Lewin,J. (eds.), *Timescales in Geomorphology*, Wiley, Chichester, 131-142.

Calvet, M.: 1993, *Crues catastrophiques et vitesse de l'érosion dans les Pyrénées Orientales*, Servei Geologie de Catalunya, 93-101.

Calvet, M.: 1994, *Morphogenèse d'une montagne méditerranéenne; les Pyrénées orientales*, Presse Universitaire de Perpignan.

Calzolari, C. and Ungaro, F.: 1998, 'Geomorphic features of a badland biancane area (Central Italy): characterisation, distribution and quantitative spatial analysis', *Catena* **31**, 237-56.

Camuffo, D.: 1987, 'Freezing of the Venetian Lagoon since the 9th century A.D. in comparison to the climate of western Europe and England', *Clim. Change* **10**, 43-66.

Camuffo, D. and Enzi, S.: 1992, 'Reconstructing the climate of northern Italy from archive sources', in Bradley, R.S. and Jones, P.D. (eds.), *Climate Since A.D. 1500*, Routledge, London, 143-54.

Camuffo, D. and Enzi, S.: 1994, 'The climate of Italy from 1675 to 1715', in Frenzel, B., Pfister, C. and Gläser, B. (eds.), *Climatic trends and anomalies in Europe 1675-1715*, Fischer, Stuttgart, 243-54.

Camuffo, D. and Enzi, S.: 1996, 'The analysis of two bi-millennial series; Tiber and Po river floods', in Jones, P.D., Bradley, R.S. and Jouzel, J. (eds.), *Climatic Variations and Forcing Mechanisms of the last 2000 Years*, NATO ASI Series 1: Global Environmental Change, 41, Springer, Berlin.

Cencini, C.: 1998, 'Physical processes and human activities in the evolution of the Po delta, Italy', *Jnl. Coastal Res.* **14**, 774-93.

Cremaschi, M., Marchetti, M. and Ravazzi, C.: 1994, 'Geomorphological evidence for land surfaces cleared from forest in the central Po plain (northern Italy) during the Roman period', in Frenzel, B., Reisch, L. and Weiß, M.M. (eds.), *Evaluation of land surfaces cleared from forests in the Mediterranean region during the time of the Roman empire*, Fischer, Stuttgart, 119 - 132.

Davidson, N.S.: 1985, 'Northern Italy in the 1590's', in Clark, P. (ed.), *The Crisis of the 1590's*, Allen and Unwin, London, 157-76.

Díez, A.G., Salas, L., Díaz de Terán, J.R. and Cendrero, A.: 1996, 'Late Quaternary climate changes and mass movement frequency and magnitude in the Cantabrian region, Spain', *Geomorph.* **15**, 291-309.

Ely, I. L.: 1997, 'Response of extreme floods in the southwestern United States to climatic variations in the late Holocene', *Geomorph.* **19**, 175-201.

Enzel, Y. & Wells, S.G.: 1997, 'Extracting Holocene palaeohydrology and palaeoclimatology information from modern extreme flood events; an example from southern California', *Geomorph.* **19**, 203-226.

Flocas, A.A.: 1988, 'Frontal depressions over the Mediterranean sea and central Southern Europe', *Méditerranée* **66**, 4, 43-52.

Font Tullot, I.: 1988, *Historia del Clima de España*, Instituto Nacional de Meteorologia, Madrid.

Gamper, M. and Suter, J.: 1982, 'Postglaziale Klimageschichte der Schweizer Alpen', *Geographica Helvetica* **37**, 105-114.

Gellatly, A.F., Grove, J.M., Bücher. A., Latham, R. and Whalley, W.B.: 1995, 'Recent historical fluctuations of the Glacier du Taillon, Pyrénées', *Phys. Geogr.* **15**, 399-413.

Grove, A.T.: 1997, 'Classic in physical geography revisited: The Mediterranean Valleys by C. Vita-Finzi', *Progr. Phys. Geogr.* **21**, 251-6.

Grove, A.T. and Rackham, O.: 2000 (in press), *The Nature of Mediterranean Europe: an ecological history*, Yale U. P.

Grove, J.M.: 1988, *The Little Ice Age,* Methuen, London.
Grove, J.M.: 1996, 'The Century Timescale'in Driver, T.S. & Chapman, G.P. (eds.), *Timescales and Environmental Change,* Routledge, London and New York, 39-87.
Grove, J.M.: 1997, 'The spatial and temporal variations of glaciers during the Holocene in the Alps, Pyrenees, Tatra and Caucasus', in Frenzel, B., Boulton, G.S., Gläser. B. and Huckriede, U. (eds.), *Glacier fluctuations during the Holocene,* Fischer, Stuttgart, 95-103.
Grove, J.M. and Conterio, A.: 1994, 'Climate in the eastern and central Mediterranean, 1675 to 1715', in Frenzel, B (ed.), *Climatic Trends and Anomalies in Europe 1675-1715,* Fischer, Stuttgart, 13, 275-85.
Grove, J.M. and Conterio, A.: 1995, 'The climate of Crete in the sixteenth and seventeenth centuries', *Clim. Change* **30**, 223-47.
Guidoboni, E.: 1998, 'Human factors, extreme events and floods in the Lower Po Plain (Northern Italy) in the 16th Century', *Environment and History* **4**, 279-308.
Guillén, J. and Palanques, A.: 1997, 'A historical perspective of the morphological evolution in the lower Ebro river', *Environ. Geol.* **30**, 174-80.
Holzhauser, H.: 1984a, 'Rekonstruktion von Gletscherschwankungen mit Hilfe fossiler Hölzer', *Geogr. Helvetica* **39**, 3-15.
Holzhauser, H.: 1984b, 'Zur Geschichte der Aletschgletscher und des Fieschgletschers', *Phys. Geogr. (Geographisches Institut der Universität Zurich)* **13**, 1-148.
Holzhauser, H.: 1997, 'Fluctuations of the Grosser Aletsch Glacier and the Gorner Glacier during the last 3200 years: new results'. In Frenzel, B., Boulton, G.S., Gläser. B. and Huckriede, U. (eds.), *Glacier fluctuations during the Holocene,* Fischer, Stuttgart, 16, 35-58.
Holzhauser, H. and Zumbühl, H.J.: 1996, 'The history of the Lower Grindelwald Glacier during the last 2800 years - paleosols, fossil wood and historical pictorial records - new results', *Zeits. Geom. NF Suppl.* **104**, 95-127.
Hurrell, J.W.: 1995, 'Decadal trends in the North Atlantic Oscillation: regional temperature and precipitation', *Science* **269**, 676-79.
Innocenti, L. and Pranzini, E.: 1993, 'Geomorphological evolution and sedimentology of the Ombrone River delta, Italy', *J. Coastal Res.* **9**, 481-93.
Jacob, N.: 1997, 'La crue d'octobre 1940 dans la basse vallée du Tech (Roussillon) d'après les dossiers des sinistrés, *Ann. de Géogr.* **596**, 414-424.
Jameson, M.H., Runnels, C.N. and van Andel, T.H.: 1994, *A Greek Countryside: the southern Argolid from prehistory to the present day.* Stanford University Press.
Jeftic, L., Milliman, J.D. and Sestini, G.: 1992, *Climatic Change and the Mediterranean,* Arnold, London.
Kington, J.: 1999. 'The severe winter of 1697/8, *Weather* **54**, 43-9.
Kushnir, Y.: 1999, 'Europe's winter prospects', *Nature* **398**, 289-91.
Lamb, H.H.: 1963, 'On the nature of certain climatic epochs which differed from the modern (1900-1939) normal', in *Changes of Climate: Proceedings of the UNESCO/WMO Rome 1961 Symposium* (UNESCO Arid Zone Research Series XX) Paris, 125-150.
Le Roy Ladurie, E.: 1965, 'Le climat des XIe et XVIe siècles: séries comparées', *Annales Économies Sociétés Civilisations* **20**, 899-922.
Lindgrén, S. and Neumann, J.: 1981, 'The cold and wet year 1695 - a contemporary German account', *Clim. Change* **3**, 173-87.
Llasat, M-C. and Rodriguez, R.: 1997, 'Towards a regionalization of extreme rainfall events in the Mediterranean area', in Gustard, E. et al. (eds.), *Friend '97 - Regional Hydrology: Concepts and Models for Sustainable Water Resource Management (Proceedings of the Postojna, Slovenia, Conference, September-October 1997)* IAHS Publ. **246**, 215-22.
Lopez Gomez, A.: 1983, 'Las lluvias catastroficas mediterraneas', *Estudios Geograficos* **44**, 11-29.
Météo-France: 1995, *Inventaire des situations à précipitations diluviennes sur le Languedoc-Roussillon, la Provence-Alpes, Côte d'Azur et la Corse, Période 1958-1994* , Service Central d'Exploitation de la Météorologie, Aix-en-Provence.
Millán, M., Estrela, M.J. and Caselles, V.: 1995, 'Torrential precipitations on the Spanish east coast: the role of the Mediterranean sea surface temperature', *Atmos. Res.* **36**, 1-16.

Molina Sempere, C., Vidal-Abarca, M.R., and Suárez, M.L.: 1992, 'Floods in arid south-east Spanish areas: a historical and environmental review', in Rossi, G. et al. (eds.), *Coping with Floods*, Kluwer Academic, Dordrecht, Boston and London, 271-8.

Obled, C. and Tourasse, P.: 1992, 'Uncertainty in flood forecasting: a French case study', in Rossi,G., et al (eds.), *Coping with Floods,* Kluwer Academic, Dordrecht, Boston and London, 473-501.

Pardé, M.: 1924-5, *Le régime du Rhône, étude hydrologique,* 2 vols. Allier, Grenoble.

Pardé, M.: 1941, 'La formidable crue d'octobre 1940 dans les Pyrénées-Orientales', *Rev. géogr. des Pyrénées et du Sud-Ouest* **12**, 237-279.

Patzelt, G.: 1996, 'Former lakes dammed by glaciers and resulting floods in the Ötztal, Tyrol', *Mountain Res. and Devt.* **16**, 298-301.

Pavese, P.M., Banzon, V. and Colacino, M.: 1992, 'Three historical data series on floods and anomalous climatic events in Italy'in Bradley, R.S. and Jones, P.D. (eds.), *Climate since A.D. 1500,* Routledge, London, 155-70.

Pfister, C.: 1994, 'Spatial patterns of climatic change in Europe A.D. 1675-1715'in Frenzel, B., Pfister, C. and Glaser, B. (eds.), *Climatic trends and anomalies in Europe 1675-1715,* Fischer, Stuttgart, 287-316.

Pfister, C., Schwarz-Zanetti, G., Hochstrasser, F. and Wegmann, M.: 1998, 'The most severe winters of the fourteenth century in Central Europe compared to some analogues in the recent past', in Wishman, E., Frenzel, B. and Weiß, M.M. (eds.), *Documentary Climatic Evidence for 1750-1850 and the fourteenth century,* Fischer, Stuttgart, 45-61.

Pichard, G.: 1995, 'Les crues sur le bas Rhône de 1500 à nos jours. Pour une histoire hydro-climatique', *Méditerranée* **80**, 105-116.

Porta, G. (ed.): 1990-1, *Giovanni Villani, Nuova Cronica,* Bk XI chapter 168, 3 vols. Parma.

Potter, T.W.: 1976, 'Valleys and settlement, some new evidence', *World Archaeology* **8**, 206-19.

Probst, J-C.: 1989, 'Hydroclimatic fluctuations of some European rivers since 1800'in Petts, G.E. with Moller, H. and Roux, A.L. (eds.), *Historical Change of Large Alluvial Rivers: Western Europe,* Wiley, Chichester, New York, Brisbane, Toronto, Singapore.

Reichenbach, P., Cardinali, M., De Vita, P. and Guzzetti, F.: 1998, 'Regional hydrological thresholds for landslides and floods in the Tiber river basin (central Italy)', *Environ. Geol.* **35**, 146-59.

Rodrigo, F.S., Esteban-Parra, M.J., Castro-Diez, Y.: 1994, 'An attempt to reconstruct the rainfall regime of Andalusia from 1601 A.D. to 1650 A.D. using historical documents', *Clim. Change* **27**, 397-418.

Rodrigo, F.S., Esteban-Parra, M.J., Castro-Diez, Y.: 1996, 'Seasonal distribution of precipitation anomalies in Andalusia during the sixteenth and seventeenth centuries: a preliminary analysis from documentary sources', *Weather* **51**, 210-9.

Rumsby, B.T. and Macklin, M.G.: 1996, 'River response to the last neoglacial (the 'Little Ice Age') in northern, western and central Europe', in Branson, J., Brown, A.G. and Gregory, K.J. (eds.), *Global Continental Changes: the Context of Palaeohydrology,* Geol. Soc. London) Sp. Publi. No. 115, 217-233.

Russo, G. and Sacchini, A.: 1994, 'Brief survey on recurrencies of extreme rainfalls in Genoa, Italy', *Idójárás (Quarterly Journal of the Hungarian Meteorological Service)* **98**, 251-60.

Stanley, D.J. and Warne, A.G.: 1994, 'Worldwide initiation of Holocene marine deltas by deceleration of sea-level rise', *Science* **265**, 228-31.

Starkel, L.: 1991, 'Fluvial environments as a source of information on climatic changes and human impact in Europe', in Frenzel, E., Pons, A. and Gläser, B. (eds.), *Evaluation of Climate Proxy Data in Relation to the European Holocene,* Fischer, Stuttgart, 241-254.

Trousset, P. (ed.): 1987, *Déplacements des Lignes de Rivage en Méditerranée d'après les données de l'archaeologie,* Editions du CNRS, Paris.

Veggiani, A.: 1983, 'Degrado ambientale e dissesti idrogeologici indotti dal deterioramento climatico nell'Alto medioevo in Italia: I casi riminesi', *Studi Romagnoli* **34**, 123-146.

Vita-Finzi, C.: 1969, *The Mediterranean valleys,* Cambridge University Press.

Wigley, T.M.L.: 1992, 'Future climate of the Mediterranean Basin with particular emphasis on changes in precipitation', in Jeftic, L., Millman, J.D. and Sestini, G. (eds.), *Climatic change and the Mediterranean*, Edward Arnold, London, 15-44.

(Received 15 June 1998; in revised form 25 September 2000)

IS THERE AN INSECT SIGNAL FOR THE "LITTLE ICE AGE"?

PAUL C. BUCKLAND and PAT E. WAGNER

Department of Archaeology and Prehistory, University of Sheffield, Sheffield, S10 2TN, UK.

Abstract. The downturn in temperature in the late medieval period is likely to have had significant impact upon insect distribution. Despite the amount of study fossil insect faunas have been afforded, there is presently little convincing evidence of climatic, rather than human impact upon faunas during the "Little Ice Age". This probably reflects as much the paucity of suitable sites, as the overarching scale of habitat destruction by Man.

Key Words: "Little Ice Age"; Fossil insects; British Isles; Greenland; Iceland; USA

1. Introduction

The use of fossil insect assemblages in the interpretation of Quaternary climate change has been an established, if somewhat infrequently used technique for nearly thirty years (Buckland and Coope, 1991; Elias, 1994). Whilst the research has been eminently successful in the Weichselian (Wisconsinan) glacial, where changes have been both large in scale and precipitate (e.g. Ashworth, 1996; Brooks et al., 1997; Lowe et al., 1995), the subtleties of Holocene climatic, as opposed to human, impact have been less successfully resolved. In 1965, Osborne pointed out that fossil evidence showed that forest clearance had had a major impact upon the British insect fauna, and that this included a wide range of extinctions, or rather extirpations (*sensu* Elias, 1994) from the British fauna and many major retractions in range. Subsequent work has considerably extended the list of species no longer found in the British Isles (e.g. Osborne, 1969; Buckland and Kenward, 1973; Whitehouse, 1997). Most lost species appear to have an association with old established woodland and the term *Urwaldtiere* has been adopted from the German entomological literature to encompass them (Buckland, 1979). Many have retracted their distribution southwards and eastwards to the less manicured forests of Central Europe, but others have gone north, and not all extirpated species have any association with woodland. In an early attempt to dissect out human from climatic factors, Osborne (1976) suggested a correlation between periods of northern hemisphere glacier expansion and the occurrence of northern species in fossil beetle assemblages from Midland England. More recently, both Osborne (1997) and Wagner (1997) have reviewed the evidence for Holocene climate, but come to slightly differing conclusions. The former sees evidence of climate change over the Holocene and the latter, concentrating on the "Little Ice Age", finds it impossible to separate any climate signal from the overwhelming impact of Man upon the British fauna.

Climatic Change **48**: 137–149, 2001.

2. The "Little Ice Age" and Faunal Change

The overall pattern of climate and weather, in the sense of day-to-day changes, over the past four hundred years is so well known for most of the developed world from both direct observational data and other proxies (Grove, 1988), that it is unlikely that insect fossils have much to add. Much of the evidence for changes in distribution over the last century, since recording began is anecdotal and occasionally doubtful (*cf.* Osborne, 1997). The impact of the general cooling trend over the late medieval period through to the late nineteenth century has been invoked to explain many changes in insect distribution, but there are few sites where the Late Holocene fossil record has been afforded the detailed study which it deserves, and for many local extinctions it is simply not known when the latest record occurred. It has to be significant that the most recent site with an *Urwald* beetle fauna, with several species now extinct in Britain and many others of Red Data Book status (cf. Shirt, 1987; Hyman, 1992; 1994), at Thorne Moors, south Yorkshire, (Buckland and Dinnin, 1993; Whitehouse, 1997) dates from *c.*3000 BP. More recent sites show faunas which approach those of the present landscape, or at least that before the impact of herbicides and insecticides on crops and ivermectin in stock. Even where woodland survived, it was so heavily managed that it provided few refuges for the old forest fauna. The process which makes the present interglacial unique, the conversion of virtually continuous forest to culture steppe (*sensu* Hammond, 1974), was well under way by the Late Bronze Age, when the Wilsford Shaft in Wiltshire, despite its interesting mix of dung beetles, shows a surprisingly modern fauna (Osborne, 1969; 1989). By the Roman period this transformation appears to have been largely complete at least in the lowlands. Changes over the last two millennia, or at least in comparison with the earliest modern records of species, might therefore have a discernible climatic component, but it is rarely that easy. As Osborne (1997) points out, several species, like the large chafer *Polyphylla fullo* (L.) and the lucanid *Platycerus caraboides* (L.), both known as Holocene fossils, the former from Roman Hibaldstow in north Lincolnshire (Girling, pers. comm.), and the latter from mid-Holocene East Yorkshire (Wagner, in Constantine, 1994), appear in collections down to the middle of the nineteenth century. *P. caraboides* is yet another casualty of the loss of old woodland, since it appears to have survived in a few isolated areas until the 1830s (Hyman 1992). Like many others, they appear to hang on through the most severe part of the "Little Ice Age" and disappear when climate begins to improve. *P. fullo* is as likely to be a victim of the ploughing up of the heaths as of climate change, but, an active flier, its appearance in Kent (Hyman, 1992) may reflect the one summer when warm south-easterly winds coincided with its mass emergence on the Continent. The case of *P. fullo* highlights the problem of differentiating between the impact of an individual year's weather, facilitating very temporary increases in range, and an overall trend in climate which allows colonisation of new areas. Whilst the annual temperature may not change, a trend towards warmer summers and colder winters would benefit species with a more continental distribution, and many of the extinctions from Britain seem to exhibit this pattern. The warm winters of the oceanic fringe may lead to premature emergence of adults and subsequent death during late frosts. In addition, cool, wet summers may accumulate insufficient warmth to allow larval development to be completed. Where habitat fragmentation is extreme and species have low dispersal potential, individual difficult years may be sufficient to eliminate permanently populations isolated by human impact. As a result, extinction is stochastic, both spatially and temporally and surviving communities may appear to have retreated northwards, southwards or eastwards, as local factors come into play in their fates. The very process of forest clearance not only destroys a range of habitats for relatively immobile taxa, but also significantly alters microclimates, leading to other less easily explained changes in the faunas. In the Alps, south-east of Grenoble, Ponel (in Ponel et al., 1992) suggested that "Little Ice Age" cooling was sufficient to bring the obligate snow patch species *Helophorus glacialis* Villa to lower altitudes, whilst his palynological colleagues (*op. cit.*) preferred an interpretation based upon forest destruction leading to changes in the albedo and longer periods of snow cover.

2.1. THE DUNG FAUNA

Dung beetles, Scarabaeidae, tend to be restricted not to the herbivore producing the dung, but to the nature of the environment in which the droppings were deposited (Landin, 1961). As occupants of ephemeral biotopes, they are often highly mobile and range widely in pursuit of their food source (*cf.* Baker et al., 1993). Theoretically at least they should provide an ideal group with which to examine climate change. Osborne (1997) has commented upon changes in the frequency of dung beetles of the genera *Onthophagus* and *Aphodius* both geographically and temporally. The latter genus tends to dominate in dung pats in the British Isles at the present day, and the numbers of *Onthophagus* increase significantly southwards and eastwards across Europe. In the Bronze Age faunas from Wilsford (Osborne, 1969) and Bidford on Avon (Osborne, 1988), individuals of *Onthophagus* species, including several which are now doubtfully part of the British fauna, roughly equal those of *Aphodius*. In addition, *A. quadriguttatus* Hbst., no longer occurring in England, is present in the well deposit at Wilsford (Osborne, 1969). This may imply a warmer or at least more continental climate *c.* 3000 BP. By the Roman period the modern pattern of low numbers of *Onthophagus* in dung pats in Britain appears to have become established. Whilst changing frequencies may be a better clue to climate change, so few sites have been studied that the evidence has to be treated with caution. *A. quadriguttatus* has declined extensively in western Europe over the last century (Horion, 1956), and this pattern has been shared by many elements in the dung fauna, even before the widespread use of ivermectin and other biocides in domestic animals (e.g. Biström et al., 1991). Lemdahl and Thelaus (1989), however, in examining material from a medieval site at Falun in southern Sweden, favour a climate change explanation for the decline of *Heptaulacus testudinarius* (F.), although it has shown a similar reduction in range in England over the past one hundred years of agricultural improvement (Hyman, 1992). In Britain, changes in agriculture within the Roman period may have been sufficient to lead to local extinctions and changes in species frequency; but the problem remains that we do not know when many species disappeared from Britain. A similar pattern of decline has been noted in other larger scarabaeids, such as *Oxythyrea funesta* (Poda) (Baraud 1977) across Europe, and whilst this decline has been most noticeable at the limits of species' range, a common factor, shared with the dung fauna, has been the disappearance through improvement of old grasslands, a habitat perhaps now more rare than old woodland.

2.2. THE ATLANTIC ISLANDS

Working with fossil assemblages in areas remote from significant human interference might offer one possibility of examining "Little Ice Age" impacts. Böcher (1998; Böcher and Bennike, 1996) has noted two probable extinctions from the beetle fauna of Greenland in regions beyond the Southwest, where modern sheep farming, and earlier Norse activity has had a significant influence on the biota (*cf.* Buckland et al., 1998). The chrysomelid *Phratora polaris* Schneider was recovered from two localities in the High Arctic of north-east Greenland in deposits of *ca.* 8000 years ago (Böcher and Bennike, 1996). Feeding on Arctic willow, *Salix herbacea* (Larsson and Gígja, 1959), the species is known from Iceland, and must have formed one of the original early Holocene colonists across the North Atlantic (Buckland, 1988). A further example from the Norse farm site at Gården under Sandet (the Farm beneath the Sand) in the Western Settlement, east of Nuuk, may have originated in peat used as bedding (*cf.* Amorosi et al., 1998), but it does at least suggest a wider distribution during the Holocene; its extinction from Greenland may have taken place at any stage during the latter part of the Postglacial. The small rove beetle, *Euaesthetus laeviusculus* Mann. occurs across much of the Palaearctic in a range of swampy and bog habitats (Koch, 1989). Its presence on the palaeo-Eskimo Saqqaq site of Asummiut, on the Davis Strait, west of Søndre Strømfjord in west Greenland in the period *ca.* 2500-500 BC (Böcher, 1998) and at GUS in deposits of *ca.* 1000-1350 AD may reflect "Little Ice Age" cooling sufficient to remove the species from Greenland. It does, however, occur on a number of Lateglacial sites in western Europe (e.g. Drumurcher, Ireland; Coope et al., 1979), where the climate is likely to have been as severe as that prevailing in the Davis Strait region, and it may equally have been overlooked in the modern fauna of Greenland. Böcher

(Böcher and Fredskild, 1993) has also commented upon a number of apparent southward retractions in range in the faunas from the Saqqaq site on Qeqertasussak in Disko Bay, west Greenland, and these presumably reflect cooling more recently than *ca.* 3900-3100 ^{14}C yr. BP. As a cautionary tale over the use of modern distribution data in isolated regions, however, McGovern et al. (1983) had suggested that the occurrence of the small ladybird *Nephus redtenbacheri* Muls. in deposits on the farm site at Niaqussat in the Western Settlement of Norse Greenland provided evidence of "Little Ice Age" cooling, since modern records did not extend outside of the very south of the country (Böcher, 1988). A subsequent extensive pitfall trapping project by P. I. Buckland in 1995 found not only this species, but also a small rove beetle, *Mycetoporus nigrans* Mækl. not previously recorded from Greenland (Buckland et al., 1998).

One piece of anecdotal evidence from Iceland may be relevant to "Little Ice Age" impact. The sheep fleece louse, *Damalinia ovis* L. was, on the fossil evidence, clearly present on the North Atlantic islands from the beginnings of European settlement (Sadler, 1991), yet it first appears in the written record under the interesting name *Hafíslus* - "sea-ice louse" - as late as 1857 (Guðmundur Ólafsson, pers. comm.). Whilst the term may long predate its written form and earlier sources may be found, it seems probable that farmers had noted the correlation between severe sea ice years, when the hay growing season was seriously curtailed (Fridriksson, 1972; Ogilvie 1984), and weakness in sheep leading to heavy lice infestation of their fleeces.

Using the mutual climatic range model (MCR), developed by Atkinson et al. (1986), Perry (in Buckland et al., 1983) was able to examine the one apparent extinction in the Icelandic beetle fauna, and relate its disappearance to cooler summers during the "Little Ice Age". The presence of frequent tephra horizons in the peat stratigraphy allows close dating of deposits (e.g. Buckland et al., 1986). *Hydraena britteni* Joy is a widespread fossil in samples from sites in Mýrdalur and east of Markarfljót in Eyjafjallasveit in southern Iceland until the late medieval period when it appears to become extinct. MCR can be played backwards such that it is possible to input temperature parameters and output a list of species which could, given free access, occur at that locality. Southern Iceland is currently warm enough for *H. britteni*, but a 1° C decline in July temperatures would appear sufficient to eliminate it, and because of the remoteness of potential populations to feed re-immigration, the beetle has been cut out of the fauna. If the model is correct and the species has not succumbed to destruction and eutrophication of wetlands, this interpretation carries an important corollary: at least in Iceland, the coldest part of the Holocene would have to have been after AD 1500.

2.3. MEDIEVAL FAUNAS FROM EUROPE

On the whole, distributional data for MCR are at far too coarse a scale to resolve Holocene climates, and do not extend to those groups which have suffered major retractions in range, simply because habitat destruction has probably been more important than climate and their modern distributions are a reflection of this. Hellqvist and Lemdahl (1996) working on urban insect faunas from Uppsala in Sweden have applied the MCR method to their lists, and suggested that summers were "slightly warmer than present" *ca.* 1400. The evidence, however, is tenuous. Of the species which imply warmer conditions, one, *Amara fusca* Dej. has only a *cf.* identification and has therefore to be discounted, and the other, the bombardier beetle, *Brachinus crepitans* (L.) has historical records as far north as Uppland (Lindroth, 1986). Its decline, as in the Czech and Slovak Republics (Hurka, 1996) is probably more recent, a further reflection of changes in agriculture, again perhaps the reduction in unimproved short grazed calcareous grassland, its preferred habitat.

Wetlands are another habitat under threat on a worldwide scale from both drainage and pollution, and it is not surprising that there have been significant retractions in range amongst its fauna. Dinnin (1991) has discussed the case of the large hydrophilid *Hydrous* (=*Hydrophilus*) *piceus* (L.), which has nineteenth century and Late Holocene fossil records as far north as Yorkshire, but is presently restricted to southern England. The same post-medieval samples from Sutton Common in south Yorkshire also include the riparian ground beetle *Odacantha melanura* (L.), which now extends no further north than Norfolk (Luff 1998). Its habitat, *Phragmites* reed swamp, remains fairly extensive in the north of Britain, and a climatic parameter in its reduced range may be involved. In the Early Holocene *O. melanura* is recorded as far north as Dumfriesshire in southern Scotland (Bishop and Coope 1977), and this may imply, along with other species (Buckland and Sadler 1997), that the beginning of the present interglacial was significantly warmer than present, although a more continental regime with warmer summers and colder winters would can the same effect on insect distribution.

Despite the amount of research carried out on closely dated faunas from archaeological sites in the British Isles, it is only Kenward's work, principally in York, where there is sufficient data to examine climatic impact on the insect faunas through the medieval period. The available evidence was reviewed in 1976 (Addyman et al., 1976). In 1983 (Hall et al., 1983), Kenward still felt that several species, including the beetles *Anthicus bifasciatus* Rossi, *Acritus homoeopathicus* Woll. and *Apion difficile* Hbst., had withdrawn southwards because of climatic deterioration since the Anglo-Scandinavian period. By 1995, however, he had become more circumspect (Kenward and Hall 1995), and only the distributional shift of the bug *Heterogaster urticae* is accorded a climatic interpretation. Archaeological faunas suffer not only from the problem of species which have been dispersed accidentally by man, best evinced by the pests of stored products (Buckland, 1990), but also from the past occurrence of urban environments which, in our modern more hygienic society, have now all but disappeared. The latter point has been discussed by Kenward (1975) in relation to the colydiid *Aglenus brunneus* Gyll., but is also relevant to *A. bifasciatus* and many other species. The deep accumulations of dung, plant and other animal debris which so characterised towns until the advent of the motor vehicle and organised refuse disposal, provided artificially warm habitats for many species north of their natural range, and it only required the right circumstances of accidental introduction in ballast or dunnage (*cf.* Sadler, 1991) to establish temporary populations. Some, like the scarabaeid *Oxyomus silvestris* (Scop.) from medieval Bessastaðir in Iceland (Amorosi et al., 1992), may appear in the fossil record, but have failed to establish themselves, and others, like *A. difficile,* which feeds upon dyer's greenweed, *Genista tinctoria* L. would have been brought into York with the plant on numerous occasions; its host has now virtually disappeared from the region (Kenward and Hall, 1995, 772-3). Of the Coleoptera which Kenward (in Hall et al., 1983) saw as possibly having a climatic component to their reduced distribution, only *A. homoeopathicus* appears to have no direct association with human activity; a further record from the Roman villa at Dalton Parlours, W. Yorkshire, (Sudell, 1990) is a misidentification of the congener *A. nigricornis* (Hoff.). This small histerid is currently recorded only from Southeast England (Hyman, 1992) and Worcestershire (Whitehead, 1996), and is frequently associated with burnt ground (Welch, 1968). On the Continent, however, Koch (1989a) notes it also from fermenting hay and mouldering stable manure, and it may have once found suitable habitats in the urban heat islands of dung and waste animal fodder.

Compared with the earlier medieval period , few later medieval and post-medieval sites have been studied, and there appear to be few examples of extirpations which could be linked to the "Little Ice Age". Girling (1984) found an elytron of the whirligig beetle *Gyrinus colymbus* Er. in medieval deposits from a ditch on the site of the Austin Friary at Leicester. The remainder of the fauna is dominated by synathropic species (Girling, 1981), and the gyrinid may have come in with other water beetles from the adjacent River Soar, but it is also possible that peat was used on the site and that the specimen had been incorporated from deposits much older than medieval. *G. colymbus*' present distribution in Europe excludes the British Isles and ranges northwards as far as southern Zealand in Denmark, and eastwards across Germany, Poland, Switzerland, Austria, Hungary. Italy, and the Balkans, to Turkey and southern Russia (Holmén, 1987). Koch (1989a) gives the most westerly records as Holstein and Bavaria, but Girling (1984) notes a single locality on the west coast of France, and if this is correct, a simple decline of approximately 1° C in July temperatures would appear sufficient to remove this species from the British fauna. The cucujid *Airaphilus elongatus* (Gyll.) is recorded by Osborne from sites of the Roman period at Stourport (Osborne 1996) and Droitwich (Osborne 1974) in Worcestershire, England. An inhabitant of wet, boggy meadows on the Continent (Koch 1989b), its post-Roman disappearance, if such it is, may reflect climatic deterioration.

Buckland (1975), in reviewing the evidence for the declining distribution of the death watch beetle *Xestobium rufovillosum* DeG., suggested that summer temperatures in northern England were sufficiently warm to enable it to disperse naturally to its host trees, largely oak, during the medieval period, and that its later restriction to structural timber in medieval buildings reflected colder summers. Kenward (Addyman et al., 1976; Hall et al., 1983; Kenward and Hall, 1995) has presented similar arguments for the southward retraction of the nettle bug *Heterogaster urticae* (F.). Buckland (1981) has argued that the rapid expansion of another bug, *Ischnodemus sabuleti* Fall./*quadratus* Fieb. during the present century may be the result of the "post-Little Ice Age" warming trend. The two species, which cannot be separated on the fossil thorax from Bronze Age Brigg in north Lincolnshire, were restricted to Surrey and Folkestone Warren, Kent, at the end of the nineteenth century, but *I. sabuleti* has since re-established its northern distribution.

2.4. NORTH AMERICA

Given the size of the continent, few Late Holocene sites for fossil insects have been examined in the USA and Canada. In the relative flatlands of Minnesota, Schwert and Ashworth (1985) found no evidence for climatic impact upon insect faunas over the last 3500 years. In contrast, Elias (1996) has argued that altitudinal changes in species distribution, apparent in fossil assemblages from the Rockies, accurately reflect the "Little Ice Age". Although the area of Colorado where most of Elias sites lie should be relatively far from large scale human disturbance, and have sufficient relief for small scale climate change to have significant impact in insect distribution, Ponel and Richoux (1997) have urged caution in the interpretation of high altitude assemblages, particularly because of the impact of local climatic factors, including those leading to the incorporation of lowland species by updraft. MCR, used by Elias (1996), is also liable to skewing of results by a single individual of a species with an apparently very limited climatic range, which may have arrived in the deposit during the one atypically cool year in an otherwise warm period, and it cannot be employed to evaluate changes in relative frequency of species. Other taphonomic considerations are also significant. Perversely, a depression of the treeline as a result of cooling may lead to increased representation of lowland taxa in fossil assemblages, simply because the filtering effect of trees has been removed. This does not, however, invalidate Elias' conclusions, but serves to show that a careful analysis of situation and site taphonomy and comparison with modern detailed collecting data are essential before the nuance of the impact of climate change on past insect faunas can be interpreted.

3. Conclusion

To the question posed as the title to this paper, 'Is there a fossil insect signal for the "Little Ice Age"?, the answer has to be a guarded, 'Yes, probably'. No site has been studied which provides a complete series of samples through the last two millennia in an environment which was sufficiently stable or free from human interference for a clear climatic signal to be evident. The few fragments available, from North America, Greenland to Sweden and southwards to England, provide several hints, but the overwhelming picture is one of the progressive fragmentation and destruction of habitats even in the most remote of places. The scale of change has been unprecedented, in that even after the creation of *tabulae rasae* by repeated glaciation there were always refuges from whence the biota returned. The loss of these places will leave the next interglacial much the poorer in species.

Acknowledgements

Research in Iceland and Greenland was funded initially by the Leverhulme Trust, UK, and later by a major research award to NABO, the North Atlantic Biocultural Organisation, by US National Science Foundation, (Grant no. OPP-9523529). This paper owes much to discussion with several colleagues in the North Atlantic region, including Kevin Edwards and Guðmundur Ólafsson. Tim Atkinson, Scott Elias, Eva Panagiotakopulu and Jon Sadler kindly commented upon the text.

References

Addyman, P. V., Hood, J. S. R., Kenward, H. K., MacGregor, A. and Williams, D.: 1976, Palaeoclimate in urban environmental archaeology at York, England : problems and potential. *World Archaeol.* **8**, 220-233.

Amorosi, T., Buckland, P. C., Ólafsson, G., Sadler, J. P., and Skidmore, P.: 1992, Site status and the palaeoecological record: a discussion of the results from Bessastaðir, Iceland, in: Morris C. D. and Rackham, D. J. (eds.) *Norse and later settlement and subsistence in the North Atlantic.* Glasgow, Dept. of Archaeology, University of Glasgow, pp. 169-192.

Amorosi, T., Buckland, P. C., Edwards, K. J., Mainland, I., McGovern, T. H., Sadler, J. P. and Skidmore, P. (1998) They did not live by grass alone: the politics and palaeoecology of fodder on the North Atlantic islands. *Environmental Archaeology* **1**, 41-54.

Ashworth, A. C.: 1996, The response of beetles to Quaternary climatic changes, in: Huntley, B., Cramer, W., Morgan, A. V., Prentice, H. C. and Allen, J R. M. (eds.) *Past and future rapid*

environmental changes : the spatial and evolutionary response of terrestrial biota, Berlin, Springer, pp. 119-127.
Atkinson, T. C., Briffa, K. R., Coope, G. R., Joachim, J. M. and Perry, D. W.: 1986, Climatic calibration of coleopteran data, in: Berglund, B. E. (ed.) *Handbook of Holocene Palaeoecology and Palaeohydrology*, Chichester , J. Wiley and Son. pp. 851-858.
Baker, R. G., Schwert, D. P., Bettis III, E. A., and Chumbley, C. A.: 1993, Impact of Euro-American settlement on a riparian landscape in northeast Iowa, midwestern USA: an integrated approach based on historical evidence, floodplain sediments, fossil pollen, plant macrofossils and insects. *The Holocene* **3**, 314-323.
Baraud, J.: 1977, Coléoptères Scarabaeoidea. Faune de l'Europe occidentale. Belgique, France, Grande Bretagne, Italie, Péninsule Ibérique. *Nouvelle Revue d'Entomologie, Suppl.* **VII, 1**. Toulouse.
Biström, O., Silfverberg, H. and Rutanen, I.: 1991, Abundance and distribution of coprophilous Histerini (Histeridae) and *Onthophagus* and *Aphodius* (Scarabaeidae) in Finland (Coleoptera). *Entomologica Fennica* **2**, 53-66.
Bishop, W. W. and Coope, G. R.: 1977, Stratigraphical and faunal Evidence for Lateglacial and Early Flandrian environments in South-West Scotland, in: Gray, J. M. and Lowe, J. J. (eds.) *Studies in the Scottish Lateglacial Environment*, Oxford, Pergamon. pp. 61-88.
Böcher, J.: 1988, The Coleoptera of Greenland. *Meddelelser om Grønland, Biosciences*, **26**.
Böcher, J.: 1998, Insect remains from Asumiut, in: Arneborg, J. and Gulløv, H. C. (eds.) *Man, Culture and Environment in Ancient Greenland*, Copenhagen , Danish National Museum and Danish Polar Centre. pp. 133-136.
Böcher, J. and Bennike, O.: 1996, Early Holocene insect and plant remains from Jameson Land, East Greenland. *Boreas* **25**, 187-193.
Böcher, J. and Fredskild, B.: 1993, Plant and arthropod remains from the palaeo-Eskimo site on Qeqertasussuk, West Greenland. *Meddelelser om Grønland, Geoscience*, **30**.
Brooks, S. J., Mayle, F. E. and Lowe, J. J.: 1997, The Late Devensian Lateglacial palaeoenvironmental record from Whitrig Bog, SE Scotland. 2. Chironomidae (Insecta: Diptera). *Boreas* **26**, 297-308.
Buckland, P. C.: 1975, Synanthropy and the deathwatch : a discussion. *Naturalist* (Hull), **100** (no. **933**) 37-42.
Buckland, P. C.: 1979, *Thorne Moors : a palaeoecological study of a Bronze Age site (a contribution to the history of the British insect fauna)*. Birmingham, University of Birmingham, Department of Geography Occasional Publication **8**.
Buckland, P. C.: 1981, Insect remains from beneath the Brigg 'Raft', in: S. McGrail (ed.) *The Brigg 'Raft' and her prehistoric environment*, Oxford, British Archaeological Reports, **89**, pp. 155-175.
Buckland, P. C.: 1988, North Atlantic faunal connections - introduction or endemics? *Entomologica Scandinavica Suppl.* **32**, 7-29.
Buckland, P. C.: 1990, Granaries stores and insects. The archaeology of insect synanthropy, in: Fournier, D. and Sigaut, F. (eds.) *La préparation alimentaire des céréales*, Rapports présentés à la Table ronde, Ravello au Centre Universitaire pour les Biens culturels, avril 1988. Rixensart, PACT. pp. 69-81.
Buckland, P. C., Buckland, P. I. and Skidmore, P.: 1998, Insect remains from GUS: an interim report, in Arneborg, J. and Gulløv, H. C. (eds.) *Man, Culture and Environment in Ancient Greenland*, Copenhagen, Danish National Museum and Danish Polar Centre. pp. 74-79.
Buckland, P. C. and Coope, G. R.: 1991, *A Bibliography and Literature Review of Quaternary Entomology*. Sheffield, John Collis Publications, University of Sheffield.
Buckland, P.C. and Dinnin, M. J.: 1993, Holocene Woodlands: the Fossil Insect Evidence, in: Kirby, K. and Drake, C. M. (eds.) *Dead wood matters: the ecology and conservation of saproxylic invertebrates in Britain*, Peterborough, English Nature Science **7**, 6-20.
Buckland, P. C. and Dinnin, M. H.: 1998, Insect faunas at Landnám : a palaeoentomological study at Tjørnuvík, Streymoy, Faroe Islands. *Fróðskaparrit*, **46**, 277-286.

Buckland, P. C., Edwards, K. J. and Sadler, J. P.: 1998, Early Holocene investigations at Saksunardalur and the origins of the Faroese biota. *Fróðskaparrit,* **46**, 259-266.

Buckland, P. C., Gerrard, A. J., Larsen, G., Perry, D. W., Savory, D. R. and Sveinbjarnardóttir, G.: 1986, Late Holocene Palaeoecology at Ketilsstaðir in Mýrdalur, South Iceland. *Jökull*, **36**, 41-55.

Buckland, P. C. and Kenward, H. K.: 1973, Thorne Moor : a palaeoecological study of a Bronze Age site. *Nature* (London) **241**, 405-406.

Buckland, P. C., Perry, D. and Sveinbjarnardóttir, G.: 1983, *Hydraena britteni* Joy (Coleoptera, Hydraenidae) fundin á Íslandi í setlögum frá því seint á nútima. *Náttúrufræðingurinn* **52**, 37-44.

Buckland, P. C. and Sadler, J. P.: 1997, Insects, in: Edwards, K.J. and Ralston, I.B.M. (eds.) *Scotland. Environment and Archaeology 8,000 BC to AD 1000*, Chichester, J. Wiley & Sons. pp. 105-108.

Constantine, B.: 1994, *Dorcus parallelipipedus* (L.) (Col., Lucanidae) in East Yorkshire. *Entomologist's Monthly Magazine* **130**, 62.

Coope, G. R., Dickson, J. H., McCutcheon, J. A. and Mitchell, G. F.: 1979, The Lateglacial and Early Postglacial Deposit at Drumurcher, Co. Monaghan. *Proc. Royal Irish Acad.* **B79**, 63-85.

Dinnin, M. H.: 1991, The sub-fossil occurrence of the Greater Silver Water-Beetle *Hydrophilus piceus* (L.) at Shirley Pool, South Yorkshire. *Naturalist* (Hull) **116**, 57-59.

Dugmore, A. J. and Buckland, P. C.: 1991, Tephrochronology and Late Holocene Soil Erosion in South Iceland, in: Maizels, J. K. and Caseldine, C. (eds.) *Environmental Change in Iceland: Past and Present.* Dordtrecht, Kluwer Academic Press. pp. 147-160.

Elias, S. A.: 1994, *Quaternary Insects and Their Environments*. Washington, Smithsonian Institution Press.

Elias, S. A.: 1996, Late Pleistocene and Holocene seasonal temperatures reconstructed from fossil beetle assemblages in the Rocky Mountains. *Quaternary Research* **46**, 311-318.

Fridriksson, S.: 1972, Grass and grass utilisation in Iceland. *Ecology* **53**, 785-796.

Girling, M. A.: 1981, The environmental evidence, in: Mellor, J. E. and Pearce, T., *The Austin Friars, Leicester*, London, Council for British Archaeology, Research Report **35**, 169-173.

Girling, M. A.: 1984, A Little Ice Age extinction of a water beetle from Britain. *Boreas*, **13**, 1-4.

Grove, J.: 1988, *The Little Ice Age.* London, Methuen.

Hall, A. R., Kenward, H. K., Williams, D. and Greig, J. R. A.: 1983, *Environment and Living Conditions at Two Anglo-Scandinavian Sites*. Archaeology of York, **14/4**. London, Council for British Archaeology for York Archaeological Trust.

Hammond, P. M.: 1974, Changes in the British coleopterous fauna, in: Hawksworth, D. L. (ed.) *The changing flora and fauna of Britain*, London, Academic Press, Systematics Association Special Volume **6**. pp. 323- 369.

Hellqvist, M. and Lemdahl, G.: 1996, Insect assemblages and local environment in the mediaeval town of Uppsala, Sweden. *J. Archaeol. Sci.* **23**, 873-881.

Holmén, M.: 1987, The Aquatic Adephaga (Coleoptera) of Fennoscandia and Denmark. 1. Gyrinidae, Haliplidae, Hygrobiidae and Noteridae. *Fauna Entomologica Scandinavica*, **20**. Leiden, E.J.Brill.

Horion, A.: 1956, *Faunistik der mitteleuropäischen Käfer. VI: Lamellicornia (Scarabaeidae-Lucanidae).* Überlingen, A. Freyel.

Hurka, K.: 1996, *Carabidae of the Czech and Slovak Republics*. Zlín, Kabourek.

Hyman, P. S.: 1992, *A review of the scarce and threatened Coleoptera of Great Britain, Part 1* (Revised and updated by M.S.Parsons). Peterborough, UK Joint Nature Conservation Committee.

Hyman, P. S.: 1994, *A review of the scarce and threatened Coleoptera of Great Britain, Part 2* (Revised and updated by M.S.Parsons). Peterborough, UK Joint Nature Conservation Committee.

Jessen, K. and Rasmussen, R.: 1922, Et profil gennem en Torvemose paa Faeroerne. *Danmarks geologiske undersøgelse, 4R.* **1**, nr. **13**.

Kenward, H. K.: 1975, The biological and archaeological implications of the beetle *Aglenus brunneus* (Gyllenhall) in ancient faunas. *J. Archaeol. Sci.* **2**, 63-69.

Kenward, H. K. and Hall, A. R.: 1995, *Biological evidence from 16-22 Coppergate*. Archaeology of York, **14/7**. York, Council for British Archaeology for York Archaeological Trust.

Koch, K.: 1989a, *Die Käfer Mitteleuropas. Ökologie*, 1. Krefeld,. Goecke and Evers.

Koch, K.: 1989b, *Die Käfer Mitteleuropas. Ökologie*, 2. Krefeld,.Goecke and Evers.

Landin, B.-O.: 1961, Ecological studies on dung-beetles. *Opuscula Entomologica*, Suppl. **19**.

Larsson, S.G. and Gígja, G.: 1959, *Coleoptera. Zoology of Iceland* **43a**. Copenhagen, Munksgaard.

Lemdahl, G. and Thelaus, M.: 1989, Subfossila skalbaggsfynd från medeltida Halmstad. *Entomologiske Tids.* **110**, 39-41.

Lindroth, C.H.: 1986, The Carabidae (Coleoptera) of Fennoscandia and Denmark. *Fauna Entomologica Scandinavica* **15,2**. Leiden/Copenhagen, E.J.Brill/Scandinavian Science Press Ltd.

Lowe, J. J., Coope, G. R., Sheldrake, C., Harkness, D. D. and Walker, M. J. C.: 1995, Direct comparison of UK temperatures and Greenland snow accumulation rates, 15000-12000 yr ago. *J. Quat. Sci.* **10**, 175-180.

Luff, M. L.: 1998, *Provisional atlas of the ground beetles (Coleoptera, Carabidae) of Britain.* Abbots Ripton, Centre for Ecology & Hydrology, Biological Records Centre.

McGovern, T. H., Buckland, P. C., Savory, D., Sveinbjarnardøttir, G., Andreasen, C. and Skidmore, P.: 1983, A Study of the Faunal and Floral Remains from Two Norse Farms in the Western Settlement, Greenland. *Arctic Anthropol.* **20**, 93-120.

Ogilvie, A. E. J.: 1984, The impact of climate on grass growth and hay yield in Iceland. A.D. 1601 to 1780, in: *Climate changes on a yearly to millennial basis* (eds Mörner N.-A. and Karlén, W.), Reidel, Dordrecht, 343-352.

Osborne, P. J.: 1965, The effect of forest clearance on the distribution of the British lnsect fauna. *Proc. XII International Cong. Entomol. London, 1964*, pp. 466-457.

Osborne, P. J.: 1969, An insect fauna of Late Bronze Age date from Wilsford, Wiltshire. *J. Anim. Ecol.* **38**, 555-566.

Osborne, P. J.: 1974, *Airaphilus elongatus* (Gyll.) (Col., Cucujidae) present in Britain in Roman times. *Entomologist's monthly Magazine* **109**, 239.

Osborne, P. J.: 1976, Evidence from the insects of climatic variations during the Flandrian period a preliminary note. *World Archaeol.* **8**, 150-158.

Osborne, P. J.: 1988, A Late Bronze Age insect fauna from the River Avon, Warwickshire, England: its implications for the terrestrial and fluvial environment and for climate. *J. Archaeol. Sci.* **15**, 715-727.

Osborne, P. J. (1989) Insects, in: Ashbee, P., Bell, M. and Proudfoot, E., *Wilsford Shaft. Excavations 1960-62*, English Heritage, London, pp. 96-99 + fiche C1-7.

Osborne, P. J.: 1996, An insect fauna of Roman date from Stourport, Worcestershire, U. K., and its environmental implications. *Circaea, The Journal of the Association for Environmental Archaeology* **2**, 183-189.

Osborne, P. J.: 1997, Insects, Man and climate in the British Holocene, in: Ashworth, A.C., Buckland, P.C. and Sadler, J.P. (eds.) *Studies in Quaternary entomology - an inordinate fondness for insects. Quat. Proc.* **5**, 193-198.

Ponel, P., De Beaulieu, J.-L. and Tobolski, K.: 1992, Holocene palaeoenvironments at the timberline in the Taillefer Massif, French Alps: a study of pollen, plant macrofossils and fossil insects. *The Holocene* **2**, 117-130.

Ponel, P. and Richoux, P.: 1997, Difficultés d'interprétation des assemblages de Coléoptères fossiles Quaternaires en milieu d'altitude. *GEOBIOS* **M.S. 21**, 213-219.

Sadler, J. P.: 1991, Beetles, boats and biogeography. *Acta Archaeologica* **61**, 199-211.

Schwert, D. P. and Ashworth, A. C.: 1985, Fossil evidence for Late Holocene faunal stability in southern Minnesota (Coleoptera). *The Coleopterists' Bulletin* **39**, 67-79.

Shirt, D.B. (ed.): 1987, *British Red Data Books. 2. Insects*. Peterborough, Nature Conservancy Council.

Sudell, T. L.: 1990, The insects from Well 1, in: Wrathmell, S. and Nicholson, A. (eds.), *Dalton Parlours: Iron Age settlement and Roman Villa Site,* Yorkshire Archaeology **3**, 267-271. Wakefield, West Yorkshire Archaeology Service.

Wagner, P. E.: 1997, Human impact or cooling climate? The "Little Ice Age" and the beetle fauna of the British Isles, in: Ashworth, A. C., Buckland, P. C. and Sadler, J. P. (eds.) *Studies in Quaternary entomology - an inordinate fondness for insects. Quat. Proc.* **5**, 269-276.

Welch, R.C.: 1968, *Acritus homeopathicus* Woll. (Col., Histeridae) in Huntingdonshire. *Entomologist's monthly Magazine* **104**, 122.

Whitehead, P.: 1996, The notable arboreal Coleoptera of Bredon Hill, Worcestershire, England. *The Coleopterist* **5**, 45-53.

Whitehouse, N. J.: 1997, Silent witnesses: an '*Urwald*' fossil insect assemblage from Thorne Moors. *Thorne and Hatfield Moors Papers* **4**, 19-54.

(Received 9 October 1998; in revised form 22 June 2000)

CAN LICHENOMETRY BE USED TO DATE THE "LITTLE ICE AGE" GLACIAL MAXIMUM IN ICELAND?

MARTIN P. KIRKBRIDE[1] and ANDREW J. DUGMORE[2]

[1]*Department of Geography, University of Dundee, Dundee DD1 4HN, Scotland*
[2]*Department of Geography, University of Edinburgh, Drummond Street, Edinburgh EH8 9XP, Scotland*

Abstract. In Iceland, there are numerous examples of glacier advances dated to the latter half of the last century. However, in marked contrast to the Alps and northern Europe, the record of historical-age moraines before ca. 1850 is rather sparse. This paper examines to what extent this pattern reflects the actual history of glacier fluctuations, and to what extent it could be a function of preservation and dating of the geomorphological record. Measurements of *Rhizocarpon geographicum* sp. lichen thalli on ice-marginal moraines and sandur form the basis for the late "Little Ice Age" glacial chronology in south Iceland. Recent studies have converged on the view of a maximum glacier extent in the late nineteenth century, and not during earlier and possibly colder parts of the "Little Ice Age". Here, results of independent dating of moraine ridges and a *jökulhlaup* deposit demonstrate that conventional lichenometric techniques tend to cluster dates of these landforms to the 1860s-1880s, underestimating tephrochronological dates on the same landforms by >100 yr in some cases. A mid-eighteenth century glacial maximum may be better represented in the landform record than hitherto thought, with implications for reconstruction of North Atlantic circulation patterns.

Key Words: glaciers; Iceland; lichenometry; tephrochronology

1. Introduction: The Timing of the "Little Ice Age" Glacial Maximum in South Iceland

Iceland's central location within the North Atlantic Ocean places it within the zone of convergence of cold polar and warm sub-tropical water and air masses. Migration of the Polar Front during the "Little Ice Age" has caused significant climatic variation in the region, as the relative effects of meridional and zonal circulation have changed over time. Consequent environmental changes are documented in records such as winter sea-ice cover (Ogilvie, 1984, 1992), the fluctuating fortunes of the cod fishery (Jonsson, 1965; Lamb, 1979), and as periods of glacial advance and retreat (Grove, 1988; Guđmundsson, 1997). Iceland's "Little Ice Age" moraine chronology is based on historical records and lichenometric dating. A notable feature of the established glacial history is the uncertainty surrounding the timing of the glacial maximum of the "Little Ice Age". At outlet glaciers of the large icecaps the "Little Ice Age" maximum apparently occurred in the mid- to late-nineteenth century, but at steeper glaciers in the mid- eighteenth century (Björnsson, 1979). However, the extent of these two maxima appears similar at many glaciers, with advances in the later period

Climatic Change **48**: 151–167, 2001.

being only slightly greater or less extensive than in the earlier period (Thórarinsson, 1943).

Recent lichenometric studies of glacier forelands have produced detailed chronologies on a ridge-by-ridge basis (Caseldine, 1991; Stötter, 1991; Kugelmann, 1991; Evans et al., 1999). This level of detail contrasts with the rather vague information on glacier terminus locations contained in documentary accounts (cf. Thórarinsson, 1943). Lichenometric studies in south Iceland place the greatest "Little Ice Age" extent of glaciers in the late nineteenth century, consequently according the status of "Little Ice Age maximum" to advances in the 1880s-1890s even at smaller, steeper glaciers. For example, Evans et al. (1999, p.38) summarise this emphasis by stating:

> Most Icelandic glaciers have been predominantly in a state of recession since their "Little Ice Age" maximum around the turn of the twentieth century, though exceptions have been documented.

Yet these advances postdate the periods of coolest "Little Ice Age" climate in the late-seventeenth and mid-eighteenth centuries (Lamb, 1979), and contrasts with Thórarinsson's (1943) recognition of major "highstands" of ice in the 1750s and 1840s, and lesser events in the decades following 1710, 1810, and around 1890. The apparent tight clustering of late-nineteenth century dates for the "Little Ice Age" maximum in the south contrasts with northern Iceland, where lichenometrically-dated "Little Ice Age" maximum moraines at 24 glaciers (Caseldine and Stötter, 1993) are spread across an interval of over a century (1813-1921).

If the existing record of glacier fluctuations in south Iceland is accepted, then a climatic explanation must be sought. In Iceland, such an explanation involves changes in ocean and atmospheric circulation associated with movements of the North Atlantic sub-polar convergence zone (Lamb, 1979). Comparison with northern Norway suggests that the lack of moraine evidence for glacier advances predating the late-nineteenth century may reflect suppression of winter precipitation during the coldest periods, and a reduced zonal influence on climate (Ballantyne, 1990). In Iceland, such a pattern would indicate isolation of the island from the warmer, humid influence of the Irminger Current by southward incursions of polar water, associated with high incidences of sea ice along the north and east coasts. Accurate dating is clearly an essential precursor to such interpretations of palaeo-circulation patterns and inter-regional comparisons.

A key question is to what extent the nineteenth century maximum is a real reflection of glacial history, and to what extent the preservation and dating of the palaeoenvironmental record may have influenced the apparent pattern of glacier fluctuations. In this paper, the established lichenometric method is scrutinised using two approaches. First, moraine and flood (*hlaup*) deposits are dated both

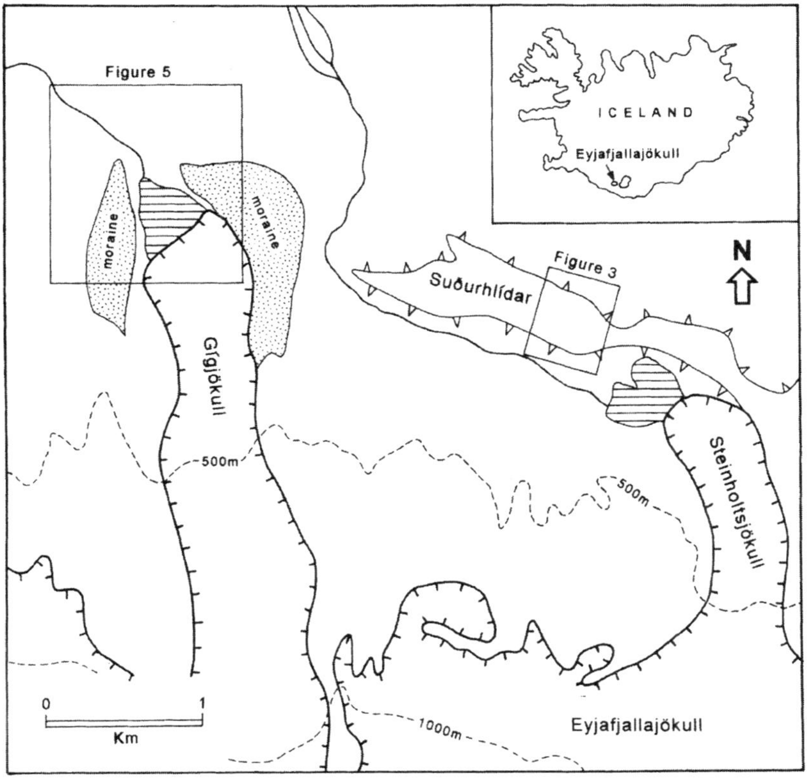

Figure 1. Location map of Steinholtsjökull and Gígjökull, northern outlets from the Eyjafjallajökull icecap. Boxes show the locations of the geomorphological maps of Figures 3 and 5.

by lichenometry and by tephrochronology, and the independent estimates of landform age used to determine the time beyond which the reliability of lichenometry becomes questionable. Second, three procedures for assessing the representative lichen size on a deposit are applied to the samples to investigate the sensitivity of the age estimates to the method employed. The results are used to review the reliability of late-nineteenth century glacier advances in the context of the "Little Ice Age", with implications for wider patterns of climatic variability in the North Atlantic region. Field work was undertaken at two small outlet glaciers in southern Iceland (Figure 1). The study exploits the opportunity presented in this area by a well-established tephrochronology (Thórarinsson, 1967; 1975; 1980; Einarsson et al., 1980; Larsen 1984; Larsen and Thórarinsson, 1977; Dugmore 1987; 1989; Haflidason et al., 1992; Larsen et al., 1999), allowing independent dating of moraines and *jökulhlaup* deposits from 1821-3 and the ca. 1740s, and forms part of a wider study of Holocene glacier fluctuations in Iceland (Kirkbride and Dugmore, submitted).

2. General Problems of Age Assignment Using Lichen Samples

There are three stages of development and application of lichenometry where problems may arise. The initial construction of a regional growth curve relies on the identification of rock surfaces of known age (Jaksch, 1970, 1975; Maizels and Dugmore, 1985; Evans et al., 1999). Various methods are then available for measuring thallus size. Some workers use the long axis of each thallus (Maizels and Dugmore, 1985; Evans et al., 1999), others the shortest axis or inscribed circle method (Locke et al., 1979; Caseldine, 1983, 1985; Gordon and Sharp, 1983; Häberle, 1991). Finally, a choice has to be made between different procedures for selecting a thallus size deemed to be representative of the deposit age. Commonly, the long axis of the single largest thallus (Maizels and Dugmore, 1985; Kugelman, 1991), or the mean of the five largest thalli found on a deposit (Evans et al., 1999) are used for speed and convenience. Most workers employing these procedures concentrate on minimum cut-off sizes during sampling, and do not record data allowing comparisons between ages derived from different procedures. Some workers advocate population-based size-frequency methods for estimating representative size (Benedict, 1985; Caseldine, 1991). Techniques for lichen measurement have been comprehensively described elsewhere (e.g. Locke et al., 1979; Worsley, 1981).

In Iceland there is a general lack of appropriate surfaces for the construction of a calibration curve. In the south, key sources of surfaces of known age are a limited number of twentieth-century gravestones and glacial landforms. Moraines have been dated by reference to stereoscopic aerial photographs of 1946 (and the later aerial coverage from 1960), and the Geodetic Survey maps of 1904 and 1935-7. Additional data have also been gathered from the few known examples of moraines and ice-marginal positions dated by historical sources to the late-nineteenth century (Thórarinsson 1943; Maizels and Dugmore, 1985; Evans et al., 1999). Dates of older glacial landforms rely on extrapolated curves. A summary of published growth curves for the region is presented by Evans et al., (1999).

Tephrochronology has provided control points for lichen growth curves in Alaska and the Yukon Territory of northwestern North America (Denton and Karlén, 1973; Begét, 1994). In Alaska, Begét, (1994) used the Jarvis Creek Ash to test the reliability of lichenometric dating at Gulkana Glacier. However, the opportunities for integrating multiple dating techniques provided by tephras within the glaciovolcanic regions of Iceland have not been exploited previously. Here, we use tephrochronology to conduct independent testing of the lichenometric ages of glacial deposits in south Iceland.

Three procedures for estimating representative lichen size are compared: the "single largest", the "mean of the five largest", and the size-frequency based "1 in 1000" methods (Caseldine, 1991). Using all three methods for comparative

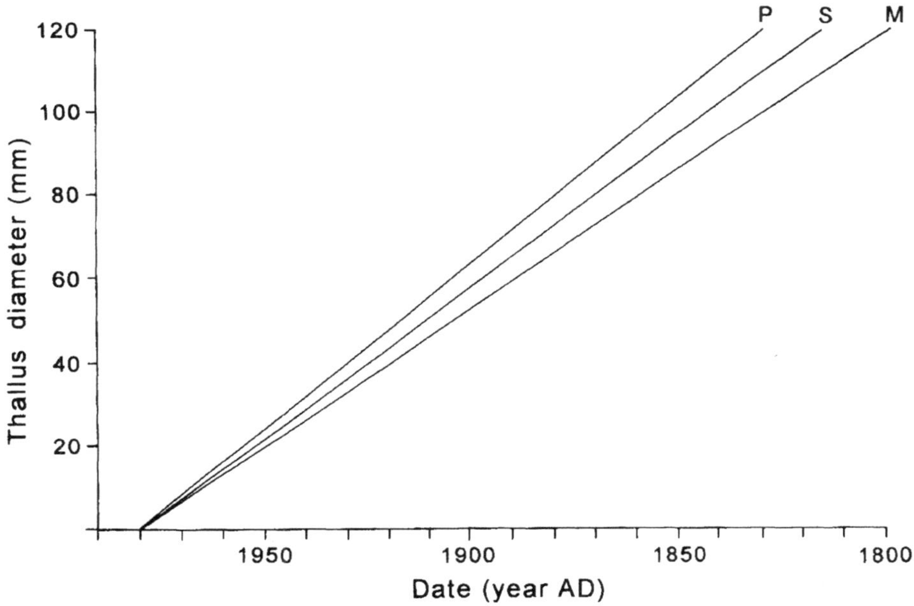

Figure 2. Lichen growth curves for southern Iceland, based on the empirical single-largest calibration of Maizels and Dugmore (1985) (Line S), with additional lines based on the mean of the five largest thalli (Line M) and the extrapolated size (Line P) methods (see text for explanation of the construction of lines M and P).

purposes requires adjustment of Maizels and Dugmore's (1985) calibrated growth curve (line S in Figure 2). Their curve was based on measures of the single largest lichen, and data from the deposits used for calibration do not exist from which to construct curves based on the other two methods. In Figure 2, lines M and P have been calculated using empirical relationships. Ten samples of between 452 and 858 thalli from deposits at Gígjökull and Steinholtsjökull were used to calculate the ratio of thallus size based on the mean of the five largest, the single largest, and the "1 in 1000" methods. Ratios of thallus diameters using each method were 0.91:1.00:1.10 respectively. These ratios were then applied to Maizels and Dugmore's curve to construct growth curves M and P, to allow ages based on each method to be estimated. The original curve uses a colonisation period of 25 years. Evans et al. (1999) suggest that lichens colonise proglacial areas along the south coast after ca. 6.5 years. We present deposit ages based on the first value, but these dates may be converted to concord with more recent research by the addition of ca.18 years. Thus, a date of 1870 based on the Maizels and Dugmore curve corresponds to a date of 1888-89 using the shorter colonisation time. The latter estimates are presented in parentheses throughout the paper, thus: 1870 (1888).

3. Verification of the Temporal Range for Lichenometric Dating

In south Iceland, rapid lichen growth and weathering rates are thought to limit the range of application to deposits younger than ca. 160 years old (Maizels and Dugmore, 1985). Evans et al. (1999) ascribe a lichenometric date of 1777 to a surge moraine at Eyjabakkajökull, suggesting a somewhat older limit of ca. 220 yr for that area. Thompson and Jones (1986) identified a range of environmental factors that may be responsible for placing the Kota fan (Öræfi), formed in 1727, beyond the range of lichenometry, indicating an effective limit of <220 years. In the north, Caseldine and Stötter (1993) apply a linear extrapolation to Kugelmann's (1991) growth curve, calibrated to 1887, to produce dates as old as 1813. These authors suggest that their extrapolation may lead to underestimates of ages beyond the calibration period. The temporal range claimed for lichenometric dating has largely been based on the availability of calibration points for the growth curves employed, but has not been tested against an independent dating technique as called for by Begét (1994). Extensive surfaces of known ages are needed for better calibration, and possibilities include both glacial landforms and deposits from catastrophic floods, or *jökulhlaups*. The established tephra stratigraphy for the region (Dugmore, 1987, 1989) provides known ages for such surfaces.

3.1. DATING THE STEINHOLTSJÖKULL MORAINES

Five moraine ridges on the Suðurhlíðar plateau (SW1 to SW5), to the south of the modern glacier terminus (Figure 3A), were investigated by lichenometric sampling of ridge crests and tephrostratigraphy within inter-moraine hollows. Age estimates from three of the moraines are given in Table I, and lichen population size distributions in Figure 4. Inter-moraine hollows form narrow troughs between moraine ridges, where meltwater from the distal slope of each inner, younger moraine has been channelled against the proximal slope of the corresponding outer, older moraine. Aeolian and volcanogenic sedimentation within the hollows therefore commenced following the deposition of the younger bounding moraine, whose deposition is bracketed by the basal date of each soil pit.

Moraine SW2 shows some consistency of dating between the two techniques. Lichenometry on SW2 (Table I) gives a date of deposition of 1886 (1904) using the single-largest method. K1918 is the oldest tephra layer in pit 30, within soil accumulated on a boulder pavement washed by meltwater from the moraine. The tephra is draped over the basal boulders, indicating moraine deposition very shortly prior to the 1918 eruption.

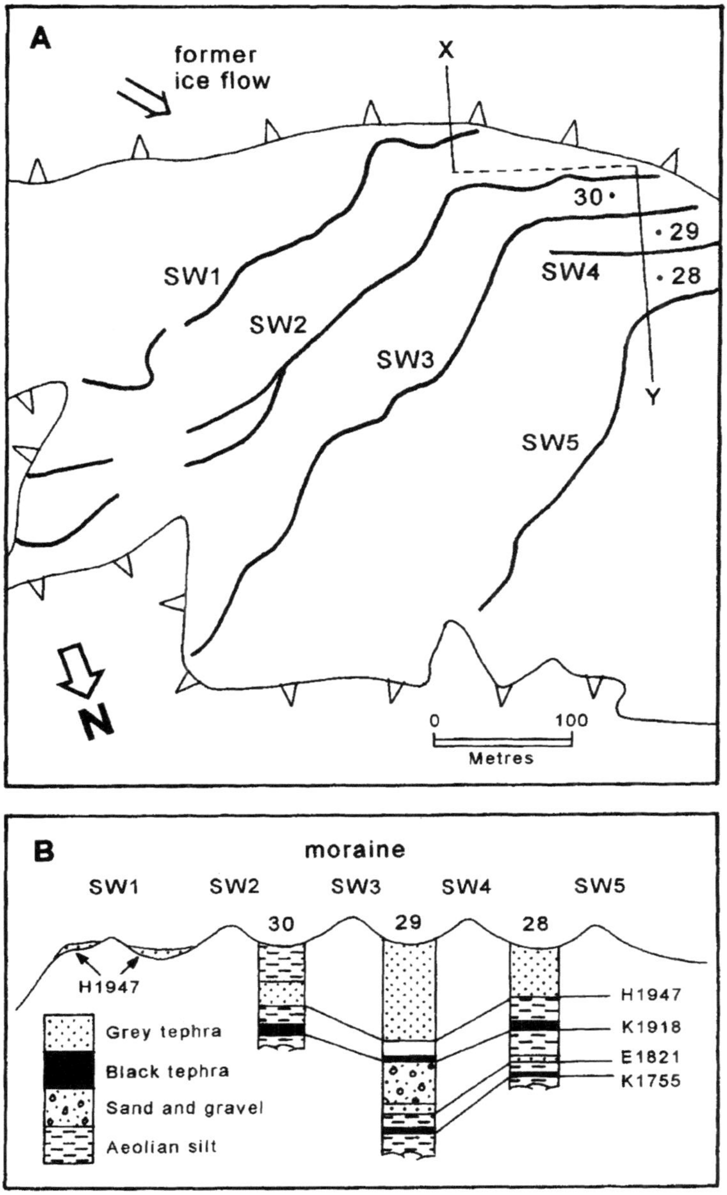

Figure 3. (A) Sample sites for lichenometry and tephrochronology on the Steinholtsjökull moraines. X-Y marks the location of the section shown in part B. Dots indicate the locations of soil pits for tephra stratigraphy. (B) Sketch section along X-Y showing tephra stratigraphy in the inter-moraine troughs. Note how troughs distal to moraines would have channelled meltwater from the adjacent ice front so that soil accumulation will postdate the advance to the younger bounding moraine. See text for detailed discussion.

TABLE I: Comparison of landform dates derived by different lichenometric procedures with tephrochronological dates. Moraines SW2-SW4 are from Steinholtsjökull, and the northern and western fans refer to the 1821 *hlaup* at Gígjökull (Figure 5). [1]Maximum limiting date given by absence of tephra layer; [2]minimum limiting date given by lowest tephra in the distal soil sequence; [3]precise date for event (see text).

Deposit code and lichen sample size	Age-limiting tephra layers		Thallus size and estimated date of deposition					
			Population		Single largest		Mean of five largest	
	Max.[1]	Min.[2]	Size (mm)	Date	Size (mm)	Date	Size (mm)	Date
SW2 (n=550)	E1821	K1918	77	1874	69	1886	63	1893
SW3 (n=717)	K1721	K1755	80	1870	80	1870	71	1883
SW4 (n=482)	K1721	K1755	91	1854	81	1868	73	1880
N. fan (n=706)	K1755	K1918	91	1854	90	1855	81	1868
W. fan (n=748)	E1821[3]		91	1854	121	1813	102	1840

For moraines SW3 and SW4, ages derived by different methods cannot be reconciled. Lichenometry on SW3 (Table 1) gives a date of moraine stabilisation of 1870 (1888) by the single-largest method, consistent with the basal K1918 tephra in Pit 30 in the proximal hollow. However, Pit 29 in the distal hollow contains the K1755 tephra as a discontinuous layer occupying the spaces between the basal boulders, but does not contain K1721. The age of SW3 is estimated on the basis that the basal pavement in distal pit 29 is associated with meltwater from the moraine, whose minimum age is therefore given by the K1755 tephra, and whose maximum age is given by the absence of the K1721 tephra. In this case, there is a discrepency of >115 years between the results of the two methods.

For moraine SW4, K1755 tephra occurs in pits 29 and 28, in both the distal and proximal hollows. In the proximal pit, the tephra occupies the infill between boulders; in the distal pit, it drapes over the boulders as a continuous layer 10-20mm above the boulder lag. K1721 tephra does not occur in either pit. The best estimate of moraine age lies between 1721 and 1755. Conversely, lichenometric dating of SW4 (Table 1) suggests stabilisation around 1868 (1886) by the single-largest method. At best, the two estimates differ by >113 years. The tephra dating at this site is unequivocal, because the two eighteenth-century Katla tephras are separated by just 34 years. Finally, the outermost moraine SW5 clearly lay beyond the lichenometric age range and was not sampled. Tephra stratigraphy provides only a minimum date of 1755.

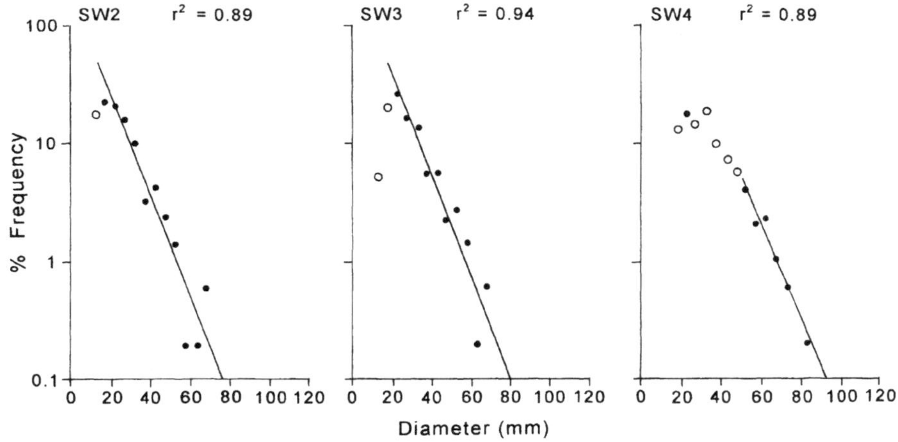

Figure 4. Log-normal percentage frequency distributions for lichen samples from the Steinholtsjökull moraines. Correlation co-efficients are calculated using the bold points only.

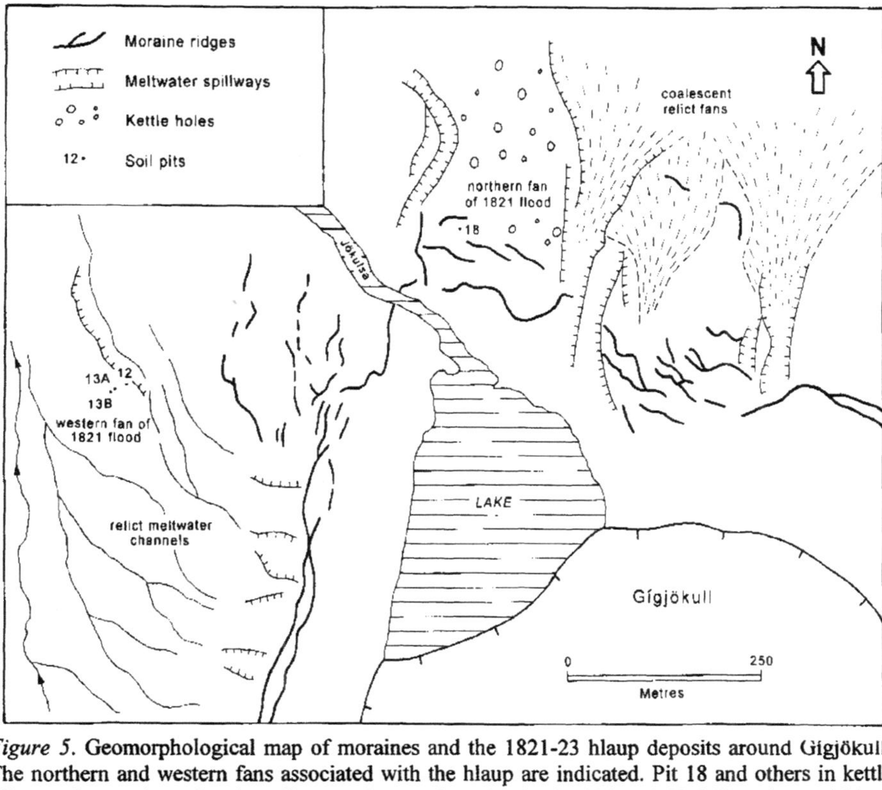

Figure 5. Geomorphological map of moraines and the 1821-23 hlaup deposits around Gígjökull. The northern and western fans associated with the hlaup are indicated. Pit 18 and others in kettle fills on the northern fan (not illustrated) contain only the K1918 and H1947 tephras within a shallow accumulation of aeolian soil.

3.2. DATING THE 1821-23 JÖKULHLAUP AT GÍGJÖKULL

The Eyjafjallajökull eruption of 1821-3 was on a small scale and restricted to the central crater of the volcano; it generated a small volume of silicic tephra (Larsen et al., 1999) and a limited flood from the snout of Gígjökull, the glacier draining from the summit crater. The tephra was produced at the start of the eruption period because at Lambafell, south of Eyjafjallajökull, the layer underlies and is stratigraphically separated from the 1823 Katla tephra by a narrow band of aeolian sediment that possibly represents 1-2 years' accumulation (Dugmore, 1987). The extent of the 1821-3 flood (*hlaup)* deposit has recently been mapped and dated (Dugmore and Kirkbride, unpublished). Two fans deposited by the flood have been identified: one to the north of the present glacier, and one to the west (Figure 5). The dating evidence for the flood deposit comes from a series of stratigraphic sections that cross the margin of the western flood route (Figure 6). The layer of fine white tephra deposited by atmospheric fallout at the start of the eruption can be traced towards the edge of the later flood, where it underlies fine sands and silts deposited by the outermost parts of the *hlaup*. Further in, the flood deposit becomes more coarse grained, and the tephra airfall is absent. The direct stratigraphic link may be continued to prove the age and origin of the boulders making up the majority of the *hlaup* deposit. These boulders provide >500 m^2 of lithologically similar rock surfaces unaffected by moss cover, burial by sediment or enhanced weathering.

Age estimates based on lichenometric samples from flood-deposited boulders on both fans are shown in Table 1. Maximum thallus sizes from each sample were 90mm and 121mm, which give apparent dates of 1855 and 1813 (1873 and 1831) by the single-largest method for the deposits of the 1821-23 flood. The latter date lies beyond the calibrated range of the technique. However, the population structure, consistent between-site results, and high correlation of size and frequency (Figure 7) may encourage some operators to accept a "1 in 1000" age estimate of 1854, which actually underestimates the date of deposition by 33 years.

3.3. COMPARISON OF LICHENOMETRIC METHODS

Estimates of the lichen thallus size which is taken to best represent the age of a deposit have been made using the three methods outlined above, associated with the three growth curves presented (Figure 2). Table 1 shows that for moraine SW2, the representative thallus size varies between 77mm and 63mm depending on the procedure employed. These sizes translate to apparent dates of between 1874 and 1893 (1892-1911), a variation of two decades or 18% of the mean age estimate. For older deposits, the differences between representative size/age estimates also vary by a similar amount. Of particular concern is that the "mean

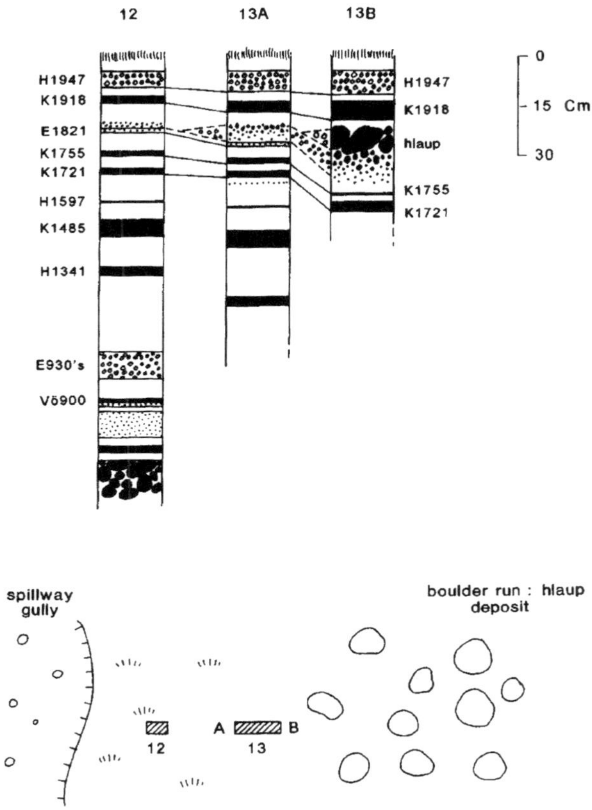

Figure 6. Interpolated tephra stratigraphy of the margin of the 1821-23 *hlaup* deposit, showing the relationship between the stratigraphy of individual soil pits and the tapering margin of overbank deposition by the *hlaup*. Note the detailed relationship between the tephra fall and flood-deposited coarse alluvium. (Soil pit locations are shown in Figure 5).

of the five largest" technique gives smaller thallus sizes and therefore is more likely to give dates which appear to fall within the age range of the technique. Moraines SW3 and SW4 would give a lichenometric dates of 1883 (1901) and 1880 (1898) by this method, whereas the "1 in 1000" method would arouse suspicion by producing dates of 1870 and 1854 (1888 and 1872).

Estimates of representative lichen size and age are therefore sensitively dependent on the estimation procedure employed. Not only does this mean that deposits datable by the technique can be expected to give results of low accuracy and poor repeatability, but that deposits older than the dating range are more likely to be mis-dated using some procedures than others. Furthermore, the use of different estimated colonisation times of 6.5 to 25 years gives a very different impression of deposit age and, significantly, of lichenometric reliability.

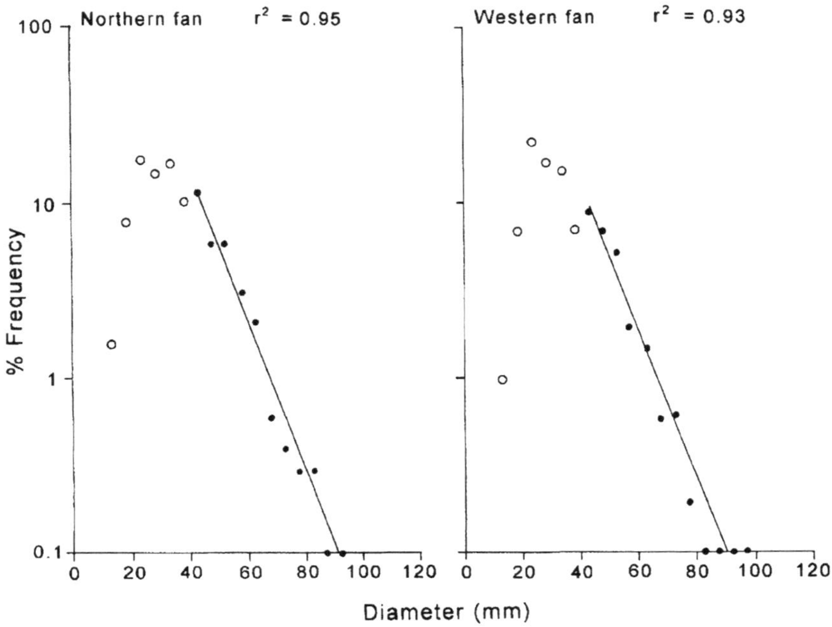

Figure 7. Log-normal percentage frequency distributions for lichen samples from large boulders in the 1821-23 *hlaup* deposit. Correlation co-efficients are calculated using the bold points only.

3.4. SUMMARY

These sites show that age estimates based on lichenometry and tephrochronology are consistent only for the youngest moraine dated by both methods (SW2), dated to the late-nineteenth century (Table 1). However, two other moraines dated by tephrochronology to the early to mid-eighteenth century also give late-nineteenth century lichenometric ages, whereas the 1821 flood deposit gives very variable lichenometric dates, apparently older than those of the eighteenth-century moraines. As a result, it is difficult to establish at what point the lichenometric ages become unreliable, because accurate and inaccurate dates of ca. 1860-1870 (by the Maizels and Dugmore curve) are indistinguishable. If the shorter colonisation time is employed, the problem is exacerbated because this "grey area" corresponds to the period claimed to be the "Little Ice Age" glacier maximum of the 1880s-1890s (Evans et al., 1999). Even the supposedly longer-range "1 in 1000" method (Caseldine, 1991) is shown to suffer the same constraints of dating range as other methods. What is certain is that lichenometric dating causes an apparent clustering of dates in the mid- to late-nineteenth century, including deposits of the mid-eighteenth and early-nineteenth centuries.

4. Discussion

To what extent is the apparent late-nineteenth century "Little Ice Age" glacial maximum a real reflection of glacier fluctuations? Or does it instead reflect poor preservation and misleading dating of landforms?

Several influences apart from climate may have affected the age distribution of Icelandic moraine sequences. As a general rule, age-distribution patterns of preserved moraines reflect not just climatic history, but also the sensitivity of response of different glaciers, the preservation of depositional landforms, and the availability of reliable dating (Kirkbride and Brazier, 1998). One possibility is that the Icelandic geomorphological record is severely truncated ("censored"), and evidence of older fluctuations has been masked by later events. The proglacial areas of ice cap outlet glaciers are particularly vulnerable to destruction or burial by later fluvioglacial and *jökulhlaup* activity. As a result, the pre-nineteenth century moraine record from the forelands of large ice-cap outlet glaciers has probably suffered severe erosion censoring. Such loss of evidence is apparent from the mapped distribution of moraines at several outlet glaciers, where moraines are increasingly fragmented with age, and the oldest preserved remnants are located marginal to the main sandur (Maizels and Dugmore, 1985; Evans et al., 1999). Icelandic glaciers do, however, terminate in a diverse range of proglacial environments, suggesting that evidence of earlier events should survive in suitable locations.

A second possibility is that the "Little Ice Age" chronology is biased towards large glaciers, whose response may differ from that of small glaciers such as Gígjökull and Steinholtsjökull. The existence of mid-eighteenth century moraines at small glaciers, but only late-nineteenth century moraines at large ones, could be interpreted as a function of a size-dependence affecting the magnitude or timing of the advances, but in practice it is difficult to see how this might work.

A third possibility is that these earlier moraines have been mapped, but incorrectly dated. Lichenometry has been applied widely, but limitations already noted may be more profound than currently accepted. Our independent dating of moraines and *hlaup* deposits demonstrates that the environmental constraints affecting lichenometry can cause eighteenth-century landforms to be ascribed mid- to late-nineteenth century dates. Several moraines which are claimed to represent the "Little Ice Age" maximum have been dated to the period 1870-1890s by the mean of the five largest lichens at Kvíárjökull, Hólárjökull, Fláajökull, Heinabergsjökull/Skálafellsjökull and Eyjabakkajökull (which surges) (Evans et al., 1999). Our tephrochronological dating at Steinholtsjökull (Table 1) indicates that many of these moraines could be at least a century older. Documentary records in support of lichenometric dates of a nineteenth-century maximum at these glaciers (Thórarinsson, 1943) are not actually very clear-cut. Only at Heinbergsjökull is a late-nineteenth century maximum documented unequivocally as being larger than earlier maxima. At Hólárjökull, the advance of 1880-90 was "as large as or larger than in 1756" (Thórarinsson, 1943, p.30), indicating a probable later maximum. At Kvíárjökull, Thórarinsson (1943) states that the maximum extent in historical time pre-dated 1794, while at Fláajökull there is no documentary evidence from before 1839, so the late-nineteenth century advance can only be assumed to be the "Little Ice Age" maximum. More widely, of the glaciers in southern Iceland for which documentary records are reviewed by Thórarinsson (1943), only twelve out of twenty-five have records dating from 1756 or earlier, and six of these show eighteenth century "Little Ice Age" maxima. Thus, the historical evidence for a dominant "Little Ice Age" maximum in the nineteenth century is actually based on a rather small sample of glaciers.

Another possibility is that the growth rate of lichens, known to vary across the Icelandic landscape (Evans et al., 1999) has also varied through time, and was significantly slower during colder, pre-twentieth century conditions. In this case, lichenometry would need to use a temporally-variable growth curve calibrated to temperature records. Currently, lichen dating could underestimate "true" ages simply because the coldest periods were characterised by slower growth. To explain the observed discrepencies on the Steinholtsjökull moraines would, however, require an extremely sensitive dependence of growth rate on air temperature.

Accepting that a mid-eighteenth century glacial maximum deserves at least equal emphasis, we now consider the glacial record in the context of other palaeoenvironmental indicators. Evidence from winter sea-ice reconstructions, fluctuations in the cod-fishing industry, written accounts, and later from instrumental temperature records (Jonsson, 1965; Lamb, 1979; Grove, 1988; Ogilvie, 1984, 1992; Eythórson and Sigtryggsson, 1971) indicate that conditions apparently favourable for glacier expansion were sustained in the late seventeenth century (particularly ca. 1675-1700), between ca. 1730 and 1760, the 1780s, the 1860s-1870s, and finally in the 1920s. According to Lamb (1979) the most severe "Little Ice Age" climate occurred in the first of these periods, but Ogilvie (1992) presents evidence that minimum temperatures were attained in the 1780s. Lamb (1979) suggests that intrusion of polar water and sea ice (the East Iceland Current) was only established around the south coast of Iceland between 1685 and 1704, and then only as a very narrow band holding warmer Atlantic water about 20km off the coast. In other periods dominated by polar weather types, the Irminger Current continued to warm the south coast (Lamb, 1979) and, by implication, zonal southwesterly winds supplied abundant winter snowfall to glacier accumulation zones. Thus, the evidence makes winter snow starvation unlikely on south Icelandic glaciers in any part of the "Little Ice Age" except perhaps the late seventeenth century.

Small outlet glaciers in south Iceland may therefore be expected to have attained their "Little Ice Age" maximum in the late-seventeenth or mid-eighteenth century, but (unlike larger outlet lobes) are unlikely to have done so in the late-nineteenth century unless winter precipitation was so much greater than in earlier colder periods that it offset longer and warmer ablation seasons. The evidence suggests that this is unlikely. Our revisions to the dating of the "Little Ice Age" maximum therefore seem to point to a chronology which is consistent with other proxy evidence. Larger outlet glaciers do not appear to be rewarding sites from which to derive a complete picture of climatic variation in the early and middle parts of the "Little Ice Age".

5. Conclusions

It has been shown that a lichen-based chronology in south Iceland spuriously clusters events within the mid- to late-nineteenth century, and underestimates the age of earlier "Little Ice Age" deposits, especially at glaciers where documentary records of earlier advances are poor or lacking. Indeed, a case could be made for lichenometry to be applied only to twentieth-century deposits as a supplement to historical records. Our tephrochronological results demonstrate that the small outlet glaciers we have studied (and probably others elsewhere in southern Iceland) were more extensive in the mid-eighteenth century than in the late-nineteenth century. A lichen-based chronology would not have demonstrated this. By implication, climate along the south coast of Iceland was significantly affected by zonal circulation from at least the mid-eighteenth century onwards. Further research is needed to reconstruct glacier extents in the late-seventeenth century/early-eighteenth century period when polar influences were greater and "Little Ice Age" temperatures at a minimum. The notion of a dominant "Little Ice Age" glacial maximum in the late-nineteenth century should at best be treated cautiously, and needs further rigorous testing at a larger number of glaciers using alternative dating techniques such as tephrochronology. There have been at least five cool periods since the late-seventeenth century when glaciers seem to have advanced to maximum positions of similar extents. Which period came to represent the "Little Ice Age" maximum at any particular glacier may have depended on local glaciological responses. Indeed, "Little Ice Age" maxima dated by methods other than lichenometry are spread over nearly three centuries.

Acknowledgements

Funding towards aspects of this project was provided by the National Geographic Society, the Leverhulme Trust, and by the Royal Society (Grant 1310.V706/LUM/TH). Field assistance from Ann Davies, Mark Skidmore and Nick Spedding is gratefully acknowledged.

References

Ballantyne, C.K.: 1990, The Holocene glacial chronology of Lyngshalvöya, northern Norway: chronology and climatic implications, *Boreas* **19**, 93-117.

Begét, J.E.: 1994, Tephrochronology, lichenometry and radiocarbon dating at Gulkana Glacier, central Alaska Range, USA., *The Holocene* **4**, 303-313.

Benedict, J.B.: 1985, Arapaho Pass: glacial geology and archaeology at the crest of the Colorado Front Range, *Centre for Mountain Archaeology, Research Report* **3**, Colorado.

Björnsson, H.: 1979, Glaciers in Iceland, *Jökull* **29**, 74-80.

Caseldine, C.J.: 1983, Resurvey of the margins of Gljúfurárjökull and the chronology of recent deglaciation, *Jökull* **33**, 111-118.

Caseldine, C.J.: 1985, The extent of some glaciers in northern Iceland during the Little Ice Age and the nature of recent deglaciation, *Geographical Journal* **151**, 215-227.

Caseldine, C.J.: 1991, Lichenometric dating, lichen population studies, and Holocene glacial history in Tröllaskagi, northern Iceland, in Maizels, J.K. and Caseldine, C. (eds.), *Environmental Change in Iceland: Past and Present*. Kluwer Academic Publishers, Dordrecht, pp. 219-236.

Caseldine, C.J and Stötter, J.: 1993, "Little Ice Age" glaciation of Tröllaskagi peninsula, northern Iceland: climatic implications for reconstructed equilibrium-line altitudes (ELAs), *The Holocene* **3**, 357-366.

Denton, G.H. and Karlén, W.: 1973, Lichenometry: its application to Holocene moraine studies in southern Alaska and Swedish Lapland, *Arctic and Alpine Research* **5**, 347-372.

Dugmore, A.J.: 1987, Holocene glacier fluctuations around Eyjafjallajökull, south Iceland: a tephrochronological study, Unpublished Ph.D thesis, University of Aberdeen, 214pp.

Dugmore, A.J.: 1989, Tephrochronological studies of Holocene glacier fluctuations in south Iceland, in Oerlemans, J. (ed.), *Glacier Fluctuations and Climatic Change*, Kluwer Academic Publishers, Dordrecht, pp. 37-55.

Einarsson, E.H., Larsen, G. and Thorarinsson, S: 1980, The Sólheimar tephra layer and the Katla eruption of c.1357AD. *Acta Naturalia Islandica* **28**, 1-24.

Evans, D.J.A., Archer, S. and Wilson, D.J.H.: 1999, A comparison of the lichenometric and Schmidt hammer dating techniques based on data from the proglacial aras of some Icelandic glaciers, *Quaternary Science Reviews* **18**, 13-41.

Eythórsson, J. And Sigtryggsson, H.: 1971, The climate and weather of Iceland, in Tuxen, S.L. (ed.), *The Zoology of Iceland* **1** (3), Ejnar Munksgaard, Copenhagen and Reykjavík.

Gordon, J.E. and Sharp, M.J.: 1983, Lichenometry in dating recent landforms and deposits, south-west Iceland, *Boreas* **12**, 191-200.

Grove, J.M.: 1988, *The Little Ice Age*, Methuen, London, 498pp.

Guðmundsson, H.J.: 1997, A review of the Holocene environmental history of Iceland, *Quaternary Science Reviews* **16**, 81-92.

Häberle, T.: 1991, Holocene glacial history of the Hörgárdalur area, Tröllaskagi, northern Iceland, in Maizels, J.K. and Caseldine, C. (eds), *Environmental Change in Iceland: Past and Present*, Kluwer Academic Publishers, Dordrecht, pp. 193-202.

Hafliðason, H., Larsen, G. And Ólafsson, G.: 1992, The recent sedimentation history of Thingvatn, Iceland, *Oikos* **64**, 80-95.

Jaksch, K.: 1970, Beobachtungen in den Gletschervorfeldern des Sólheima- und Sidujökull in sommer 1970, *Jökull* **20**, 45-49.

Jaksch, K.: 1975, Das Gletschervorfeld des Sólheimajökull, *Jökull* **25**, 34-38.

Jonsson, J.: 1965, Temperature and growth of cod in Icelandic waters. *ICNAF Special Publications* **6**, 537-539.

Kirkbride, M.P. and Brazier, V.: 1998, A critical evaluation of the use of glacial chronologies in climatic reconstruction, with reference to New Zealand, in Owen, L.A. (ed.), *Mountain Glaciation. Quaternary Proceedings* **6**, 55-64.

Kirkbride, M.P. and Dugmore, A.J.: in press, The timing and significance of mid-Holocene glacier advances in northern and central Iceland, *Journal of Quaternary Science* **15**.

Kugelmann, O.: 1991, Dating recent glacier advances in the Svarfaðardalur-Skíðadalur area of northern Iceland by means of a new lichen curve, in Maizels, J.K. and Caseldine, C., (eds.), *Environmental Change in Iceland: Past and Present*, Kluwer Academic Publishers, Dordrecht, pp. 203-217.

Lamb, H.H.: 1979, Climatic variation and changes in the wind and ocean circulation: the Little Ice Age in the northeast Atlantic, *Quaternary Research* **11**, 1-20.

Larsen, G.: 1984, Recent volcanic activity of the Veidivotn fissure swarm, southern Iceland: an approach to volcanic risk assessment, *Journal of Volcanology and Geothermal Research* **22**, 33-58.

Larsen, G. and Thórarinsson, S.: 1977, H4 and other acidic Hekla tephra layers, *Jökull* **27**, 28-46.

Larsen, G.; Dugmore, A.J. and Newton, A.J.: 1999, Geochemistry of historical age silicic tephras in Iceland, *The Holocene* **9**, 463-471.

Locke, W.W., Andrews, J.T. and Webber, P.J.: 1979, A manual for lichenometry, *British Geomorphological Research Group Technical Bulletin* **26**, 47pp.

Maizels, J.K. and Dugmore, A.J.: 1985, Lichenometric dating and tephrochronology of sandur deposits, Sólheimajökull area, southern Iceland, *Jökull* **35**, 69-77.

Ogilvie, A.E.J.: 1984, The past climate and sea-ice record from Iceland, *Climatic Change* 6, 131-152.

Ogilvie, A.E.J.: 1992, Documentary evidence for changes in the climate of Iceland, AD 1500 to 1800, in Bradley, R.S. and Jones, P.D., (eds.), *Climate since A.D. 1500*, Routledge, London and New York, pp.92-117.

Stötter, J.: 1991, New observations on the postglacial glacial history of Tröllaskagi, northern Iceland, in Maizels, J.K. and Caseldine, C., (eds.), *Environmental Change in Iceland: Past and Present*, Kluwer Academic Publishers, Dordrecht, pp. 181-192.

Thomson, A. and Jones, A.: 1986, Rates and causes of proglacial river terrace formation in southeast Iceland: an application of lichenometric dating techniques, *Boreas* **15**, 231-246.

Thórarinsson, S.: 1943, Vatnajökull. Scientific results of the Swedish-Icelandic investigations, 1936-37-38. Oscillations of the Icelandic glaciers in the last 250yr, *Geografiska Annaler* **25**, 1-54.

Thórarinsson, S.: 1967, The eruptions of Hekla in historical times, *The Eruption of Hekla 1947-1948* **1**, 1-170.

Thórarinsson, S.:1975, Katla og annáll Kötlugosa (Katla and the annals of Katla tephras), *Árbók ferðafélags Íslands* **1975**, 125-149.

Thórarinsson, S.: 1980, Langleiðir gjósku úr Þremur Kötlugosum (English Abstract: Distant transport of tephra in three Katla eruptions and one Grímsvötn (?) eruption), *Jökull* **30**, 65-73.

Worsley, P.: 1981, Lichenometry, in Goudie, A., (ed.), *Geomorphological Techniques*, Allen and Unwin, London, pp.302-305.

(Received 16 February 1999; in revised form 12 May 2000)

EARLY INSTRUMENTAL METEOROLOGICAL OBSERVATIONS IN ICELAND

TRAUSTI JÓNSSON and HILMAR GARÐARSSON

Icelandic Meteorological Office, Bústaðavegur 9, IS-150 Reykjavík, Iceland

Abstract. When the Danish Meteorological Institute (DMI) was established in 1872 it took over the responsibility of meteorological observations in Iceland. Prior to that time, the observations were in the hands of enthusiastic individuals, in some cases in close contact with the international scientific community. The original data compiled by these latter observers are kept at the Manuscript Section of the Icelandic National Library and in the Archives of the Icelandic Meteorological Office. Much of this material was gradually delivered to Iceland (along with other documents from official Danish archives) after the formal independence of Iceland in 1918. This paper presents an overview of instrumental observations available from the pre-1872 era. The oldest such observations that have been found were made in 1749 to 1751 and in 1763. The temporal density increases after 1775, and from 1822 there is at least one instrumental observation to be found on each and every day, somewhere in Iceland. During the period 1776 to 1822 there are many gaps, some only short, others long; with the longest from late 1814 to mid 1820. During this last gap almost no instrumental records have been found. The quality of the observations is very variable, but generally they seem to be at least internally consistent, and will certainly become an important addition to the knowledge of the climate and climatic variability of the Northeast Atlantic region. A large part of the data are available in digital format. The present study gives a richer and more comprehensive view of the meteorological activities in Iceland during the "Early Instrumental Period" (EIP), than previously available. It includes a summary of the early history of meteorological measurements in Iceland. Weather diaries without quantitative measurements are outside the scope of this paper.

Key Words. Early instrumental period; meteorological observations and measurements; Iceland

1. Introduction

The surge of interest in the natural sciences in Europe in the seventeenth and eighteenth centuries also spread to Iceland, and a few Icelanders acquired some academic training in the natural sciences in Copenhagen (Guðmundsson, 1998). Nevertheless, no meteorological measurements of any substance survive from the very early part of this period. The only known temperature measurements of any kind in Iceland prior to the middle of the eighteenth century deserve a mention, however. In 1664, the Royal Society of London sent a thermometer to Páll Björnsson (1621-1706) in Selárdalur in Iceland (Seaton, 1935). No information has been found on any readings of the instrument, save one, which has survived in a letter from Björnsson to the Society. During the seventeenth and eighteenth centuries Iceland was severely affected by famines and other hardships. It is likely that these were exacerbated by unfavourable weather and the presence of

Climatic Change **48**: 169–187, 2001.

sea ice off the coasts (Ogilvie, 1982, 1984, 1992; Ogilvie and Jónsson, this volume). The Danish authorities subsequently became aware of the need for an objective assessment of the state of the country as well as information on its geographical position, environmental conditions, natural resources and climate (Sigurðsson, 1978). The Royal Danish Scientific Society (*Det Kongelige danske Videnskabernes Selskab*) was established in 1742 and it soon initiated regular observations at the Rundetaarn observatory in Copenhagen, then led by the well known Danish scientist Ole Röhmer (Skovmand, 1986a). The Icelandic census of 1703 (Magnússon and Vídalín, 1924-1947), the reports by Horrebow (1752) and Ólafsson (1974, 1975), the efforts of the Danish Scientific Society and the Danish Society of Natural history late in the eighteenth century were all important landmarks in the scientific investigation of the Icelandic environment. This interest was just one of the many manifestations of the Enlightenment movement in Europe that influenced both individual scientists and literary figures as well as many of the European governments of the time. A firm belief in general progress and the benevolent effect of knowledge on man and society stimulated scientific research and the investigation of nature and natural phenomena (Sigurðsson, 1990). The sudden interest in the climate of Iceland in the late-eighteenth century must be seen in light of this general trend.

In the late-eighteenth century the interest for climatic mapping culminated in the effort of the *Societas Meteorologica Palatina* based in Mannheim (Kington, 1974). The Danish Scientific Society established a "permanent" meteorological and astronomical observatory at Lambhús near Reykjavík in 1779. Formally it was not a part of the Mannheim effort, but it is very likely that at least some of the instruments were from the Mannheim stocks, as Thomas Bugge (1740-1815), of the Danish Scientific Society, was cooperating closely with the Palatinean Society and his observatory at Rundetaarn was a formal member of the Mannheim network from 1782 to 1788 (Skovmand, 1986a,b). Only a part of the observational data from the Lambhús observatory has been found. During the period up to 1807, other observations were made by enthusiastic individuals, most notably the district doctor and naturalist Sveinn Pálsson (1762-1840) and Sheriff Magnús Ketilsson (1732-1803). In 1807, a Danish coastal surveying team established a base in Akureyri in northern Iceland. Its main task was to establish a long baseline in Iceland and make triangular measurements of the Icelandic coastal area (Thoroddsen, 1892-1904). At the Akureyri base, the team made meteorological measurements for seven years.

In 1820 the, *Landlæknir* of Iceland (national medical officer, the formal title of the head of the nascent health service), Jón Thorsteinsson (1794-1855), began observations in Reykjavík that were to continue for more than 30 years. This was done in cooperation with the Danish Scientific Society. A few other stations were established later in the 1820s but these lasted only a few years. A major effort was initiated in 1840 by the Icelandic Society of Letters at the urging of the natural scientist and poet Jónas Hallgrímsson (1807-1845). The Danish Scientific Society donated 45 thermometers, and these were sent to parish ministers all over the country. Most of these priests then made observations and sent reports to the Society of Letters for a few years and some even continued for more than 30 years. One of the thermometers eventually came into the hands of Árni Thorlacius (1802-1891), a merchant in Stykkishólmur. Árni had both the means and interest to embark on what was to become a major undertaking, making observations many times a day with a set of thermometers, barometers and a precipitation gauge (Bergþórsson, 1957). Thorlacius later cooperated with the Scottish Meteorological Society, and after 1872 with the then new Danish Meteorological Institute. The observations at Stykkishólmur have continued from their beginning in 1845 to this day, except for a 5 months break in 1919.

2. Search for Material

For the present study, a comprehensive search for weather diaries containing instrumental measurements was made at the Manuscript Section of the National Library of Iceland (*Handritadeild, Landsbókasafn Íslands*) and at the National Archives (*Þjóðskjalasafn*). The result of this search was reported by Hávarðsson (1997). Some of the series have been known for a long time and some background material had been collected and annotated in Icelandic publications. In this connection, the seminal works of Thoroddsen (1892-1904 and 1916-1917) as well as papers by Eyþórsson (1956, 1962, 1963 and 1964), Kington (1972) and Sigfúsdóttir (1969) must be mentioned. Furthermore, Guðmundsson (1989) gives an account of the astronomical activities in Iceland in the eighteenth century and provides some additional information not included in the Eyþórsson papers, especially regarding the activity in the 1770's. The present study gives a more detailed and comprehensive view of the meteorological activities in Iceland during the "Early Instrumental Period" (EIP) than previously available. It includes a summary of the early history of meteorological measurements in Iceland. Weather diaries without instrumental observations are not systematically listed and the reader in search of those is referred to Hávarðsson (1997). Another search was made for official correspondence located in the National Archives relating to the eighteenth-century instrumental observations. A number of documents were found and most of these were also transcribed. These documents contain only very little meteorological information. In spite of this they offer an interesting view on financial and practical matters, for example details of the building costs of the Lambhús observatory, and the house of the observer. These documents are not referenced directly in the present paper, but the transcribed copies are accessible in a report (Þorbjarnardóttir et al., 1998). Another report (Garðarsson, 1998) includes short biographical entries for the main individuals involved in the eighteenth and early-nineteenth century observations. A search was also made in Copenhagen in the Archives of the Danish Scientific Society and the Danish National Archives, as discussed below.

A number of the series have been transcribed and digitised as indicated below. In most cases it was not possible to digitize directly from the manuscripts. It is fairly straightforward to digitize numbers from instrumental observations, but the visual (subjective) parts of the observations (for example wind direction, speed, clouds and weather) had to be transcribed and coded before the actual digitization. In some cases this was both a tedious and strenuous procedure as the manuscripts are difficult to read. A transcription of subjective observations always includes a compromise; a slight loss of information must be expected. A decision was made to concentrate the digitizing effort on the series that included daily or semi-daily instrumental observations, especially those containing pressure data. All the digitized material is available upon request. It must be emphasized that much material still remains to be digitised. For station location, see Figure 1.

3. On the Individual Series of Measurements

The individual series are presented below in an approximate chronological order. Only the Lievog series (not all) and a part of the Jón Thorsteinsson series were available in digital format on a daily basis prior to this study. References to manuscript sources (if any) are given below, in italics, at the end of the discussion of each series. For a more detailed listing see Hávarðsson (1997). Some of the manuscripts were hard to find. The Horrebow series had been published previously, in an eighteenth century book (Horrebow, 1752). Monthly averages for the temperature and pressure in the Stykkishólmur series have been available, but not the daily values.

3.1. THE OBSERVATIONS OF NIELS HORREBOW AT BESSASTAÐIR

The observations of Niels Horrebow (1712-1760) at Bessastaðir, in southwest Iceland, extend from 1 August 1749 to 30 July 1751. This is the oldest known instrumental series in Iceland. The observations were made as part of a general survey of Iceland, formally under the Danish Scientific Society. However, it seems that Horrebow was more or less on his own in rather primitive circumstances in Iceland (Sigurðsson, 1990). The observations were printed at the end of his long and generally balanced report on Iceland (Horrebow, 1752). In the opinion of Horrebow himself the data revealed a much more benign climate than had been assumed before to prevail in Iceland. He was especially surprised by the mild winters. The observations of pressure (French inches), temperature (early Reaumur) and wind direction have been digitized. The temperature was measured in an unheated house, but not outdoors. It is not unlikely that Horrebow left the instru-

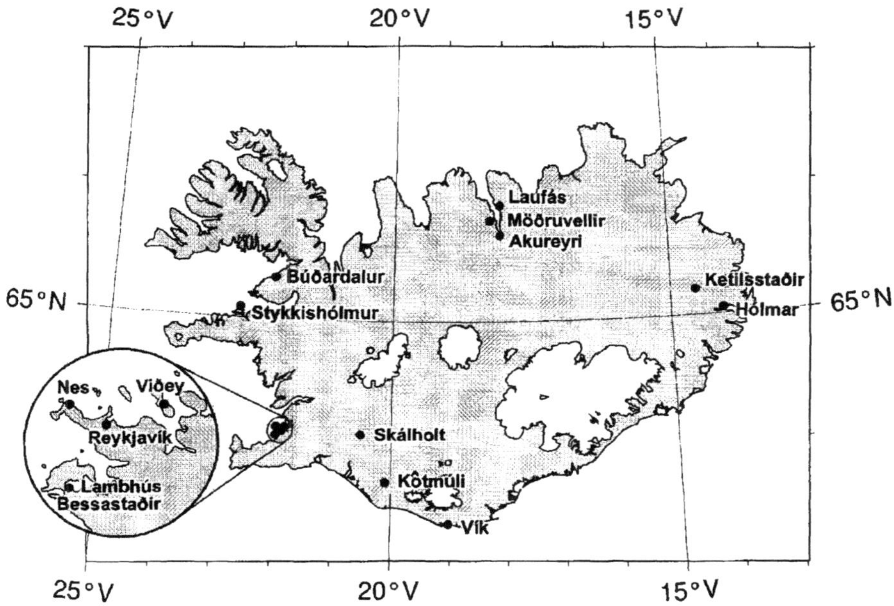

Figure 1. The location of most of the stations mentioned in the text.

ments in Iceland. The original observation lists have not been systematically searched for. Niels Horrebow was the brother of Christian Horrebow the younger (1718-1776), professor in Copenhagen, and from 1751 the director of the Copenhagen observatory and Peder Horrebow (1728-1814), who was for many years the main observer there. Source: *Horrebow (1752)*.

3.2. THE OBSERVATIONS OF EGGERT ÓLAFSSON AND BJARNI PÁLSSON

The Danish Scientific Society supported a major survey of the geography of Iceland which was carried out during 1752-1757 by Eggert Ólafsson (1726-1768) and Bjarni Pálsson (1719-1779). The team was in the possession of thermometers and a barometer. An extensive travel journal was originally published in Danish (Olafsen, 1772). It was subsequently translated into German, French, English (in 1805) and Icelandic. A second edition of the Icelandic version was published in 1974 (Ólafsson, 1974) and a new English edition in 1975 (Ólafsson, 1975). A collection of notes and documents relevant to the travels has survived. It is clear that regular observations of temperature, pressure and wind were made, both on the journeys during the summer and also at the winter quarters, evidently throughout the years 1752-1757. These observations have not been found although they were sent annually to the Danish Scientific Society (Ólafsson 1974, p.181). Temperature measurements were made with their instruments in Húnavatnssýsla in northern Iceland for at least two years (Ólafsson 1974, p.16). Which years, however, is not known (1758 and 1759?). Ólafsson also made observations during 1760 to 1764 in Barðarstrandarsýsla county in northwestern Iceland, but nothing has been found. An unsuccessful attempt was made to find some traces of this information at the National Libraries of Iceland and Denmark, and in the collections of the Danish Scientific Society. In spite of this it is still possible that at least some of these observations might be found in the future. Ólafsson and his wife disappeared tragically at sea in a shipwreck during a change of residence in 1768. It is possible that the 1760-1764 observations were also permanently lost. Guðlaugur Þorgeirsson (1711-1789), a priest at Garðar, near Reykjavík, made some observations of weather and pressure by using a barometer sent to him by Ólafsson or Pálsson (Ólafsson 1974, p.182). In the letters of a Swedish gentleman travelling with the well-known Englishman Sir Joseph Banks in 1772, (von Troil, 1961) it is noted that Þorgeirsson had taken over as a weather observer immediately after the departure of Horrebow in 1751.

3.3. THE OBSERVATIONS OF 1763

A series of observations exists from the year 1763. The location is not known, but it is probably either Garðar or Viðey (a small island off the coast near Reykjavík). The observations have been attributed to the naturalist Eggert Ólafsson and would then be a part of his 1760-1764 series of observations mentioned above, but this is highly uncertain. It is also possible that they were made by the previously mentioned Guðlaugur Þorgeirsson at Garðar. The pressure observations have been digitized, but there are no temperature observations. Source: *National Library, ÍB 5. fol.*

3.4. THE INITIATIVE OF THE "COUNTRY COMMISSION"

A further economic survey was initiated in 1770, carried out by the so-called *Landsnefndin* (the Country Commission). One of the first formally educated Icelandic natural scientists, Eyjólfur Jónsson (1735-1775), was appointed secretary of the commission. From his letter of appointment in June 1770, it is clear that he was also expected to make astronomical and geographical measurements. It was stated that he was to get three barometers and three thermometers in order to measure the state of the atmosphere daily, and also the heights of mountains by the help of barometric measurements (Eyþórsson, 1962). It seems that Jónsson never made the extensive measurements he was supposed to undertake. It is, however, mentioned in von Troil's travel letters (1961, p 55), that he had been making meteorological observations at Arnarhóll in Reykjavík since 1769. Jónsson was also appointed a priest designate at Staðarstaður in western Iceland with a side-duty as a meteorological observer. The *Stiftamtmaður* (the highest government official living in Iceland) was supposed to make sure that he would report the results of the measurements every year to the Danish Scientific Society. A special house (observatory) was to be erected at Staðarstaður. Unfortunately, Jónsson never made it to his new post; he was in poor health and died at the age of forty in 1775. Before he died, the plans for an observatory at Staðarstaður were abandoned and a new site selected at Lambhús, a small subsidiary farm close to Bessastaðir. Regarding this, nothing occurred for a year, and then it happened that Professor Christian Horrebow in Copenhagen died. At first it was expected that his brother, Peder Horrebow, who had been a professor designate would take over. He was instructed that he could take the student Eyjólfur Jónsson Moe (or Moh), the name-sake of the first Jónsson, as an apprentice observer. However, at the end of 1776, Peder Horrebow was forced to retire (at the unusually early age of fifty) and Thomas Bugge took over as the new Professor (Skovmand 1986a and 1986b). The student Eyjólfur Jónsson Moe (1745-?) never took up his position, and in 1778 Bugge asked for permission to take another student, Rasmus Lievog (1738 or 1739-1811) as an apprentice for the vacant post in Iceland. It is not known if Eyjólfur was simply disqualified because of his affiliation with Peder Horrebow, but Rasmus Lievog did indeed go to Iceland in 1779 as an observer (see below). Additional information on this episode in the history of meteorological observations in Iceland is found in Eyþórsson (1962), in Guðmundsson (1989), and a transcription of the official correspondence can be found in Þorbjarnardóttir et al. (1998).

3.5. THE OBSERVATIONS OF HANNES FINNSSON AT SKÁLHOLT

The Bishop of Skálholt, Hannes Finnsson (1739-1796), was a multi-disciplinarian and had a great interest in the workings of nature. His work on the demographical constraints of the Icelandic natural environment is still a classic of its time (Finnsson, 1796, 1970; Hugason,1990; Ogilvie and Jónsson, this volume). At least a part of his weather diary made at Skálholt has survived. The surviving part covers the period 9 December 1777 to 2 March 1781 and includes both pressure and temperature. It has not yet been digitized. Source: *National Library, LBS. 46. fol., pp. 190-211.*

3.6. THE OBSERVATIONS OF HELGI SIVERTSEN AT SKÁLHOLT

Observations were made at Skálholt during the years 1779-1780 by Helgi Sivertsen (1743-1819, also known as Sigurðsson). The temperature measurements were made in an unheated room on the north side of a house. There is a note on the weather and wind of every day. The instrumental observations of pressure (French inches) have been digitized. The observations have not yet been compared to the Hannes Finnson journal, allegedly made at the same location. Source: *National Library, ÍB 234, a-b, 4to.*

3.7. OBSERVATIONS AT AN UNKNOWN LOCATION 1776 TO 1779

Observations were also undertaken at an unknown location for 1776 to 1777 and 1779. There are no temperature measurements except in January 1779. At the start of the 1779 observation lists there is a reference to southern Iceland. (*Observationes meteorologicæ in australi qvadrante Islandiæ*). It is likely that this refers to Reykjavík or its vicinity (Garðar or Viðey?) and could have been made by Pastor Guðlaugur Þorgeirsson, mentioned above. The name of Pastor Þorgeirsson and his alleged observations surface repeatedly in sources during the 1751 to 1779 period. It is thus tempting to attribute these and other unlocated observations to him. It must be emphasized that this is of course far from certain. The pressure readings (French inches) of this series have been digitized. There are notes on the weather and wind every day. Source: *National Library. Óskráð veðurgögn, kassi II.*

3.8. THE RASMUS LIEVOG SERIES

Rasmus Lievog was appointed as a Royal Astronomical Observer in Iceland on 21 April 1779. According to the letter of appointment, his duties included observations of the weather, and measurements of temperature and pressure at least three times per day and, further, to report the results regularly to Thomas Bugge at the Rundetaarn observatory and the Danish Scientific Society (Eyþórsson, 1962). Lievog came to Iceland in early August and began his observations almost immediately, on 10 August. The so-called observatory had not been built, so until Christmas 1779 the observations were made at Bessastaðir, a few hundred meters away. There is some information on the dismal conditions at Lambhús during the first few years, and the almost always futile requests of Lievog to the bureaucracy both in Iceland and Denmark (Þorbjarnardóttir et.al., 1998).

After a few years, information on the station conditions and also the observations, gradually become more sporadic and it is not even known with certainty when Lievog left the country. His salary receipts for the years 1794 to 1802 have survived and he was certainly on the payroll at least to the end of this period. It is likely that he left sometime during the years 1802 to 1805. He died in poverty in Denmark in 1811. His belongings were auctioned, and the famous linguistic pioneer, Rasmus Christian Rask, bought at least some of the observations still existing and later bestowed them to the then newly formed Icelandic Society of Letters. Some of the extant data have thus come to us via this society. Other data came from the archives of the Danish Scientific Society (via the Danish Meteorological Institute which delivered most of its material related to Iceland to the

Figure 2. Lambhús in 1792, the larger building is almost certainly the Astronomical Observatory (Icelandic National Library, LBS 890 fol)

Icelandic Meteorological Office and the National Library in 1954). The following Lievog data have survived: a) complete listings from 10 August 1779 to 30 June 1785 and the whole year 1789; b) annual summaries (tables) of weather frequencies from 1 July 1787 to 30 June 1788 and the whole year of 1792. (Some comments are present in the manuscripts, but the original lists are missing); c) a summary for the period 1 July 1785 to 30 June 1787 was found during the search conducted at the Copenhagen archives during the present research, but not the original observation lists. The complete observation lists consist of three to four readings per day and include temperature, pressure, wind direction, wind speed and weather. The 1789 lists contain precipitation measurements as well. The pressure was registered in French inches (Paris lines). The uncorrected pressure values are generally on the low side (even after reasonable height and gravity corrections have been added). A likely reason for this is entrapped air above the mercury column. Some work has been done in a correction evaluation of the temperature (measured in Reaumur). It seems that the temperature series of the dataset is quite inhomogenous, but it can probably be corrected at least to the level of internal consistency. In spite of this problem, cold and warm periods are easily detected. The wind frequency seems to be reliable as the local wind effects present in all recent series from this part of the country are also present in the Lievog-series. The frequency of weather element occurrences also seem reasonable. The weather has been digitized in a code similar to the present day standard SYNOPtic weather message (see Jónsson and Kristjánsson, 1998). Some additional information on the history of the Lambhús observatory is found in Eyþórsson (1962, 1963). Sources: *National Library, ÍB 234a-b, 4to* and *Óskráð veðurgögn, kassi II*. Summaries from 1785-87 in *Kongelig bibliotek, NKS 2749 quarto*.

3.9. THE OBSERVATIONS OF MAGNÚS KETILSSON

The Sheriff of the Dalir district in western Iceland, Magnús Ketilsson, who lived at Búðardalur at Skarðsströnd, observed the weather in his diary, starting 24 February 1779, and ending in 1802. When he was at home he also made pressure observations and some temperature measurements. The instrumental series, however, ends in 1797 and there are many gaps in it. The temperature series is very irregular, the notation is unusual and has not successfully been deciphered. The pressure readings (French inches) have been digitized. Parts of the observations are contemporary with the surviving part of the Lievog series. These two series have been compared and are very well correlated on a day-to-day basis, but substantial index adjustments are needed. A transcription has been made of the daily weather during 1786 to 1791. Source: *National Library LBS 573, 4to.*

3.10. THE OBSERVATIONS OF SVEINN PÁLSSON

The Sveinn Pálsson southern Iceland weather observation diaries covering the period January 1798 to December 1812 have been digitized. No part of this series has been available before. The series consists of temperature (Reaumur) and pressure (French inches) measurements, wind direction information, as well as the cloud drift direction, wind speed and weather. There are large gaps in both the temperature and pressure measurements, but in spite of this, the series gives a reasonable overview of the weather in Iceland during this period. The largest gaps occur during summer and autumn when Pálsson was traveling extensively. He was a district medical doctor in southern Iceland as well as an ardent naturalist and an unusually keen observer of physical processes in the environment (Ingólfsson, 1991).

For most of the period (1798 to 1809) Pálsson was living at Kotmúli in the southern lowlands of Iceland, but in 1809 he moved to Vík in Mýrdalur (at the extreme southern tip of the country). A comparison has been made with the contemporary von Scheel series (see below) during an overlapping period. This comparison is promising as a tool in an overall quality assessment of both pressure series. It reveals that the observers are clearly observing the same weather systems at a similar magnitude, but it has proven difficult to estimate likely index adjustments for the barometers. (This matter will be pursued further in a later paper.) It is thus advisable that one takes great care when using the daily instrumental data. The pressure and temperature series start at the end of January 1798, the pressure measurements end in late 1809, and the temperature measurements by the end of 1812. An interesting part of the series is the inclusion of the direction of cloud movement. All the measurements and visual observations are available in digital format. An effort was made to locate instrumental observations by Sveinn Pálsson prior to 1798 without great success. Some of his visual observations from the mid-1780s were found, however. (Present source: A detailed hand-made transcription by Haraldur Jónasson, kept in the library of the Manuscript Department, *Stofnun Árna Magnússonar*, University of Iceland). There are also some temperature measurements during this period. These could conceivably be used in the day-to-day analysis of selected periods. Pálsson's travel journal during 1792 to 1794 has been published (Pálsson, 1983). During these years he was surveying the nature and environment of the country with the support of the Danish Society of Natural History (*Naturhistorie Selskabet*). This society supplied the instruments. The monthly overviews Pálsson made in 1792 to 1795 have been tabulated and are available as a part of the travel journal.

An interesting manuscript was found, a sort of a manual for meteorological observers, written by Pálsson in 1792. It is evident that he was interested in establishing a kind of meteorological observation network in Iceland (National Library, Lbs. 306, 4to). He specifically mentions three priests as potential or actual observers (Salomon Björnsson at Berufjörður, Guðlaugur Sveinsson in Vatnsfjörður and Pétur Pétursson at Miklibær). It is clear from other sources (Espólín, 1855, p.8) that at least the last mentioned actually made some measurements. No trace has been found of any observation lists. As mentioned earlier, Sveinn Pálsson's temperature measurements ended in 1812. The weather diaries continue until 1840, almost devoid of measurements, however, (National Library, ÍB 22-23, 4to). During this latter period from 1812 to 1840 he lived in Vík. During the winter of 1803 to 1804, Pálsson was employed as the *Landlæknir* (national medical officer) of Iceland. During this time his diaries contain measurements from his residence in Reykjavík and his home at Kotmúli as well. Source: *National Library ÍB.5 fol, (1798-1811), ÍB. 22, 4to (1812), LBS. 306.*

3.11. THE VON SCHEEL SERIES

The von Scheel series at Akureyri, northern Iceland, from September 1 1807 to 18 August 1814 has been digitized. Sometime in the late-nineteenth century the von Scheel observations were copied and slightly simplified at the Danish Meteorological Institute. This copy was used in the digitisation of the series. Hans Jacob von Scheel (1779-1851) was a member of a Danish coastal surveying team and made observations three times per day during the above-mentioned period with his co-worker Captain Frisak (1773-1834) (Thoroddsen, 1892-1904, III, p. 265ff.). There are both pressure and temperature measurements. The pressure series partly overlaps with the Sveinn Pálsson series. As stated above a comparison was thus possible, but a need for adjustments of the absolute values of both series became evident during the comparison process. There are notes on the instruments and their use included in the manuscript, but this description seems to have been written by von Scheel much later, in 1833. According to this description, there were originally five thermometers. During the first year, two of them were regularly compared. Some of the thermometers were in use on the travels during the survey. The observations made during the survey journeys have not been found. Source: *National Library LBS. Óskráð veðurgögn, kassi V and kassi VI.*

3.12. THE JÓN THORSTEINSSON SERIES

The daily observations of Jón Thorsteinsson were made in Reykjavík and at Nes from 1 August 1820 to 28 February 1854. The pressure series (in French inches) was checked for both spurious errors as well as systematic ones. This series has been used in the compilation of a homogenous southwest Iceland monthly pressure extending from the early 1820s up to the present time (Jones et al., 1997). The earliest part of the temperature series should be used with some care. There are precipitation measurements from 1829 onwards. The visual observations (wind and weather) were transcribed and coded as well. Additional observations of temperature, humidity and pressure during the summers of 1843 to 1852 were found during the present data search. These additions are particularly welcome as they give better information about the diurnal temperature variations at the station and thus indirectly give a better indication of the actual instrumentation arrangements than was available before. It has not, however, been possible to digitise this voluminous addition at the present time. The observations were made in cooperation with the Danish Scientific Society, which supplied the instruments. The history of the station is rather complicated and only the briefest outline may be given here. After Thorsteinsson arrived in Iceland in 1820, he at first made his quarters in Reykjavík. However, the exact location in the town is not known. The first entry in the journal is 1 August 1820. In July 1821, he moved to Nes, a site a few miles west of Reykjavík, but on 18 October 1833 he moved again into Reykjavík, almost certainly to a location that now has the street address Ránargata 13. A six-month period, 1 September 1821 to 28 February 1822, is lost from the series; the reason is unknown. In 1839 the Danish Scientific Society published a part of the series, 1823 to 1837 *in extenso* in a book format (Holmsted, 1839). Source: *National Library, LBS Óskráð veðurgögn, kassi I and kassi VI.*

3.13. OTHER OBSERVATIONS DURING THE PERIOD 1824 TO 1837

As mentioned above the Danish Scientific Society sent instruments to other observers during the 1820s. It is known that Gísli Brynjólfsson (1794-1827), doctor of Philosophy and priest at Hólmar in eastern Iceland, made observations from the middle of 1824 to 10 July 1825. Brynjólfsson drowned in 1827. It is not clear if he continued observing up to the time of his death. Grímur Jónsson (1785-1849), the *Amtmaður* (the highest official of northern Iceland) made pressure and temperature observations during the years 1826 to 1833, albeit with two large gaps before November 1829. Páll Melsted (1791-1861), the Sheriff at Ketilsstaðir in the east made pressure and temperature observations from 5 August 1828 to 8 July 1834. The daily pressure of one year (1831) at Möðruvellir and Ketilsstaðir has been digitized for comparison purposes. Halldór Einarsson (1796-1846), Sheriff at Akranes in western Iceland measured temperature from 1 December 1835 to 31 July 1837 (not digitised). Source: *National Library. Óskráð veðurgögn, kassi II.*

3.14. THE INITIATIVE OF THE ICELANDIC SOCIETY OF LETTERS IN 1840

Jónas Hallgrímsson, mentioned earlier, was a poet of major stature and one of the very few Icelandic academics of his time specialising in the natural sciences. In 1840, he urged the Icelandic Society of Letters (*Hið íslenska bókmenntafélag*) to launch a general survey of the Icelandic environment, with the cooperation of local ministers and the members of the society. The project had two main components: a) a description of the natural and cultural aspects of each and every parish in the country; and b) meteorological observations at 50 sites all over the country. The Society actually did initiate this major undertaking and sent detailed instructions on meteorological observations to 49 ministers all over Iceland. The majority of these did begin observing, and thus a large set of temperature measurements exists from Iceland during the early part of the 1840s. A few of the ministers kept the effort going for more than 30 years (for example at Hvammur in western Iceland during 1841 to 1878). A large part of the parish descriptions have been published, and in many cases accounts of the peculiarities of local climate are included. These accounts also give valuable information for climatic assessments of a general nature. A detailed account of the project and a list of observers is found in Sigurðsson (1867). The project was also supported by the Danish Scientific Society which donated 45 thermometers, and the observations were made as specified in the guidelines of the society. Sources: The original proposal is found in the *National Library in ÍB. 73, fol.* and the observation lists mainly in the sets *LBS 73, fol* and *Óskráð veðurgögn, kassi III og kassi IV*. A few of the lists are to be found in the National Archives. More information on the bibliography of this large effort is found in Hávarðsson (1997).

3.15. THE STYKKISHÓLMUR SERIES

This series encompasses observations from Stykkishólmur from 1 November 1845 to the present day. The station history is a complicated one, and only a brief discussion can be given here. One of the thermometers of the Society of Letters effort went to Jón Guðmundsson, parish minister at Helgafell in the vicinity of Stykkishólmur. For reasons that are not known Guðmundsson asked Árni Thorlacius, a merchant in the village of Stykkishólmur to take over as an observer. Thorlacius, who was rather well off by the standards of the day, took the case to his heart, bought more instruments and started to record the weather in an unusually detailed manner. (The reference to this information is a letter to the Icelandic

Society of Letters, in the archives of the Icelandic Meteorological Office, written by Thorlacius in 1849). There are many temperature and pressure readings per day during some parts of the period before 1872. For comparison purposes, it is also fortunate that there is also more than one thermometer present, and after 1856 there are two different barometers. Precipitation measurements started in September 1856. Sometime after 1860, Thorlacius began a cooperation with the Scottish Meteorological Society which for a period supplied calibrated instruments for the station (Thorlacius and Buchan, 1869). These early observations from Stykkishólmur are available in a digital format, without any corrections. The Danish Meteorological Institute formally took over the observations at Stykkishólmur in 1872, but Thorlacius continued as the observer until his death in 1891 (Bergþórsson, 1957). Source: Manuscripts in the archives of the *Icelandic Meteorological Office, Gamalt 178*. Another copy is found in the *National Library, LBS 206-218.*

3.16. THE AKUREYRI SERIES OF 1846 TO 1854

Monthly averages from a series of observations made in Akureyri in northern Iceland during the period 1846 to 1854 have been known for some time, and were used for comparison purposes in the construction of the southwest Iceland pressure series mentioned above. The original observation lists were located during the present search initiative. They have not been digitised. The observations start on 18 August 1846 and end on 23 October 1854. There is a gap from 10 February 1847 to 22 August 1847. There are measurements of temperature and pressure during the whole period, humidity from 22 August 1847 and precipitation from 28 May 1851. The humidity measurements are irregular during the wintertime. The observer was Dr Eggert Johnsen (1798-1855), the district physician. Source: *National Library: LBS. Óskráð veðurgögn, kassi V.*

3.17. THE LAUFÁS SERIES OF 1854 TO 1882

This takes the form of a weather diary written by Björn Halldórsson (1823-1882), a parish priest at Laufás in the central north, during the period 1854 through 1882. The pressure (French inches) was recorded once or twice per day and the temperature occasionally, especially during the latter part of the period. The pressure readings have been digitised, but it is apparent that a considerable index adjustment is needed in addition to the height and gravity corrections. Source: *National Library, LBS 326, fol.*

3.18. REYKJAVÍK OBSERVATIONS DURING 1857 TO 1880

After the death of Jón Thorsteinsson in 1855, the Danish Scientific Society in 1857 recruited Þórður Jónasson (1800-1880), a judge at the high court, for the continuation of the Reykjavík observations. Slightly more than two years (1857-1859) of observations made by him, or his son Jónas, have been found, but include only temperature once a day, and a short description of the wind and weather. It is not clear if any observations were made in Reykjavík during 1859 to 1865. In 1865 Jónasson or Pastor Ólafur Pálsson (1814-1876) started a collaboration with the Scottish Meteorological Society (as Árni Thorlacius in Stykkishólmur did at the same time). Lists of some monthly means and extremes of temperature survive in the archives of the Icelandic Meteorological Office. An observation list, allegedly in the handwriting of Þórður Jónasson, covering the period 20 October 1865 to 30 June 1874 was found at the National Library. Other original observation lists from Reykjavík for the period 1865–1870 have not been found, so far. In September 1871, a teacher at the Reykjavík Grammar School, Jón Árnason (1819-1888), began observing regularly at the school in cooperation with the Danish Scientific Society. Correspondence between Árnason and the Society is to be found in the archives of the Icelandic Meteorological Office. The Danish Meteorological Institute formally took over the regular Reykjavík observations in 1878, but the station did not become an official part of the institute's observation network until May 1880. A new observer, Sigurður Sigurðsson then replaced Árnason. There are many uncertainties concerning the measurements in Reykjavík during the slightly more than 26-year period separating the measurements of Jón Thorsteinsson and the start of the formal Danish Meteorological Institute takeover. The Sources: *National Library LBS 488. fol.* (The "Þórður Jónsson" diary), *National Library, LBS Óskráð veðurgögn, kassi I and kassi VI* (Jónas Jónasson, 1857-1859), Manuscripts in the archives of the *Icelandic Meteorological Office, Gamalt 001* (Jón Árnason observations).

3.19. OTHER SERIES

A few other series including instrumental observations dating from the 1840s and later, have been checked, but not digitised. Most are of a rather fragmentary nature. An extensive list of these instrumental observations has been published in the report by Hávarðsson (1997).

4. Conclusion and Further Prospects

A comprehensive search has been made in Icelandic and Danish archives for pre-1872 meteorological measurements. An inventory has been made, a substantial part has been digitised and background information has been assessed. It is beyond the scope of this article to include a detailed discussion on the climate of Iceland during this period. However, a few comments may be made. The eighteenth and nineteenth centuries appear to have been generally difficult climatically, with lower temperatures and heavier sea-ice incidence than experienced in this area during the twentieth century. Milder periods are found, however. It seems likely that the seasonal variation of temperature was somewhat larger than now, and the temperature gradient between southern and northern Iceland during winter was also larger than now. In certain years the persistent sea ice during summer also had a deleterious effect on the temperature in some of the coastal areas of the north and east (see also Ogilvie, 1992; Ogilvie and Jónsson, this volume).

In spite of the extensive search for data already done, there is still hope of more "lost" instrumental data to be found. The search has revealed a number of references to observations that have not yet been found. Most of the important ones have been mentioned. Many weather diaries without any instrumental observations were checked and these will undoubtedly be very useful in assessing the weather conditions in the gaps still remaining in the records of the period (1749 to 1872). This work is the prerequisite of a comprehensive study of the weather conditions in Iceland during this period. Both the search and the digitisation will continue in the coming years.

Acknowledgments

This work was primarily sponsored as a part of the EU Fourth framework program: ADVICE (ENV4-CT95-0129), coordinated by Trevor Davies. A 1989 grant from the Icelandic Science Fund for the initiation of the Lievog digitising work and support from the Leverhulme Trust, UK (Grant no. E94M) are gratefully acknowledged. Thanks are also due to the Icelandic Meteorological Office and the staff of the Manuscipt section of the Icelandic National Library. Special thanks go to Astrid Ogilvie, Sjöfn Kristjánsdóttir and Svavar Hávarðsson for their invaluable comments and help. We also thank Guðrún Þorbjarnardóttir BA, who carried out a search for material held at the National Archives, the historian Ragnheiður Mósesdóttir who searched for material in some of the Copenhagen manuscript collections, Gísli Krisjánsson for his data management and not least, the digitising team at the Icelandic Meteorological Office. Other members of the ADVICE team have contributed to this work both directly and indirectly, and the generosity of the staff at the Climatic Research Unit in Norwich is especially appreciated and the useful comments of anonymous reviewers as well.

References

Bergþórsson, P.: 1957, *Loftin blá* (Fair skies), Heimskringla, Reykjavík.

Espólín, J.: 1855, *Íslands Árbækr í sögu formi. 12. deild* (Icelandic yearbooks in the form of history), Hið íslenska bókmenntafélag, København.

Eyþórsson, J.: 1956 'Elztu veðurathuganir með mælitækjum á Íslandi' (The earliest instrumental meteorological observations in Iceland), *Veðrið* 1, 27-28.

Eyþórsson, J.: 1962, 'Stjörnuturninn á Lambhúsum og Rasmus Lievog' (The astronomical observatory at Lambhús and Rasmus Lievog), *Veðrið* **7**, 39-49.

Eyþórsson, J.: 1963, 'Rasmus Lievog og veðurathuganir hans' (Rasmus Lievog and his meteorological observations), *Veðrið* **8**, 3-8.

Eyþórsson, J.: 1964, 'Jón Þorsteinsson landlæknir og veðurathuganir hans', (Jón Thorsteinsson physician and his meteorological observations), *Veðrið* **9**, 43-49.

Finnsson, H.: 1796, 'Um Mannfæckun af Hallærum á Íslandi' (Depopulation and Famine in Iceland), *Rit Lærdómslistafélags* **14**, 30-226.

Finnsson, H.: 1970, *Mannfækkun af hallærum*, Eyþórsson, J. and Nordal, J. (eds.), (A second annotated ed. of Finnsson (1796)), Almenna bókafélagið, Reykjavík

Garðarsson, H.: 1998, *Æviatriði nokkurra manna sem komu við sögu íslenskra veðurathugana fyrir 1872* (Meteorological observers in Iceland before 1872. Biographical notes), Icelandic Meteorological Office, **VÍ-G98029-ÚR24**.

Guðmundsson, E. H.: 1989, 'Johnsonius og Lievog, konunglegir stjörnumeistarar á Íslandi á 18. öld' (Johnsonius og Lievog, royal astronomical observers in Iceland in the eighteenth century), in *Eðlisfræði á Íslandi* **IV** (Physics in Iceland IV), Icelandic Physical Society 110-125.

Guðmundsson, E. H.: 1998, 'Gísli Einarsson skólameistari og vísindaáhugi á Íslandi á 17.öld' (Rector Gísli Einarsson and the scientific interest in Iceland in the seventeenth century), *Saga* **XXXVI**, 185-231.

Hávarðsson, S.: 1997, 'Veðurhandrit. Könnun og skráning veðurfræðilegra gagna á Handritadeild Landsbókasafns og Þjóðskjalasafni Íslands. With an English Summary and explanations'. (Includes details and commentaries on the observation manuscript source references), *Icelandic Meteorological Office*, **VÍ-G97039-ÚR30**.

Holmsted, S. (ed.): 1839, *Observationes Meteorologicae A. 1.Jan 1823 ad 1 Aug 1837 in Islandia factae a Thorstensenio medico,* København.

Horrebow, N.: 1752, *Tilforladelige Efterretninger om Island*, København..

Horrebow, N.: 1966, *Frásagnir um Ísland* (A report on Iceland, an Icelandic translation of Horrebow (1752), translated by Steindórsson, S.), Bókfellsútgáfan, Reykjavík.

Hugason, H.: 1990, 'Guðfræði og trúarlíf' (Theology and religious life) in Sigurðsson, I. (ed.), *Upplýsingin á Íslandi* (The Icelandic Enlightenment), Hið íslenska bókmenntafélag, Reykjavík, 320pp.

Ingólfsson, Ó.: 1991 '"On glaciers in general and particular ..." The life and works of an Icelandic pioneer in glacial research', *Boreas Pioneers series* **20**, 79-84.

Jones, P., Jónsson, T., and Wheeler, D.: 1997, 'Extension to the North Atlantic Oscillation using early instrumental pressure observations from Gibraltar and South-West Iceland', *International Journal of Climatology* **17**, 1433-1450.

Jónsson, T. and Kristjánsson G.: 1998, 'Instrumental meteorological observations in Iceland in the early instrumental period. A data companion', *Icelandic Meteorological Office*, **VÍ-G98026-ÚR21**.

Kington, J.A.: 1972, 'Meteorological observing in Scandinavia and Iceland during the eighteenth century', *Weather* **27**, 222-233.

Kington, J.A.: 1974, 'The Societas Meteorologica Palatina: An eighteenth-century meteorological society', *Weather* **29**, 416-426.

Magnússon, Á., and Vídalín, P.: 1924-1947, *Manntal á Íslandi árið 1703. Tekið að tilhlutan Árna Magnússonar og Páls Vídalín, ásamt manntali 1729 í þrem sýslum* (A general census of Iceland 1703, done by Árni Magnússon and Páll Vídalín, and a three county census of 1729), Hagstofa Íslands, Reykjavík.

Ogilvie, A. E. J.: 1982, *Climate and society in Iceland from the medieval period to the late eighteenth century.* A Doctoral Dissertation, Climatic Research Unit, School of Environmental Series, University of East Anglia, Norwich.

Ogilvie, A. E. J.: 1984, 'The impact of climate on grass growth and hay yield in Iceland' in N.-A. Mörner and W. Karlén (eds), *Climatic change on a yearly to Millenial Basis*, Reidel, Dortrecht, 343-352.

Ogilvie, A. E. J.: 1992, 'Documentary evidence for changes in the climate of Iceland, A.D. 1500 to 1800', in *Climate Since A.D. 1500*, in Bradley, R.S. and Jones, P.D. (eds), Routledge, London, 92-117.

Ogilvie, A. E. J. and Jónsson, T.: 2001, '"Little Ice Age" Research: A Perspective from Iceland', this volume.

Olafsen, E.: 1772, *Vice-Lavman Eggert Olafsens og Land-Physici Bierne Povlsens Reise igennem Island*, 1-2b, Sorø.

Ólafsson, E.: 1974, *Ferðabók Eggerts Ólafssonar og Bjarna Pálssonar um ferðir þeirra á Íslandi 1752-1757*, second Icelandic edition (The travels of Eggert Ólafsson and Bjarni Pálsson in Iceland), a translation of the Danish version of 1772 (Olafsen, 1772), 2 volumes, Örn og Örlygur, Reykjavík.

Ólafsson, E.: 1975, *Travels in Iceland by Eggert Ólafsson and Bjarni Pálsson: performed in 1752-1757*, second English edition, Örn og Örlygur, Reykjavík

Pálsson, S.: 1983, *Ferðabók Sveins Pálssonar* (2nd edition), vol. I - II, (The travel journal of Sveinn Pálsson), Örn and Örlygur, Reykjavík.

Seaton, E.: 1935, *Literary Relations of England and Scandinavia in the seventeenth century*, Oxford at the Claredon Press, 184-187.

Sigfúsdóttir, A. B.: 1969, 'Hitabreytingar á Íslandi 1846 - 1968' (Temperature changes in Iceland 1846-1968), in Einarsson, M.Á.(ed.), *Hafísinn* (The Sea Ice), Almenna bókafélagið, Reykjavík, 70-79.

Sigurðsson, H.: 1978, *Kortasaga Íslands I* (The mapping of Iceland, volume I), Reykjavík, 203-210.

Sigurðsson, H.: 1990, 'Náttúruvísindi og landafræði' (Natural science and geography), in Sigurðsson, I. (ed.), *Upplýsingin á Íslandi* (The Icelandic Enlightenment), Menningarsjóður, Reykjavík.

Sigurðsson, J.: 1867, *Hið íslenska bókmenntafélag. Stofnan félagsins og athafnir um fyrstu fimmtíu árin 1816-1866* (The Icelandic Society of Letters, a monograph. The establishment of the society and the activities for the first fifty years 1816-1866), Hið íslenska bókmenntafélag, Kaupmannahöfn, pp. 43-44 and 86-94.

Skovmand, P.A.: 1986a, 'Den meteorologiske aktivitet på Rundetårn i årene 1777-1781 (I)' (The meteorological activity at Rundetaarn 1777-1781), *Vejret* **28**, 15-18.

Skovmand, P.A.: 1986b, 'Den meteorologiske aktivitet på Rundetårn i årene 1777-1781. (II)' (The meteorological activity at Rundetaarn 1777-1781), *Vejret* **29**, 11-18.

Thorlacius, A.O. and Buchan, A.: 1869, 'On the meteorology of Iceland', *J. Scot. Met. Soc.* **Vol II, new series**, 285-292.

Þorbjarnardóttir, G., Jónsson, T. and Kristjánsson, G.: 1998, *Uppskrift af nokkrum skjölum á Þjóðskjalasafni varðandi íslenskar veðurathuganir á 18. öld* (Transcriptions of eighteenth century documents in the Icelandic National Arhives related to weather observations in Iceland in the eighteenth century), Icelandic Meteorological Office, **VÍ-G98027-ÚR22**.

Thoroddsen, Th.: 1892-1904 *Landfræðisaga Íslands. Hugmyndir manna um Ísland, náttúruskoðun og rannsóknir fyrr og síðar. I-IV.* (The historiography of the natural sciences in Iceland vol, I - IV), Hið íslenska bókmenntafélag, Reykjavík 1892-1896, Copenhagen 1989-1904.

Thoroddsen, Th.: 1916-17, *Árferði á Íslandi í 1000 ár* (The climate of Iceland through one thousand years), Hið íslenska fræðafélag, Copenhagen.

von Troil, U.: 1961 *Bréf frá Íslandi* ("Letters from Iceland", an Icelandic translation of von Troil: "Brev om Island", Stockholm, 1933), Reykjavík.

Archival Material

National Library of Iceland
(Landsbókasafn Íslands, handritadeild)

Source	*Name of observer*	*Location*
ÍB. 5, fol.	Guðlaugur Þorgeirsson?	Garðar?
Lbs. 46, fol., pp. 190-211.	Hannes Finnsson	Skálholt
ÍB. 234a-b, 4to.	Helgi Sivertsen	Skálholt
Lbs. Óskráð veðurgögn, kassi II.	Guðlaugur Þorgeirsson	Garðar
ÍB. 224a-b, 4to.	Rasmus Lievog	Lambhús
Lbs. Óskráð veðurgögn, kassi II.	Rasmus Lievog	Lambhús
Lbs. 573, 4to.	Magnús Ketilsson	Búðardalur
ÍB. 5, fol. (1798-1811).	Sveinn Pálsson	Kotmúli/Vík
Lbs. 306, 4to.	Sveinn Pálsson	
Lbs. Óskráð veðurgögn, kassi V og VI.	Hans Jacob von Scheel	Akureyri
Lbs. Óskráð veðurgögn, kassi I og VI.	Jón Þorsteinsson	Nes
Lbs. Óskráð veðurgögn, kassi II.	Gísli Brynjólfsson	Hólmar
Lbs. Óskráð veðurgögn, kassi II.	Grímur Jónsson	Möðruvellir
Lbs. Óskráð veðurgögn, kassi II.	Páll Melsteð	Ketilsstaðir
Lbs. Óskráð veðurgögn, kassi II.	Halldór Einarsson	Akranes
ÍB. 73, fol.	Parish ministers at many locations	
Lbs. Óskráð veðurgögn kassi III og IV.	Parish ministers at many locations	
Lbs. 206-218.	Árni Thorlacius	Stykkishólmur
Lbs. 326, fol.	Björn Halldórsson	Laufás
Lbs. 488, fol.	Þórður Jónasson	Reykjavík
Lbs. Óskráð veðurgögn, kassi I.	Jónas Jónasson	Reykjavík

Det Kongelige Bibliotek, Copenhagen -
Håndskriftsafdelingen

Source	*Name of observer*	*Location*
NKS 2749 quarto.	Rasmus Lievog	Lambhús

Icelandic Meteorological Office

Source	*Name of observer*	*Location*
Gamalt 178.	Árni Thorlacius	Stykkishólmur
Gamalt 001.	Jón Árnason	Reykjavík

(Received 11 February 1999; in revised form 22 September 2000)

THE TALLINN TEMPERATURE SERIES RECONSTRUCTED BACK HALF A MILLENNIUM BY USE OF PROXY DATA

A. TARAND[1] and P. Ø. NORDLI[2]

[1] *Estonian Parliament address*
[2] *The Norwegian meteorological institute, Box 43 Blindern, N-0311 Oslo*

Abstract. In order to extend the Tallinn temperature series backward in time, three different climate proxies were used. These were: the first day of ice break-up in Tallinn port, a proxy for the mean winter air temperature (December to March); the first day of ice break-up on the rivers in northern Estonia, a proxy for the beginning of spring; and, the first day of the rye harvest, a proxy for the mean air temperature in spring and summer (April to July). On the basis of these proxies the mean winter temperature could be extended back to the year AD 1500, and the spring and summer temperature back to 1731. The series of winter temperatures was analysed for long-term trends and variations on different timescales. The most striking feature is the warming of the winters from about the mid nineteenth century to the present. The warming is especially noticeable over the latest decades. The climate from the start of the series (AD 1500) to the mid nineteenth century was in general somewhat colder, and should be recognised as a part of the Little Ice Age, though the period was intercepted by warmer winters in the first half of the eighteenth century.

Key Words. Little Ice Age, temperature time series, ice break-up, rye harvest.

1. Introduction

During the last decade, many long time series of air temperature have been revised by homogeneity checking (Brazdil and Dobrovolný, 1993; Moberg and Alexandersson, 1997). The goal of the reconstruction of such long time series is to increase knowledge of changing climate at certain fixed points. Tallinn may serve as one point in a larger network of long temperature time series (see Figure 1). The homogenisation of the Tallinn air temperature time series started in the 1970s (Tarand, 1976). Not only were the observations studied, but also their history (metadata). Several early, previously unknown observations were discovered in the archives, and subsequently used in the work on the series. Papers describing this work were mostly published in Estonian or in Russian (Tarand, 1976, 1982, 1992b), and remained unknown to the international climate change research community.

The use of proxy data to reconstruct air temperature in pre-instrumental time, the main topic of this article, may be said to have begun in Estonia with a publication by Kirde (1924). His studies focused on the relationship between the ice cover at the port of Tallinn and air temperature. These data were limited to a

Climatic Change **48**: 189–199, 2001.

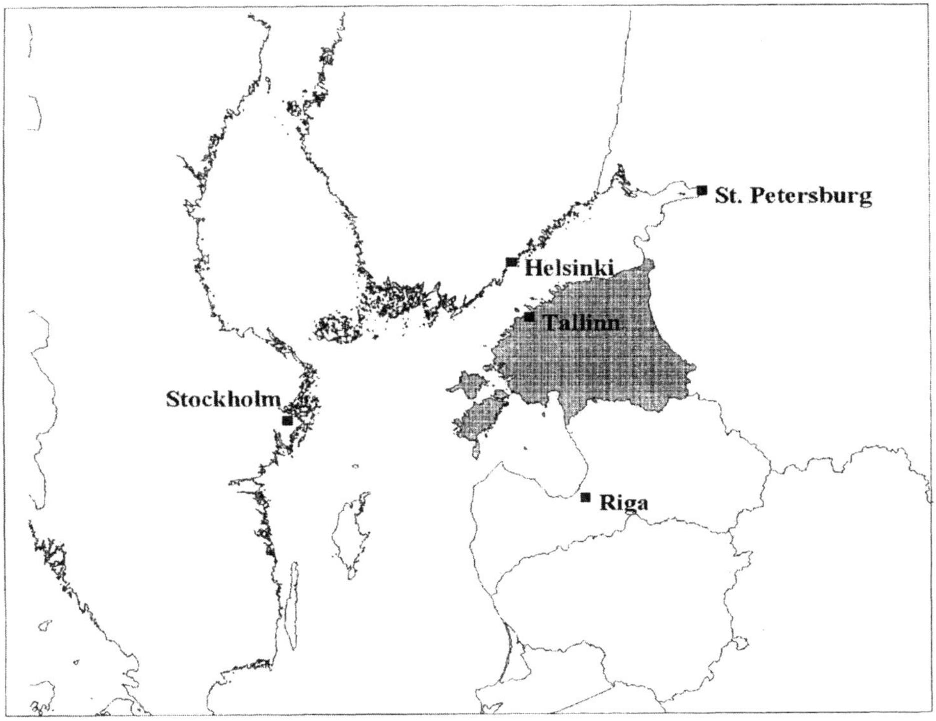

Figure 1. Estonia, the adjacent part of the Baltic Sea and the neighbour area.

period of twenty years just before World War I. More recently, proxy data have been used by Tarand (1992b) and by Tarand and Kuiv (1992). However, all earlier published data have been revised in connection with the project "A Country Case Study on Climate Change Impacts and Adaptation Assessments in the Republic of Estonia" (Tarrand and Kallaste, 1998).

In choosing the methods to be used, the main aim was to obtain the best possible results for the study of climate trends, i.e. to minimise errors in the monthly or seasonal averages.

2. Data

2.1. MODERN INSTRUMENTAL DATA

In 1849, the "Estonian Literary Society" was established (Jordan, 1889). This conducted a standardisation of the network of observations, and has shown to be of great importance regarding the quality of the observations. It seems reasonable to consider the starting year of the modern observations at Tallinn to be in 1850.

The modern observations have been homogenised for local changes at the observing sites and for relocations (Tarand, 1976, 1982, 1992b). In the period 1908 to 1946, 62 monthly means are missing, mainly caused by activities during World War I and World War II, but also in connection with a reorganisation of the station network. By comparison with rural stations, urban trends in the Tallinn series are removed (Tarand, 1974, 1975, 1986).

2.2. EARLY INSTRUMENTAL DATA

In Tallinn the first known instrumental observations date back to 1774. The homogenised series consists of several shorter series originating from different observing sites (Tarand, 1992b). From the start of observations up to 1805, only seasonal or monthly averages by Julian calendar are partly available. Recalculation to the Gregorian calendar is not straightforward as original data are lacking, and searches in archives for original journals have so far not been successful. Fortunately, there exist parallel measurements for large parts of the early instrumental period within Estonia that may be used as a control. However, no records from the year 1814 have been found, and the gap has therefore been filled with the help of the series from Stockholm (Moberg, 1996) and St Petersburg (Wild, 1881).

2.3. PROXY DATA

For the purpose of temperature reconstruction in pre-instrumental time, three main proxies of air temperature are available: 1) The day of ice break-up in the port of Tallinn, a proxy for the mean winter air temperature (December to March); 2) The day of ice break-up on the rivers in northern Estonia, a proxy for the beginning of spring; and 3) The first day of the rye harvest, a proxy for the mean air temperature in spring and summer (April to July).

A description of an Estonian time series of ice-data was first published by Tarand (1992a). The longest of the time series is that of average winter air temperature reconstructed on the basis of ice break-up in the port of Tallinn (Tallinn Road). From AD 1600 onwards there are no gaps, and it is also almost complete in the sixteenth century as only nine years are missing. In the fourteenth and fifteenth centuries a few data also exist, 11 and 29 respectively, but that is not enough to allow reconstruction of average air temperature for those centuries. The first proxy in this series dates back to 1349.

The main sources of ice break-up dates are port journals, custom books, town council minutes, letters, trade companies' documents, newspapers, periodicals, and memoirs. As a member of the Hanseatic League, Tallinn has a long history of trading, and the most valuable material up to the eighteenth century is found in the Tallinn town archives with a rich collection of port journals and custom books. Parts of these records were evacuated to Germany in 1944 and were unavailable until 1992, when the evacuated relevant documents were returned to Estonia. However, the exact year of the collapse of the Soviet Union could not be foreseen twenty years ago, and therefore systematic efforts were made to find other relevant sources. This proved to be a success as the important Clayhills Storehouse documents were detected in the "Estonian Historical Archive" at Tartu. Thus, independent sources exist that make it possible to evaluate the deviations in indirect determined ice break-up dates. Even as early as in the eighteenth century, scattered extra data exist, and in the seventeenth century, custom books and port journals for Tallinn in the Swedish National Archives are a valuable additional source of data.

Tallinn has suitable natural conditions for the use of ice as a proxy for air temperature. The port of Tallinn lies close to the average western limit of the ice cover in the Gulf of Finland, and therefore it is sensitive to ice-cover variations. Moreover, drift ice is brought in to the port by cool winds. Because of this, the occurrence of drift ice does not necessarily lead to a reduction of the correlation between winter temperature and the day of ice break-up. This correlation is reduced in other locations. In Helsinki, for example, drift ice comes in to the port by warm winds from the southern sector. Tallinn thus has a unique position among the coastal towns in the northern part of the Baltic Sea, where the freezing of seawater happens often enough to have indicator value.

For the winter (December to March) mean air temperature was reconstructed by simple, linear regression with the day of ice break-up in the port of Tallinn as predictor. The series of winter temperatures was extended back to 1757 by the observations in Stockholm and St Petersburg. A combination of these series was adjusted to be valid for Tallinn, and the following regression equation was calculated:

$$T_m = -0.0980N_d + 5.69 \qquad (1)$$

where,
T_m – mean temperature for Tallinn in winter (December-March)
N_d – ice break-up in Tallinn port (number of days from 1 January)

In unusually mild winters there is no ice in the port of Tallinn. This situation occurs when the mean winter temperature is -2.8°C or higher. Equation (1) cannot be used for the reconstruction of the highest winter temperatures because of lacking predictor values. An alternative method could be to look for narrative sources that give direct information concerning air temperature, but for some winters no documents exist, and another difficulty is that the travellers often were foreigners who had different criteria for what was meant by "mild" and "cold" than the local residents. The high winter temperatures were therefore reconstructed by use of a Monte-Carlo method. Obviously, producing extreme temperatures by chance makes them incomparable with systematically observed temperatures from recent centuries. However, used on timescales of decades to centuries the standard error reduces and no bias in the series is introduced.

The start of spring is defined as the day when the smoothed air temperature curve passes 0 °C. A proxy for this day is the ice-drift on rivers (Tarand, 1992a).

The best-known time-series of ice-drift in the region was published by Lamb (1977) for the Daugava river, which flows through Riga. This series goes back to 1530, but there are many gaps before 1709. After the 1950s, human impact has, to a large extent, disturbed the homogeneity of the series. To overcome this difficulty, the series was furnished with ice-drift data from a few Estonian rivers during the last decades. In our compilation, the river Neva at St Petersburg, and the rivers Aura and Kokemäki in Finland are also included. All series are adjusted to be valid for the small river Palmse by the method of average differences. Palmse crosses the Lahemaa National Park only 70 km east of Tallinn, and the oldest known ice-drift data in Estonia originates from this river. The series starts in the eighteenth century.

Tarand and Kuiv (1992) have shown that the mean spring and summer temperature (April to July) may be calculated by a combination of two proxies, the day of ice break-up on the rivers and the first day of rye harvest. The method is formulated by equation (2).

$$T_m = 1400/D - T_{dr} + T_{mr} \qquad (2)$$

where

T_m - mean air temperature of April to July of individual years

D - number of days between ice-drift on the rivers and first day of harvest

T_{dr} - air temperature of the reference period during the period D

T_{mr} - average air temperature of April to July of the reference period

The data and methods used are described in detail in Tarand and Kuiv (1992). They used harvest dates from 25 locations, of which 20 were located in Estonia, 3 in Finland, and 2 in Latvia. All data were adjusted to be valid for Tallinn by the method of average differences.

Most of the early harvest dates (75-80 percent) are connected with dry and warm Junes, and the late ones with rainy weather in August or during the summer. Summers with much precipitation are usually cold. These factors lead to a decrease of the error of temperature reconstruction by equation (2), and the equation is valid also for the extreme cases.

3. Error Estimation

3.1. ICE BREAK-UP IN THE PORT OF TALLINN

In most cases, the day of ice break-up is usually the day for the arrival of the first ship in spring. In general, ships were unable to pass through the ice, and stayed often somewhere nearby waiting for the ice to break up. The dating of ice break-up may thus be one or two days too late. This is not significant compared to the relativity of the term "ice break-up", which is a process that continues for several days. In some years, the first ship is registered before the final opening of the port, if the first ship was able to slip through although the drift ice stayed for some more weeks. Sometimes the ship was registered by the Captain, having walked across the ice to the port, before the ship had got through. On the other hand, the port could be open, but no trade was imminent. Background information of on-going war or blockade was also helpful in estimating the day of ice break-up, especially when the dating was made on the basis of correspondence or the handing over of passports.

The dates were checked against other independent eighteenth century sources. Analysing the data of 57 years from which two or more independent sources exist, the mean difference is only 1.3 days with a standard deviation of 8.5 days. By use of equation (1) this corresponds to about 1 °C deviation for individual winters. The maximum error for any winter as a result of late drift-ice or a cold winter is about 3°C.

3.2. THE FIRST DAY OF THE RYE HARVEST

The degree days for rye to ripen were determined with an error of ±50 degree days, which lead to an error in average April - July temperature of ±0.4 °C. In about 30 percent of the cases, the dates had to be determined indirectly from stacking in the drying barn, beginning of threshing, or carting into the granary. Due to this, the mean error is somewhat larger, and is estimated to about ±0.6 °C.

4. Results

The reconstruction of the mean winter temperatures (December–March) is shown in Figure 2 through Gaussian low pass filters with standard deviations in its distribution of 3, 9 and 30 years, corresponding to variation on timescales of about 10, 30 and 100 years respectively. Severe winters occurred around AD 1600, where the 500-year minimum values are located for the curves representing timescales of 30 as well as 100 years. Around AD 1800 there are also severe winters, and analysed on a decadal timescale the absolute minimum value occurs in the 1780s. This decade was followed by a period of milder winters, but a cluster of cold winters reoccurred in the 1800s, among these the winters 1800, 1803 and 1809, which were the most severe ones. The time interval from the 1780s to the 1820s is characterised by extraordinary large variability.

Analysed on a 100-year scale the temperature has increased the last 200 years (since around 1800) and at the end of the series it reached its absolute maximum value. A warming also took place through the seventeenth century, which resulted in a local maximum in the first part of the eighteenth century. However, already around 1860 the temperature had reached the level of the local maximum of the eighteenth century.

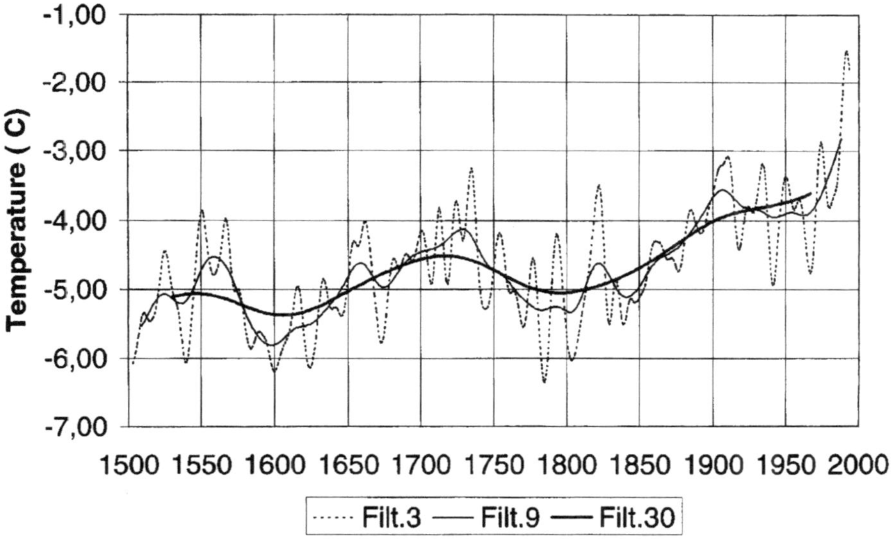

Figure 2. Winter temperature (December–March) at Tallinn since 1500 based on ice break-up dates in Tallinn port. The series is smoothed by Gaussian filters of 3, 9 and 30 years as standard deviations in the Gaussian distribution.

Testing the series for linear trends, an increasing trend for the whole period of 500 years was detected. Most of the increase has occurred during the last 150 years, i.e. during the modern instrumental period. The linear trend amounts to +1.8°C for all five centuries, while for the latest 150 years it amounts to +1.6°C. The trend over the last three decades is the most striking feature within the decadal temperature curve, Figure 2. The mean annual temperature increase corresponds to an annual decrease of the maximum ice cover in the Baltic Sea by 185 km^2 (Seinä and Palosuo, 1996; and personal correspondence).

The recent warming shown by the Tallinn series is supported by a large number of ice-free winters in Tallinn port. In the sixteenth to the nineteenth century only six to eight winters were without ice, while in the twentieth century (1999 included) the number has been 16.

The dates of ice break-up in the rivers have a highly significant trend during the last 150 years as demonstrated in Figure 3. The ice break-ups nowadays are 15 days earlier than in the first 150 years of the series. In 1990 and 1992, for the first time in the nearly 300-year record, no ice formed on the rivers (not shown in Figure 3), and the two early ice break-ups that occurred in 1989 and 1995 are many standard deviations from the mean value. In the early twentieth century, the ice break-up came about 10 days earlier than in the eighteenth century.

The average air temperature of April-July, Figure 4, also shows a positive trend, 1.5°C, for the whole period that starts in 1731. Further analysis is under-

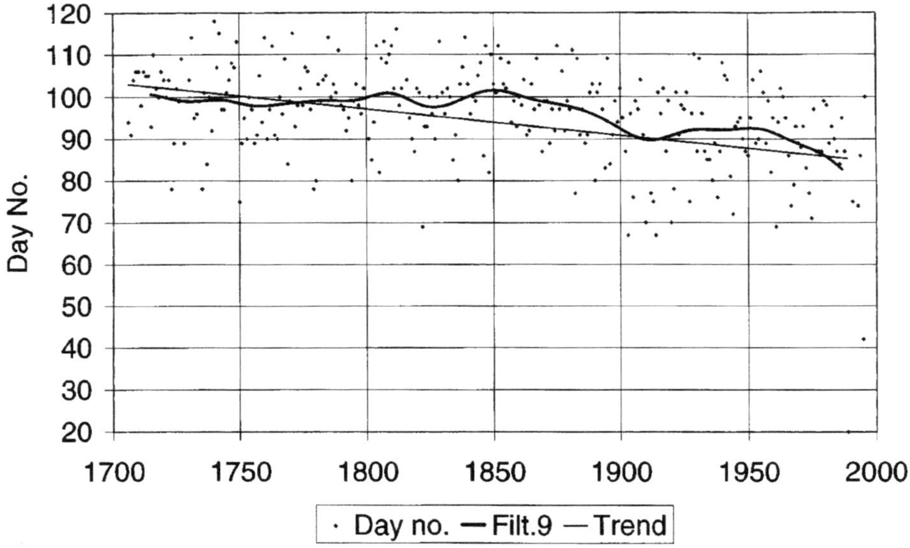

Figure 3. Ice break-up dates for Estonian rivers and the rivers Aura and Kokemäki in Finland, and Neva in Russia, all adjusted to be valid for the river Palmse near Tallinn (day No. counted from 1 January). Each spring is represented by a dot, and to ease the study of variations on a timescale of approximately 30 years, the observations are smoothed (curve) by a Gaussian filter with standard deviation 9 years in its distribution. A trend line for the whole period is also shown. In 1990 and 1992 no ice cover was established on Palmse (not marked).

taken by use of a Gaussian low pass filter designed for the study of variations on a timescale of about 30 years. During the eighteenth century the increasing trend is absent, and the curve reaches an absolute minimum value in the 1800s and a local minimum is present in the 1900s. In the 1850s there exists a local maximum while the absolute maximum occurs at the end of the curve. The absolute extremes of individual summers both occurred in the twentieth century, maximum in 1937 (12.9°C), and minimum in 1902 (7.2°C).

In the modern instrumental period (the last 150 years) the warming has taken place in March, April and May, while in June and July the temperature trend has been very small or negative.

5. Discussion and Conclusion

In areas without glaciers, like Estonia, the Little Ice Age has been defined by the presence of cold winters. In climatological literature the severity of the winters in Northern Europe has been described by mean and minimum temperature, duration of the winter and the extent of the snow cover. To a certain extent the start and end of the Little Ice Age depends on its definition. Caused by different definitions

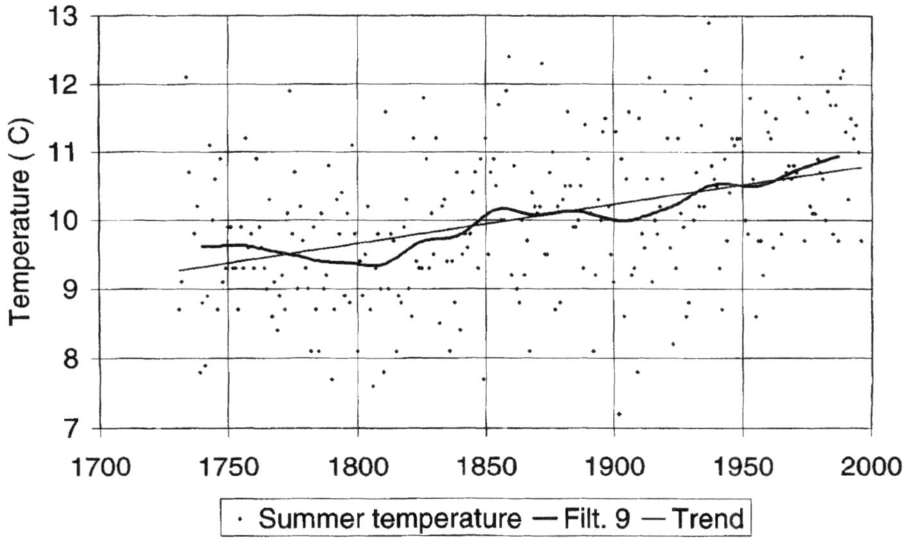

Figure 4. Summer temperature (April–July) for Tallinn based on ice break-up and rye harvest data, and of instrumental observations. To easy the study of variations on a timescale of approximately 30 years, the observations are smoothed by a Gaussian filter with standard deviation 9 years in its distribution (curve). A trend line for the whole period is also shown.

and by differences in climate between regions, the Little Ice Age is after all a relative term. Data presented in this paper should characterise the Eastern coastal zone of the Baltic Sea that includes Estonia, Southern Finland, and Latvia.

The most striking feature in our analysis is the warming of the winters from about the mid nineteenth century to present. The warming is especially noticeable during the latest decades. The climate from the start of the series (AD 1500) to the mid nineteenth century was in general somewhat colder, and should be recognised as a part of the Little Ice Age, though the period was intercepted by warmer winters in the first half of the eighteenth century.

During the whole period of 500 years, four winters occurred with mean temperatures below -10°C. These are the winters of 1541, 1809, 1829 and 1942. In the first three decades of the nineteenth century a cluster of very cold winters occurred: 1800, 1803, 1809, 1820, and 1829.

The temperatures during the season April – July since 1731 have been reconstructed by use of ice break-up dates of the river Palmse and of dates for the beginning of rye harvest. For the whole period a significant, positive, linear trend of 1.5°C was detected. In the modern instrumental period (last 150 years) this warming has taken place in March, April and May, while in June and July the temperature trend has been very small or negative. The absence of positive trend in high summer may be connected to increased cyclonic activity over the North Atlantic that generates cool summers in the Baltic region.

Acknowledgements

First of all we would like to express our gratitude to Leida Tarand, whose skill as a trained historian helped us to solve many problems connected with archive work. During the research in Tallinn Botanical Garden, Anu Joon, Tiiu Relve, Helve Meitern, Paavo Kuiv and Agu Eensaar were of great help. Their work is highly appreciated.

Due to constructive remarks the present paper is much different from that first proposed, and we want to express our gratitude to all helpers. We also want to thank Trausti Jónsson for his valuable comments in a late face of the manuscript and Astrid Ogilvie for style and language corrections.

References

Brazdil, R. and Dobrovolný, P.: 1993, 'Utilization of long temperature series for studying climatic fluctuations in the central Europe', in Obrebska-Starkel, B. (ed.): *Early meteorological instrumental records in Europe - Methods and results*. Jagellonian University, Kraków, pp. 151-162.

Jordan, P.: 1889, 'Beiträge zur Geographie und Statistik des Gouvernements Estland'. *The Estonian Government's Contributions to Geography and Statistics*, Reval.

Kirde, K.: 1924, 'Die Eisverhältnisse der Revaler Reede und ihre Beziehungen zur Lufttemperatur und Wassertemperatur der Ostsee' (The ice conditions in the Tallinn Port and its relationship to air and water temperature in the Baltic Sea), *Sitzungsberichten den Naturforschende Gesellschaft bei der Universität Dorpat*, **XXX**, 3-4, p. 8.

Lamb, H.: 1977, *Climate. Present, Past and Future*, Vol. 2, London-N.Y. p. 835.

Moberg, A.: 1996, 'Temperature variations in Sweden since the 18th century', *Dep. of Physical Geogr. Stockholm University Dissertation Series*, No. **5**.

Moberg, A. and Alexandersson, H.: 1997, 'Homogenization of Swedish temperature data. 2. Homogenized gridded air temperature compared with a subset of global gridded air temperature since 1861', *International Journal of Climatology* **17**, 35-54.

Seinä, A. and Palosuo, E.: 1996, 'The classification of the maximum annual extent of ice cover in the Baltic Sea', *Rep. Series of the Finnish Institute of Marine Resarch*, pp. 79-91.

Tarand, A.: 1974, 'The sea-land winds near Tallinn' (in Estonian), *Yearbook of Est. Geogr. So.*, 1971/72, pp. 54-62.

Tarand, A.: 1975, 'Impact of the Baltic Sea to the meteorological elements in the coastal zone'. *Water-dynamics in the Baltic Sea. Publ. of Est. Geogr. So.*, pp. 103-113.

Tarand, A.: 1976, 'Climate fluctuations in Tallinn during two centuries' (in Estonian), *Estonian Nature* **1**, 19-25.

Tarand, A.: 1982, 'Some remarks about descriptions of Estonian climate and the early instrumental meteorological observations' (in Russian), *Acta et commentationes Universitatis Tartuensis* **563**, 109-117.

Tarand, A.: 1986, 'Distribution of air temperature and precipitation in Tallinn' (in Estonian), *Tallinn Botanical Garden, preprint* No **3**.

Tarand, A.: 1992a, 'Ice-cover in the Baltic region and the air temperature of the Little Ice Age', Mikami T. (ed.), *Proceedings of the International Symposium on the Little Ice Age Climate, Tokyo, Japan 1991*, Editorial Committee of LIAC Proceedings, Department of Geography, Tokyo Metropolitan University, Tokyo, pp. 94-100.

Tarand, A.: 1992b, 'Meteorological observations in Estonia before 1850' (in Estonian), *Folia on science history in Estonia* **VIII**, Tallinn, pp. 30-50.

Tarand, A. and Kallaste T. (eds): 1998, Country Case Study on Climate Change. Impacts and Adaptation Assessments in the Republic Estonia. Tallin.

Tarand, A. and Kuiv, P.: 1992, 'The beginning of the rye harvest - a proxy indicator of summer climate in the Baltic Area', in Frenzel, B. (ed.) *Climatic trends and anomalies in Europe 1675-1715, Paläoklimaforschung **13**, Special issue: ESF project "European Paleoclimate and Man"* **8**, European Science Foundation, Strasbourg, Akademie den Wissenshaften und en Litteratur, Mainz, pp. 61-72.

Wild, H.: 1881, '*Die Temperatur-Verhältnisse des Russischen Reiches*' (The temperature conditions in Russia), St. Petersburg.

(Received 9 October 1998; in revised form 22 September 2000)

RECONSTRUCTION OF NINETEENTH CENTURY SUMMER TEMPERATURES IN NORWAY BY PROXY DATA FROM FARMERS' DIARIES.

P. Ø. NORDLI
The Norwegian Meteorological Institute, Box 43 Blindern, N-0311 Oslo

Abstract. Proxy data from five farmers' diaries in the Møre, Dovre and Trøndelag regions in central Norway were used for climatic reconstruction purposes. The method chosen was "simple linear regression analysis" with the start of the grain harvest (barley or oats) as predictor and summer temperature (May - August) as predictand. Overlapping periods with modern instrumental observations (starting 1858 or later) were used for calibration of the model. The model was tested on independent data by establishing the regression on one half of the overlapping period and applying the regression on the other half. The standard deviation in the residuals varied from 0.3°C to 0.7°C and the biases of the mean values from -0.3°C to +0.3°C. Climatic reconstructions were established for the early- and mid-nineteenth century summer temperature, i.e. during the last part of what has come to be regarded as the "Little Ice Age", in this article considered to end around 1880.

By use of the proxy data model, huge inhomogeneities of the "classical" Trondheim series were detected, the early nineteenth century part of the series evidently being too warm. The inhomogeneity was removed by use of adjustment terms. The adjusted series indicates that in the Trondheim region the summer temperature during the last part of the "Little Ice Age" phase was about 1°C lower than the latest 60 years. This is in serious contradiction to the classical Trondheim series.

Key Words. Little Ice Age, farmers' diaries, temperature time series.

1. Introduction

Instrumental observations are sparse in Norway before the Norwegian Meteorological Institute (DNMI) was founded in 1866. It is, however, possible to use proxy data for climate reconstruction in order to amplify our knowledge of climate variations and long-term trends. One promising source is farmers' diaries, an, until recently, unused source of information on Norwegian climate.

The diaries contain direct weather descriptions which may be classified in an *index system*, in which written weather information is shown converted to a relative scale, see for example, Pfister (1992) and for Norwegian data, Kastellet et al. (1998). Historical documents may also contain biological information hereafter called *historical proxy data*. These may be related directly to temperature by regression analysis. Examples are climate reconstructions based on the wine harvest (Le Roy Ladurie and Baulant, 1980), and the ripening time of rye (Tarand and Kuiv, 1994). In Norway, the most common cereals are

Climatic Change **48**: 201–218, 2001.

barley and oats. Indices based on these will be used in the present paper. Historical proxy data may also be used in combination with the index method (Yan et al., 1997).

Provided that there exist overlapping proxy and instrumental data, a long proxy series may be used to extend the length of an instrumental series backward in time. Proxy data may also be used to check the homogeneity of early meteorological observations which may be of dubious quality. In both cases it is required that the proxy and instrumental series originate from the same climate region.

The temperature reconstructions cover the last part of the "Little Ice Age". The great majority of Scandinavian glaciers reached their maximum size in the mid-eighteenth century. Waning after this was at first slow and broken by halts and minor re-advances (Grove 1988). Many of them advanced in the 1870s. Since 1880 annual mean temperature increased in Norway until the beginning of the 1930s (Hanssen-Bauer and Nordli, 1998) and the glaciers retreated during this period. Based on Scandinavian data it therefore makes sense to let the "Little Ice Age" end around 1880. The start of the "Little Ice Age" will not be addressed in this article as no diary is older than the nineteenth century.

2. Data

The study region is situated in central and southern Norway in Trøndelag and Møre and in the Dovre Mountain region. Five nineteenth century farmers' diaries from places within this region are available, as well as the longest of the Norwegian instrumental series, that from Trondheim (Birkeland, 1949). The diaries, shown in Table I, are referred to by the names of the farms where they were written, cf. figure 1.

TABLE I

The table shows the name of the farm and the height of the grain field, the period of the diary (not necessary complete), the nearest overlapping instrumental series, the overlapping period of proxy and instrumental data and the distance between the sites representing the two data sources.

Farm	Height of the field (m)	Period	Instrumental series	Overlapping period	Distance (km)
Frøystad	ca. 100	1843-1875	Ålesund	1861-1875	26
Synstbø	600-620	1813-1874	Kjøremsgrendi[1]	1865-1874	4
Simenrud	ca. 580	1874-1917	Kjøremsgrendi[1]	1874-1917	3
Kjellen	115	1805-1875	Trondheim	1858-1876	47
Brunstad	ca. 120	1859-1875	Trondheim	1859-1875	102

[1] The period 1864-1976 originates from Dombås, 5 km from Kjøremsgrendi.

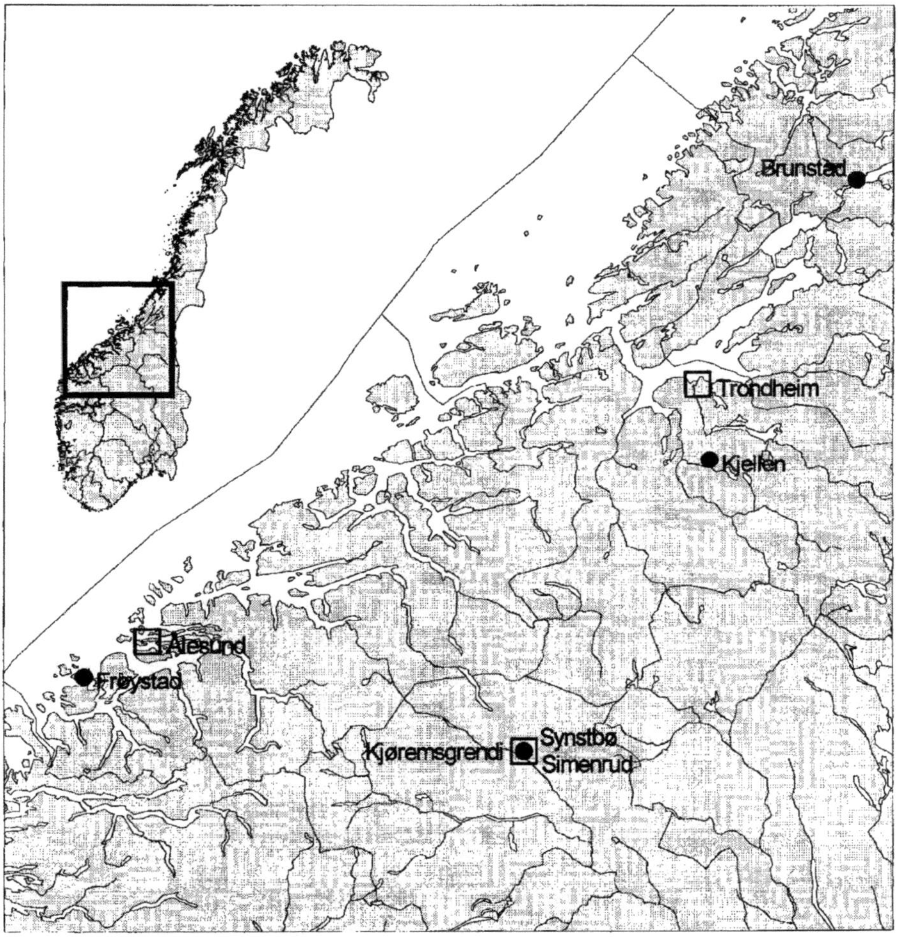

Figure 1. The Dovre, Møre and Trøndelag areas. Diary sources are marked with • and overlapping temperature series with □. When diary source and temperature series are too close to be distinguished on the map, the dot is placed inside the square.

The Frøystad farm, 62°N 6°E, at about 100 m a.s.l, is situated in the district of Møre on the south side of the Leinøya island. The fields are south-facing, with upper parts sloping steeply, and the lower parts almost flat, with wet soil. In the upper parts the soil is dryer, sandy and stony. For further details see Nordli (1997a) or Herøy sogelag (1978). The Synstbø (or Systugu Synstbø) and Simenrud farms are situated in the mountain village of Kjøremsgrendi, *62°N 9°E*, at about 600 m a.s.l. The area is known to be rather arid with yearly precipitation of only 400 mm. The grain fields are southfacing with inclinations of 25% at Synstbø and 10% at Simenrud. Due to this difference the grain ripens earlier at Synstbø than at Simenrud, figure 2b. Kjellen, *63°N 11°E*, lies on the flat banks of the river Gaula, at 115 m a.s.l. It is separated from Kjøremsgrendi by the Dovre mountain ridge. Brunstad, *64°N 12°E*, is situated near the lake Snåsavatnet with its grain fields of about 120 m a.s.l.

The diaries reveal several weather-related proxies which may be used for climatic reconstruction, such as the first day of spring work, day of moving the livestock to the mountain outfarm, first days of haymaking and of the grain harvest. The best correlations with the instrumental series were obtained by using the first day of grain harvest (Nordli, 1997b). This predictor also provides reconstructions of seasonal averages, representing longer parts of the year than any of the other proxies.

For calibration purposes the nearest possible instrumental series are used, and the distances being given in Table I. These are the *Ålesund* series (1861-1903) and the *Kjøremsgrendi* series (from 1864) both established by the Norwegian telegraph authority. The Ålesund station has been relocated once within the town, while Kjøremsgrendi was relocated by 5 km in 1976, from Dombås, which was the former name of the series. Thorough homogeneity testing of the series reveals no shift in summer temperature caused by the relocation and Kjøremsgrendi is considered to be a high quality series. The *Trondheim* series comprises data from the classical Trondheim series (1762-1935), Trondheim - Voll (1936-1956) and Værnes (Trondheim Airport) since 1957.

3. Method of temperature reconstruction

The chosen method was simple regression analysis, with the start of the grain harvest as predictor and mean summer temperature as predictand. Mean summer temperature was calculated from measurements at meteorological stations in the same regions as the diaries, and the regressions were established during overlapping periods varying from 10 to 39 years. With the restriction that mean temperatures were available for calendar months only, best possible correlations were obtained by using the mean temperature in the period May - August. This period was therefore adopted in the final regressions. However, by substituting May - August by April - August, the correlations drop only slightly (Nordli, 1997b). Scatter diagrams, regression lines and correlation coefficients are shown in figure 2.

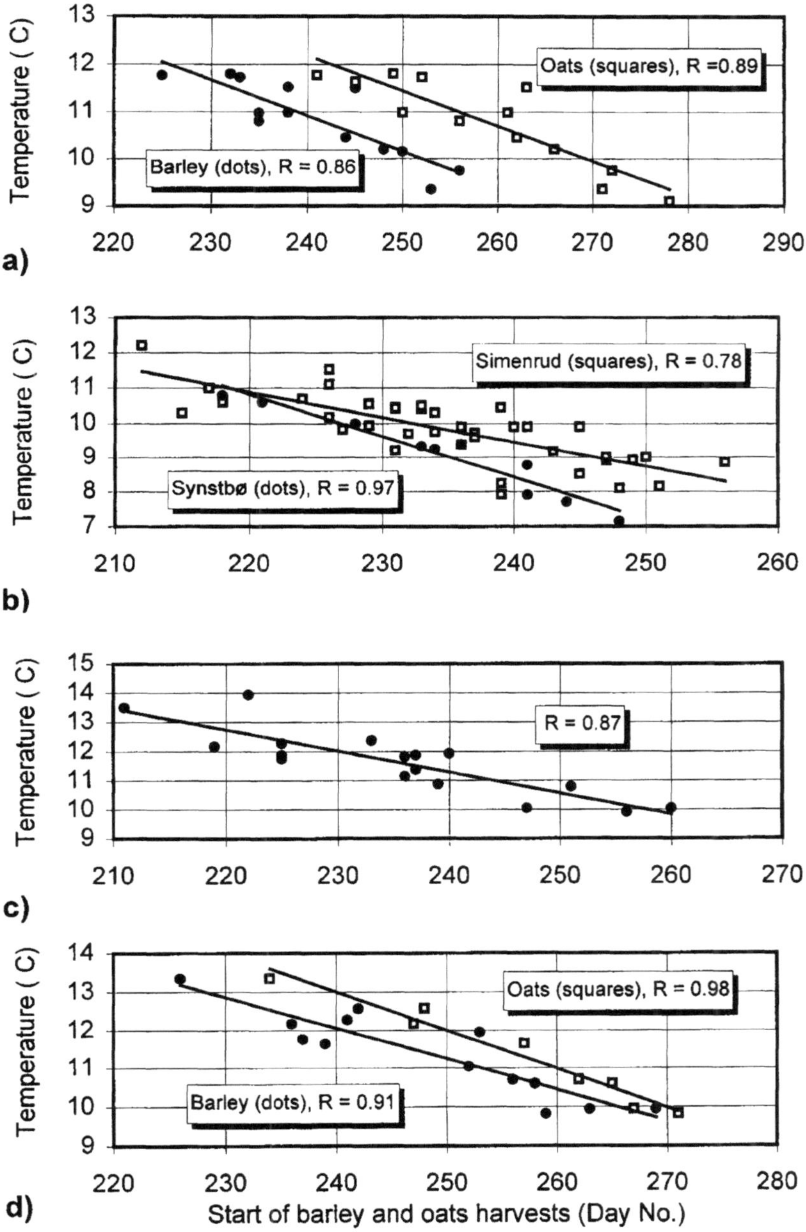

Figure 2. The first days of barley and oat harvests plotted against mean air temperature May - August. a) Frøystad diary and the Ålesund observations (1861-1875). b) Synstbø and Simenrud diaries and the Kjøremsgrendi observations (1865-1874). c) Kjellen diary and Trondheim observations (1858-1876). d) Brunstad diary and Trondheim observations (1859-1875)

The correlation coefficients vary from 0.78 at Simenrud to 0.97 at Synstbø, i.e. 60 - 93% of the variance is explained by the regressions. For climate reconstruction purposes only the diaries from Frøystad, Synstbø and Kjellen are used. Based on these 74%, 93% and 75% of the variance is explained respectively. All regressions seem to fit nicely into the linear model, figure 2.

In every proxy data method there is always a possibility of caveats. Inhomogeneity of the harvest data might arise from: 1) Change of grain field: 2) Change of the barley or oat types (in these cases the reconstructions will be biased): 3) Biased reconstructions might also be due to the period of overlapping with instrumental observations being too short. Before undertaking the climate reconstructions, the sources of potential errors should be addressed.

Change of grain field. The importance of using the same grain field seems evident, see the demonstration in figure 2b. The steeper slope at Synstbø gains more insolation than the more moderate slope at Simenrud and the grain ripens earlier at Synstbø. At Synstbø and also to a certain extent at Kjellen, the diaries ensure that the same grain fields were used throughout the period. At Frøystad, however, this is not quite clear. On the other hand, taking into account the importance of grain production, the best field is very likely to have been chosen so that the harvesting was started every year in the most favourable place.

Change of the barley or oat types. Barley, in particular, has changed significantly with the introduction of many new early-ripening varieties (Tarand and Kuiv, 1994). But this was probably not the case before 1900 as barley breeding started in Norway at the beginning of this century (Ringlund pers. comm.). At Frøystad, barley and oats were growing simultaneously, so it is possible to check the reconstructions from the two cereals against each other. Barley and oats have different ripening times. The difference in the period 1843-75 was 18 ± 5 days. This is enough time for the barley harvest to be finished before the start of the oat harvest. Thus, the beginnings of the two harvests can be considered independent of each other. The difference between the two temperature reconstructions was examined, and no abrupt shifts nor significant trends were revealed that could indicate that the cereals have changed through the period.

Biases in the reconstructed series. If the overlapping periods are too short, there might be large errors in the coefficients of the regression lines leading to biases in the reconstructed series. The robustness of the length of the overlapping period was tested by using the first (second) half of the overlapping period in establishing the regression line, while the regression was applied to the second (first) half, see Table II column **II** (**I**).

TABLE II

Regression analysis established on the basis of one half of the overlapping series, and applied to the other half. The standard deviations of the residuals (s), the biases of the mean values, the correlation coefficient, r, and the number of observations, N, are shown.

Farms	I		II		Entire series		
	s	bias	s	bias	s	r	N
Synstbø (1865-1874)	0.3	0.1	0.3	-0.1	0.3	0.97	10
Kjellen (1858-1876)	0.6	-0.3	0.5	0.3	0.5	0.87	17
Brunstad (1859-1875)	0.3	0.6	0.4	-0.6	0.4	0.91	13
Frøystad*(1861-1875)	0.5	0.0	0.5	0.3	0.4	0.89	15
Simenrud (1874-1917)	0.5	-0.2	0.7	0.1	0.6	0.78	39

* Combined barley and oat series.

The bias in the reconstructed temperature based on Brunstad seems to be aberrant. Being twice as large as the others, it was not used in the reconstructions. The other biases are ±0.3°C or less. In the final reconstructions the entire overlapping periods are twice as large as those used in the tests. This further reduces the chances of biases. Moreover, ±0.3°C is less than many of the inhomogeneities in the nineteenth-century instrumental series, caused by e.g. inadequate formulas for mean temperature (Heino, 1994), zero point attraction of the thermometer glass (Middleton, 1966) or relocations of meteorological stations (Nordli, 1997c).

The standard deviations of the residuals vary from 0.3 to 0.6°C, see Table II. Thus, summer mean temperature derived from reconstructions are more noisy than those derived from instrumental observations. However, under the assumption of randomness, the standard deviation of the mean temperature of a decade reduces to only 0.1 - 0.2°C. Thus, the proxy-data method seems to be highly suitable for the study of long-term variations and trends in summer temperature.

4. Climate reconstruction, the Kjøremsgrendi composite series

Among the proxy series in Table I, the Frøystad and Synstbø diaries are older than the nearby meteorological stations of Ålesund and Kjøremsgrendi. Therefore, the instrumental series can be extended further backwards in time. This is particularly valuable for Kjøremsgrendi, which is a high quality rural series not yet terminated. The series was analysed for variations on time scales of 10 and 30 years by a Gaussian low pass filter (shown as curves in figure 3) and also by rectangular running mean filters of 5, 10, 30, 50 and 100 years (not shown) and the trends were studied by the Mann-Kendall test. A significant long-term trend terminating before the 1930s was detected. Thus, the highest mean temperature during 5, 10, 30 and 50 years periods were all located after 1929, and the coldest periods were located before 1930.

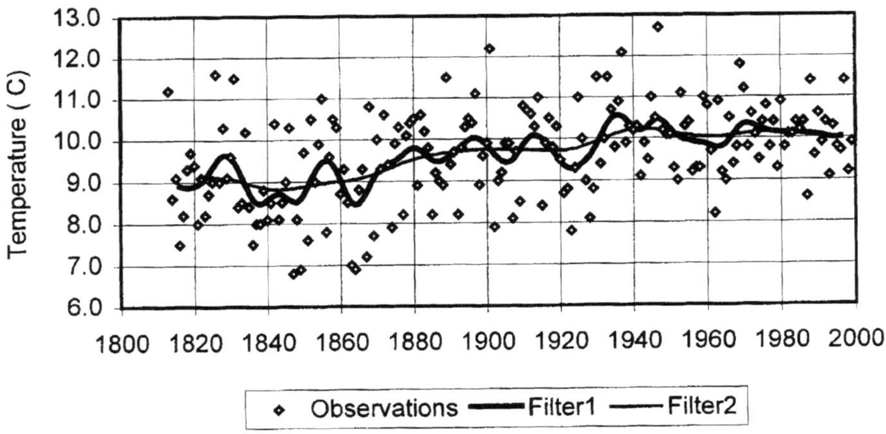

Figure 3. The Kjøremsgrendi composite series of mean summer temperature (1813 - 1999), actual values (dots) and long-term variations (curves). Gaussian filters are used with standard deviations in the distributions of 3 (Filter 1) and 9 (Filter 2) years.

The last parts of the 1830s and the 1840s were especially cold. In particular, the seven years 1835-1841 are still known as the seven bad years, which led to starvation in the Dovre mountain district. None of the single years was among the most extreme ones, but the period was not interrupted by any summer temperature above the mean for the whole series (9.6°C) and in 1836 and 1838 the grain was totally damaged by early frosts. However, in the 1850s, there was a cluster of above normal years. But in the 1860s another cluster of bad harvests occurred, and it may be noted that some of the springs were extremely cold.

The filter 2 curve in figure 3 shows that since 1930 the summer temperature has been in a steady state, about 1°C higher than during the early part of the series 1813 - 1880. The so-called "Little Ice Age" should therefore be extended at least to the end of the 1860s. However, to include the advance of glaciers in the 1870s in southern Norway, the "Little Ice Age", as it is used in this article, ends around 1880.

5. Early instrumental observations tested by proxy data

Kjellen in Midtre Gauldal is situated only 47 km to the south of the Trondheim meteorological station located in the centre of the town. The two sites lie in the same climatic region and it is possible to check the homogeneity of the early instrumental observations by the Kjellen proxy series. Nowadays no observations are made in the centre of Trondheim. However, the classical Trondheim series has been adjusted to be valid for Værnes, the Trondheim Airport, where observations continue. Although valid for Værnes, the series will in this article for simplicity be called the Trondheim series. Recently the series has been homogenised since 1870, for earlier years the homogenisation of Birkeland (1949) has so far been accepted (Nordli, 1997c).

The Trondheim series starts in 1762, but there is a gap from 1803 - 1817 that was interpolated by Birkeland, using the series from Stockholm and Edinburgh. Since 1818 the series has come from several places in the inner town. The observers of the principal series were: Vibe (1818-34, before 1928 only one observation a day), Møllerup (1835-51), Balsløw (1852-54), Rosenvinge (1855[1]-69).

According to Birkeland (1949) Rosenvinge's observations "convey a very reliable impression", and as a first approach the Rosenvinge/DNMI series (1858[1]-1875) was used for the establishing of the regression with the Kjellen proxy series, figure 2c. For control purposes the regression with the Rosenvinge/DNMI and Brunstad proxy series was also established, figure 2d. The residuals of both regressions were tested for trends by the Mann-Kendall test. Trends were detected nearly significant at the 0.05 level (not shown), but, they tilted at different directions. According to these results, the Rosenvinge/DNMI-series was considered to be homogenous. Prior to 1875 a Trondheim series was reconstructed based on the Kjellen proxies. In figure 4 filtered values of the Trondheim instrumental and proxy series as well as the Kjøremsgrendi and Ålesund composite series are shown.

Figure 4 reveals that the Trondheim instrumental series is much higher than the other series prior to Rosenvinge's observations compared to the common reference period. This indicates strongly that the Trondheim instrumental series is inhomogenous. The significance of the inhomogeneity was tested by applying the Pettitt test on the difference between the Trondheim instrumental and proxy series. The periods chosen for testing covered the time interval of one observer and his successor and the procedure was carried out stepwise from Birkeland's interpolations to Rosenvinge's observations. Balsløw's interval 1852-55 was considered to be too short and was for that reason not included in the test. The results are given in Table III, the use of the Pettitt test is shown in figure 5a.

---

[1] There has been some doubt about what Birkeland used as the principal series in 1855-57. Therefore these years were omitted as a basis for the regression.

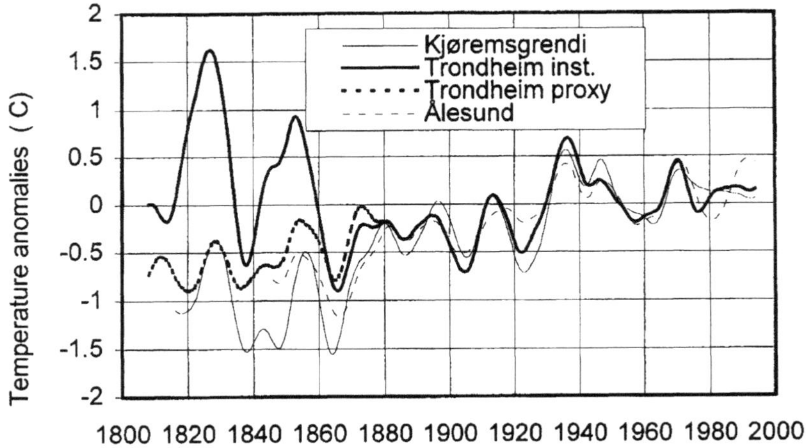

Figure 4. Trondheim instrumental and proxy series and the composite series of Kjøremsgrendi and Ålesund. Gaussian filter 1 is used, i.e. σ=3 in the Gaussian distribution. The series are shown as anomalies from the 1901-1990 mean.

TABLE III

Significance tests applied to the difference: Trondheim proxy series - Trondheim instrumental series. Significant results are marked with bold numbers, level 0.05.
1 row: Years of abrupt changes in the Trondheim series detected by the Pettitt test.
2 row: Years of relocation of the series in connection with change of observer.
3 row: Adjustment terms (°C) required for the observations of observer I to be homogenous with those of observer II. Significance (level 0.05) tested by Student's t-test.

Observer I	Interpolations	Vibe I	Vibe	Møllerup	Balsløw	Møllerup[1]
Observer II	Vibe	Vibe II	Møllerup	Balsløw	Rosenvinge	Rosenvinge
Period[2]	1805-1834	1818-1834	1818-1855	1835-1855	1852-1876	1835-1876
1 Pettitt test	**1817**	1827	**1834**			**1860**
2 Year (archive)	1817	1827	1834	1851	1854[3]	1851
3 Diff. I - II	**1.50**	**-0.47**	**-1.45**	**1.54**	**-2.08**	**-0.54**

[1]Tested without Balsløw's observations. [2]There are some missing years. [3] Some doubt about the year.

The Pettitt test points to the most likely year for a shift in the series and the significance of the shift. Shifts were detected in connection with the change of observer or observational procedure. The year was pointed out correctly in the case of the change from interpolations to Vibe's observations (1817) and from Vibe to Møllerup (1834), see figure 5a. In the case of the change from Møllerup to Rosenvinge, a significant abrupt shift was detected, but the shift came five years too early compared with the information from archives. The year (1827) of change from one to two observations a day was pointed out correctly, but the result was not significant. With the years of change predetermined, the significance was also tested by Student's t-test that found all shifts to be significant, Table III, row 3.

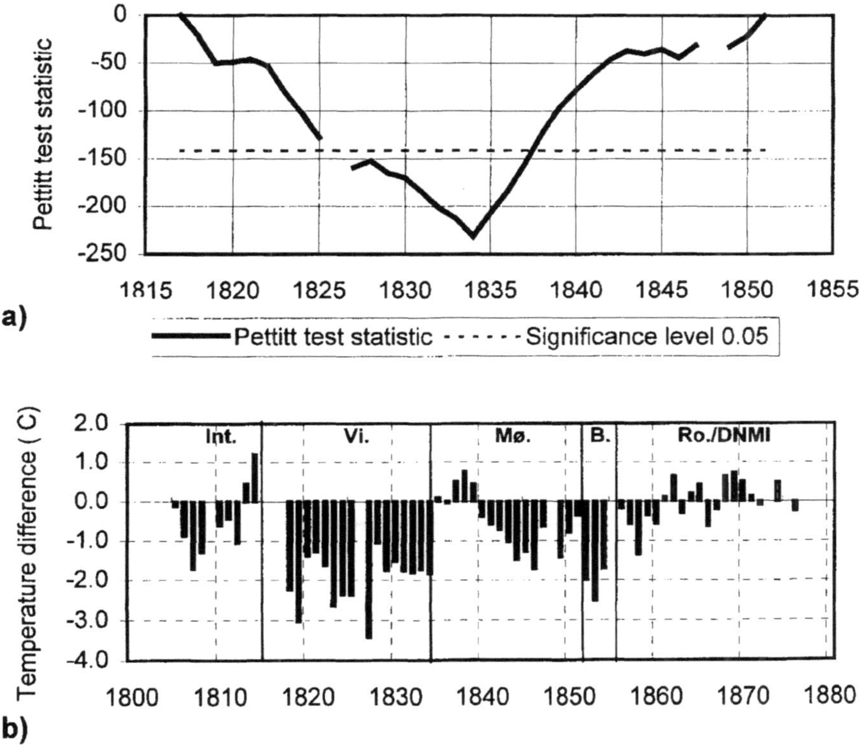

Figure 5. a) Time evolution of the Pettitt test statistic during the period of Vibe and Møllerup's observations (1818 – 1855). b) Differences of the Trondheim proxy and instrumental series (proxy - instrumental) 1805 - 1876. The partition of the diagram represents different subseries of the Trondheim instrumental series. These are denoted by the first letters of the observers, see Table III.

Before adjusting the classical Trondheim series it was split up into the original series, and taking into account Vibe's observations, a further split was found necessary to cope with the change from one to two observations a day. Adjustment terms were calculated which should be added to the shorter series to make these homogenous with the Rosenvinge/DNMI part of the series. Birkeland (1949) also used adjustment terms varying from -0.2 to -0.5 in the finishing work of the series. These were removed from the classical series and incorporated in the total adjustment terms for summer temperature presented in

Table IV. The homogenised Trondheim series is shown in figure 6a together with the nearest, classical instrumental series. The Bergen and Oslo series are homogenised by Birkeland (1823, 1928) and the Uppsala series by Moberg and Bergström (1996).

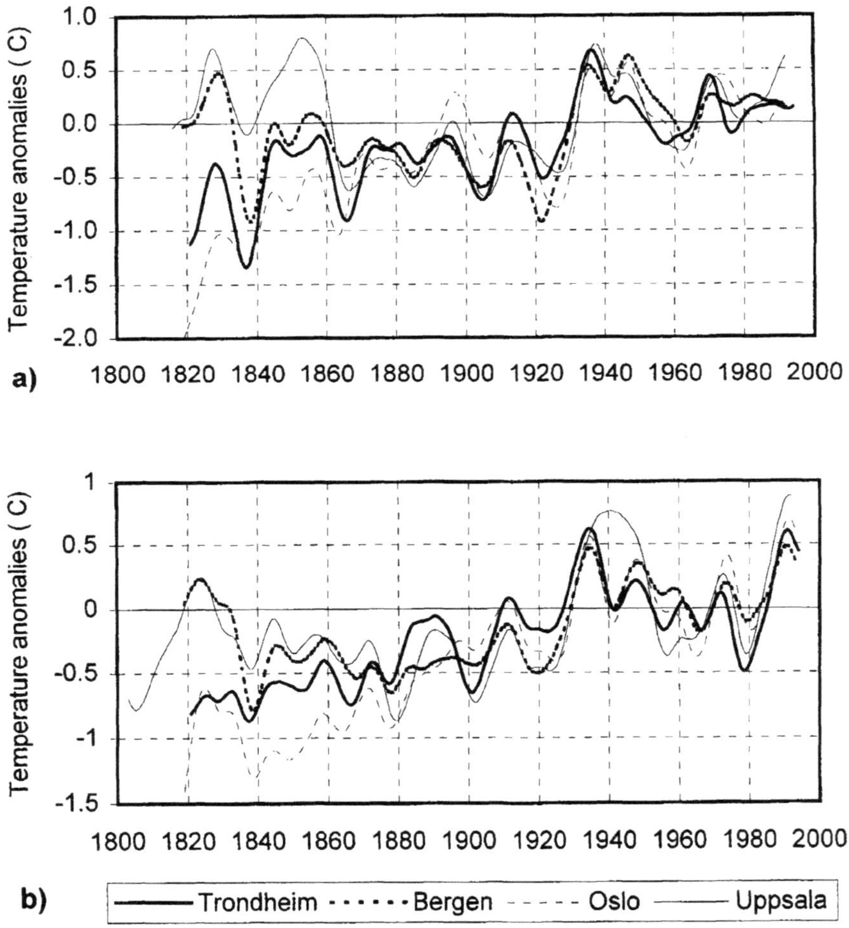

Figure 6. The newly homogenised Trondheim series and other Norwegian and Swedish long-term series. All series are smoothed by a Gaussian filter (σ = 3 years) that shows variations on time scales of about 10 years. **a**) Summer temperature (May - August). **b**) Annual temperature.

The diagram reveals an essential feature: Since about 1870 all series show quite similar long-term trends. However, in earlier years the curves differ substantially. As the adjusted Trondheim and Oslo series have more or less increasing trends from the beginning of the series up to the 1930s, no such trend is present in the Uppsala series. It should by no means be taken for granted that diverging curves in the early part of the series are caused by different regional climatic trends. It should be noted that 1870 is near the cross-roads between the early meteorological observations and the standardised networks of the national meteorological services. The different trends could simply result from inhomogeneities in the series. The increasing trend up to the 1930s of the homogenised Trondheim series is steeper than the Bergen series, but more gentle than the Oslo series. However, it may be significant that homogenisation of the early Oslo and Bergen series have not been performed since Birkeland (1923, 1928).

Without adjustments the trend of the Trondheim series would have been negative. A main problem of observing true air temperature is the overheating of the thermometers by direct or indirect short-wave radiation. Early observations are especially vulnerable to this error because of inefficient screening (Middleton, 1966; Nordli et al., 1997). As all adjustments are negative, overheating seems to be the most likely reason for the inhomogeneities of the classical Trondheim series. Based on this assumption, the following adjustments are suggested for the rest of the year: There should be no adjustments in the period October - February. Spring and autumn are considered as transition zones between summer and winter. Because of more frequent clear sky in spring than in autumn, the transition zones suggested are of different length, for details see Table IV.

TABLE IV

Adjustment terms applied to the Trondheim series derived from the Kjellen temperature reconstructions. The adjustments are: May - August: x, March: x/3, April: 2x/3, September: x/2. In the period October - February there are no adjustments.

Monthly and annual	Vibe I 1818-1827	Vibe II 1828-1834	Møllerup 1835-1851	Balsløw 1852-1854
1 Summer (May-Aug.)	-2.6	-2.0	-0.8	-2.3
2 March	-0.9	-0.7	-0.3	-0.8
3 April	-1.7	-1.3	-0.5	-1.5
4 September	-1.3	-1.0	-0.4	-1.2
5 Year	-1.2	-0.9	-0.4	-1.1
6 Former annual adjustments	-0.3	-0.3	-0.3	-0.2

The newly established monthly adjustments result in larger mean annual adjustment than those used by Birkeland, except within the observational period of Møllerup where the difference between them is negligible (compare rows 5 and 6 in Table IV). With the annual adjustments applied to the Trondheim series, filtered values are shown in figure 6b. The trend in the annual values from the beginning of the series up to the 1930s lies somewhat in between the trends of the Bergen and Oslo series. But the trends do not differ so much as in summer.

Although the homogenised Trondheim series seems to give reliable results compared with the neighbouring stations, it has not been proved that there are no inhomogeneities in winter. It is therefore essential to develop some winter proxies for temperature as well. In the absence of such a proxy, the reliability of the annual series remains unproven.

6. Discussion

There has been some confusion concerning the summer temperatures as revealed by the Norwegian long-term series during the last part of the "Little Ice Age". While the Bergen, and especially the Oslo series, showed large, positive temperature trends through the nineteenth century, the corresponding trend of the classical Trondheim series was negative. In the Trondheim series, summer temperatures in the 1820s and 1850s were even higher than those in the 1930s. Thus, there were good reasons for questioning the homogeneity of the Trondheim series.

Early temperature series may describe variations over shorter time intervals (days, months and a few years for example) very well, but studies of long-term trends are often hampered by inhomogeneities. These might be overcome by adjustments, but in the period of early instrumental observations, lack of reference stations in the surrounding regions is a rule rather than an exception. Using reference stations far away, the spatial variations could be smoothed out during the homogenisation process and an important climatic feature lost. The method is especially risky in the case of Trondheim as there are no reference stations to the north of the town. Harvest data therefore turned out to be a very valuable reference for use in the homogenisation process. Differences between regions are preserved and the homogenised series enable the identification and study of spatial variations.

The regression analysis might also be used directly to extend existing series backwards in time. But, the standard deviations of the residuals varied from 0.3°C to 0.6°C which is a drawback for the study of year-to-year variations. However, assuming randomness, the standard deviation of the mean value of a decade is reduced to 0.1 - 0.2°C which is less than the errors in the nineteenth-century thermometers. It follows that the reconstructed parts of the series facilitate trend studies on decadal or longer time scales.

The importance of local or micro climate for ripening of barley was demonstrated by two diaries from the same village. The distance between the grain fields was less than 1 km, both of them were southfacing, but one was steeper than the other and gained more insolation. Comparison with meteorological observations revealed significantly different regression lines. If the climate reconstructions had been based on a mixture of harvest data from the two fields, reconstructed summer temperature would have been biased. It should be ensured that the same grain field is used for the same temperature reconstruction. The ideal situation would be that many overlapping series were available in a given region so that the homogeneity of each series could be tested against a group of reference series, e.g. by the standard normal homogeneity test (Alexandersson, 1986). Hopefully, this will be the case in some regions if a systematic search for unregistered diaries is performed. It should be worthwhile as the knowledge of reading and writing was fairly common among farmers in Norway in the eighteenth and nineteenth centuries (Fet, 1996).

A certain impression of the homogeneity of the proxy series is given in figure 4 where the reconstructions of the three long proxy series are plotted. Local maxima and minima are present in the same decades, but the cold decades of the Kjøremsgrendi series seem to be somewhat exaggerated compared with the reconstructed Trondheim and Ålesund series. The reason for this is not clear. It remains to be seen whether this is a climatic reality or merely a consequence of inhomogenous harvest data. A climatological explanation could be that the more continental climate of Synstbø (Kjøremsgrendi), is being less influenced by conservative sea-surface temperature during extremely warm/cold summers than Frøystad (Ålesund) and Kjellen (Trondheim)

After adjustment of the Trondheim series, the climatic trends revealed by the Norwegian series become more consistent, see figure 5. The long-term trends in the nineteenth century are all positive leading up to a maximum value in the 1930s. But, the Swedish temperature series from Uppsala and Stockholm (not shown) reveal somewhat higher temperature during the early and middle nineteenth century than the ones in southern Norway compared with the common reference period. Homogenisation of the other long-term Scandinavian series by use of proxy data is therefore of great potential interest.

The Norwegian series show that the summer temperature during the last part of the "Little Ice Age" was low compared with the climate in the twentieth century. A climatic deterioration took place in the middle of the nineteenth century, for example the coldest 30-year period in the Kjøremsgrendi reconstruction occurred in the years 1835-64. The low temperatures have apparently been important for the general advance of glaciers that took place in the last part of the 1860s and the 1870s. The historically dated 1875 moraine of Nigardsbreen (a tongue of the Jostedalsbreen ice cap in southern Norway) is the most well-marked since the glacier's large advance in 1748 (Andersen and Sollid, 1971).

7. Conclusions

Overlapping periods of grain-harvest data and modern instrumental observations were used for calibration of a simple, linear, regression model for climate reconstructions. Proxy climate data in the form of dates for the start of grain harvests were extracted from farm diaries. The model was tested on independent data by establishing the regression on one half of the overlapping period and applying the regression on the other half. The standard deviation in the residuals varied from 0.3°C to 0.7°C and the biases of the mean values from -0.3°C to +0.3°[2]. Under the assumption of randomness, the standard decadal deviation reduces to 0.1° - 0.2°C used on a decadal scale. Thus, the reconstructed temperature series enable the study of long-term variations with a resolution of one decade.

At one farm, barley and oats were cultivated simultaneously, providing the possibility of comparing two independent reconstructions of the same climate. No significant difference was detected in the reconstructed summer temperatures series, thus indicating that there have been no changes in ripening abilities or in growing conditions of the cereals during the period of reconstruction.

Climatic reconstructions were established for the early-nineteenth-century summer (May - August) temperature, during the "Little Ice Age". The last part of the 1830s and the 1840s were especially cold periods. In particular, the seven years 1835-1841 are still known as the seven bad years which led to starvation in the Dovre mountain district. Also in the 1860s there is a cluster of very low summer temperatures, especially in spring. Analysing the series on a decadal scale, the 1830s/1840s and the 1860s were the coldest periods during the nineteenth and twentieth centuries. The warmest decade of the series is the 1930s.

---

2 One series had biases of 0.6°C, but this series was not used in the climate reconstructions

By use of the historical proxy data, several inhomogeneities were detected in the classical Trondheim instrumental series. These were either caused by changes of observer (thereby also location) or observational procedures. The classical series were adjusted by constant adjustment terms so as to form an approximately homogenous series 1818 - 1997. Parts of the series were subject to huge adjustments, up to ca. 2°C. This had implications for the interpretation of the summer temperature during the "Little Ice Age". In the classical series the local maxima of the 1820s and 1850s were higher than the maximum in the 1930s, while they were more than 1°C lower in the adjusted series. Compared with a 1901-1990 reference period, the adjusted Trondheim series revealed nineteenth-century temperatures somewhat higher than the Oslo series, but lower than the Bergen series.

Acknowledgements

Eirik J. Førland (DNMI) and the two anonymous reviewers are gratefully acknowledged for helpful comments on the manuscript. The author would also like to thank Astrid Ogilvie for style corrections on the manuscript and Sigmund Høgåsen (Vågå) for initiating climatic reconstructions based on Norwegian farmers' diaries.

References

Alexandersson, H.: 1986, 'A homogeneity test applied to precipitation data'. *Journal of climatology*, **6**, 661-675.

Andersen, J.L. and Sollid, J.L.: 1971, 'Glacial chronology and glacial geomorphology in the marginal zones of the glaciers, Midtdalsbreen and Nigardsbreen, South Norway'. *Norsk Geografisk Tidsskrift* **25**, 1-38.

Birkeland, B.J.: 1923, 'Ältere Meteorologische Beobachtungen in Oslo (Kristiania). Luftdruck und Temperatur seit 100 Jahren'. Old meteorological observations at Oslo. Atmospheric Pressure and Temperature since 100 years (in German). *Geofysiske Publikasjoner*. **III**, 1-56.

Birkeland, B.J.: 1928, 'Ältere Meteorologische Beobachtungen in Bergen. Luftdruck und Temperatur seit 100 Jahren'. Old meteorological observations at Bergen. Atmospheric Pressure and Temperature since 100 years (in German). *Geofysiske Publikasjoner*. **V**, 1-56.

Birkeland, B.J.: 1949, 'Old meteorological observations at Trondheim. Atmospheric Pressure and Temperature'. *Geofysiske Publikasjoner* **XV**, 1-38.

Fet, J.: 1996, 'Lesande bønder på 1700-talet'. Reading farmers in the eighteenth century (in Norwegian). *Syn og Segn* **102**, 260-268.

Grove, J.M.: 1988, *The Little Ice Age*. Routedge, London and New York, p. 498.

Hanssen-Bauer, I. and Nordli, P.Ø.: 1998, 'Annual and seasonal temperature variations in Norway 1876 – 1997'. *DNMI-klima*, Report 25/98, p. 29.

Heino, R. 1994: 'Climate in Finland during the period of meteorological observations'. *Finnish Meteorological Institute Contributions* 12, p 209.

Herøy sogelag: 1978, *Ingebrigt Frøystad. En dagbok (A diary) 1843 - 1875, 2nd edition* (in Norwegian), p. 113.

Kastellet, E., Nesje, A., Solheim-Pedersen, E.: 1998, 'Reconstructing the palaeoclimate of Jæren, southwestern Norway, for the period 1821 - 1850, from historical documentary records'. *Geografiska Annaler.* **80A**, 51-65.

Le Roy Ladurie, E and Baulant, M.: 1980, 'Grape Harvests from the Fifteenth trough the Nineteenth Centuries'. *Journal of Interdisciplinary History*, **10**, 839 - 849.

Middleton, W.E.K.: 1966 *A history of the thermometer and its use in Meteorology.* The John Hopkins Press, Baltimore, Maryland.

Moberg, A. and Bergström, H.: 1996, 'Homogenization of Swedish Temperature Data. Part III: The Long Temperature Records from Uppsala and Stockholm'. In: Moberg: Temperature variations in Sweden Since the 18th Century. The Department of Physical Geography Stockholm University, Dissertation No. 5., p. 34.

Nordli, P.Ø.: 1997a, 'The Herøy series. A composite series of summer temperatures (1843-1903)'. *DNMI-Report*, No. 26/97 Klima, p. 8.

Nordli, P.Ø.: 1997b, 'The Kjøremsgrendi series. A composite series of summer temperatures (1813 - 1997)'. *DNMI-Report*, 27/97 Klima, p. 22.

Nordli, P.Ø.: 1997c, 'Homogenitetstesting av norske temperaturseriar II'. Homogeneity testing of Norwegian temperature series II (in Norwegian). *DNMI-Report*, 29/97Klima, p. 43.

Nordli, P.Ø., Alexandersson, H, Frich, P, Førland, E., Heino, R, Jónsson, T, Tveito, O.E.: 1997, 'The effect of radiation screens on Nordic time series of mean temperature'. *Int. Journal of Climatology.* **17**, 1667-1681.

Pfister, C.: 1992, 'Monthly temperature and precipitation in Central Europe 1525 - 1979: Quantifying documentary evidence on weather and its effects'. *In* Bradley R.S. and Jones P.D. (eds.): *Climate since AD 1500,* 118-143. Routledge, London.

Ringlund, K.: 1998. Pers. comm. Department of horticulture and crop sciences, Agricultural University of Norway.

Tarand A. and Kuiv P.: 1994, 'The beginning of the rye harvest - a proxy indicator of summer climate in the Baltic Area'. *Paleoclimatic Research.* **13** (Special issue ESF project «European Palaeoclimate and man» 8 1994.), 61-72.

Yan, Z. , Alexandre, P., Demaree, G.: 1997, 'Narrative warm/cold variations in continental western Europe, AD 708 – 1426'. *Science in China (series D),* **40**, 5, 509-517.

(Received 11 February 1999; in revised form 25 September 2000)

ZONAL INDICES FOR EUROPE 1780-1995 AND RUNNING CORRELATIONS WITH TEMPERATURE

JUCUNDUS JACOBEIT[1], PETER JÖNSSON[2], LARS BÄRRING[3], CHRISTOPH BECK[1], MARIE EKSTRÖM[3]

[1]*Geographical Institute, University of Würzburg, Germany*
[2]*Malmö University, Sweden*
[3]*Department of Physical Geography, Lund University, Sweden*

Abstract. Zonal circulation indices with monthly and seasonal resolutions are calculated based on gridded monthly mean sea-level pressure (SLP) reconstructed back to 1780 by Jones et al. (1999): an overall zonal index for the whole European area between 30°W and 40°E, a normalized index for the North Atlantic Oscillation (NAO), and a similar index for Central Europe. For most of the early time up to the mid–nineteenth century we get preferred negative anomalies in the NAO index for winter and preferred positive ones for summer. The turning points in cumulative anomalies - during the 1850s for winter and during the 1870s for summer - indicate a transition period in circulation modes from the "Little Ice Age" to the recent climate in Europe. Running correlations (time windows of 21 years with time steps of one year) between zonal indices and regional temperature time series from Central England, Stockholm and two Central European regions are all indicating major instationarities in these relationships with a particular decline in winter correlations around the turn from the nineteenth to the twentieth centuries. Aspects of different circulation patterns linked with these variabilites are discussed.

Key Words: zonal indices; running correlations; temperature; Europe

1. Introduction

The atmospheric circulation of the mid latitudes is often described by so-called zonal indices measuring the meridional pressure gradient and hence the strength of the extratropical westerlies at different temporal and spatial scales. Within the North-Atlantic-European region the pressure gradient between the two semi-permanent centres of action, the Iceland Low and the Azores High, is of particular importance. This gradient represents one of the most robust recurrent modes of atmospheric behaviour (Barnston and Livezey, 1987), the North Atlantic Oscillation (Lamb and Peppler, 1987). It accounts for a large proportion of the interannual SLP variability (Rogers, 1990), controls the atmospheric factors leading to changes in the ocean (Dickson et al., 1996), and significantly accounts for hemispheric (Hurrell, 1996) and regional climatic variances (Hurrell, 1995; Hurrell and van Loon, 1997). Zonal indices, however, may also be defined for other regions in Europe (e.g. Jönsson and Bärring, 1994; Jacobeit et al., 1998) or for greater longitudinal sections comprising both the North Atlantic and the European area (Steinrücke, 1999). In any case, particular modes of zonal indices are indicators of specific circulation and climatic modes: In general terms, a large latitudinal

Climatic Change **48**: 219–241, 2001.

pressure difference (high zonality with higher pressure situated in the south) means that great parts of Europe are affected by a vigorous flow of winds from the westerly sector accompanied by high cyclonic activity and a maritime regime over the northern half of Europe. Low zonality, on the other hand, is linked to a small latitudinal pressure difference. If this gradient is reversed - for example higher pressure over Iceland than over the Azores (see Moses et al., 1987) - we get a meridional or even an easterly flow pattern with strong blocking of the westerlies and large latitudinal excursions of subtropical and polar air masses. Thus, weakened or absent westerly flow over Europe is typical for the low zonality mode representing a continental regime over the northern half of Europe.

Thus, the intensity of the westerlies is a major indicator of the atmospheric circulation and plays a key-role for the climate of Europe. To put the present climatic situation into a long-term perspective it is therefore of great importance to obtain reliable time series as long as possible. Investigations further back in time, before the end of the nineteenth century, have to be concentrated to areas that contain an adequate coverage of stations with long records of barometric pressure. Europe is favoured in that respect, since a rather dense distribution of meteorological stations already existed during the nineteenth century. This makes it feasible to reconstruct a grid point pressure data set for a time period longer than two hundred years covering Europe and the eastern part of the North Atlantic Ocean (Jones et al., 1999). From these gridded data it is possible to calculate monthly zonal indices for Europe from A.D. 1780 to get information about the atmospheric circulation for a period well extending into the "Little Ice Age" lasting from the 1560s to the 1850s (Flohn, 1993). Thus, the study period since 1780 goes back to times without particular anthropogenic forcing of the climate system and includes the transition from the "Little Ice Age" to subsequently warmer conditions. Time series of the zonal indices should indicate some circulation changes in this climatic transition period.

In the second part of this paper, relationships between the zonal indices and long European temperature time series will be investigated. For the recent century such relationships have been established in various studies, for instance by Malberg and Bökens (1997) concerning positive links between the NAO and Berlin winter temperatures, or by Jönsson and Bärring (1994) showing that the pressure difference between latitudes 45°N and 65°N has a strong average correlation with southern Scandinavian temperatures in winter ($r = +0.7$) and - with opposite sign - in summer ($r = -0.6$). As an important indication, however, the latter study using running correlations also revealed non-stationary features of these long-term average correlations. Similarly, Rogers (1985) concludes that the association between indices of the European westerly wind strength and the mean winter temperatures in southern Norway is not strictly consistent over time. These phenomena might be linked to varying air mass characteristics (Corti et al., 1999) or to frequency changes among different circulation patterns associated with a particular mode of a zonal circulation index. The gridded SLP time series back to 1780 (Jones et al., 1999) make it possible to calculate running correlations between zonal circulation indices and long European temperature time series over roughly 200 years and to investigate the consistency of these well-known circulation-climate relationships over this extended period of time. Thus, we will be able to infer whether commonly used associations between zonal index modes and European temperatures are valid beyond the frame of our recent century or, in contrast, are non-stationary and thus reflect conditions of particular sub-periods.

2. Data

The monthly mean grid point data of sea-level pressure reconstructed by Jones et al. (1999) back to 1780 for the North-Atlantic-European region are based on homogenized long-term pressure time series from stations with the oldest instrumental records in Europe. The reconstruction is based exclusively on series of measured pressure. The data base starts in 1780 with 10 continuous series, including Lund (Bärring et al., 1999) and Barcelona (Rodriguez et al., submitted). In the 1820s the network is increased to 20 stations, including the Reykjavik area and Gibraltar (Jones et al., 1997), and finally it reaches a set of 51 stations at the middle of the nineteenth century.

The historical SLP data were reconstructed on a 5° latitude by 10° longitude grid from 35 to 70°N and from 30°W to 40°E based on EOF-regression models calibrated over the period 1936-1995 and verified for the period 1881-1935. Explained variances reach above 90% around the central grid points, but decrease towards the periphery, especially for the earliest period with few station pressure series. Generally, explained variances are higher for the winter months than during summer, and this is also true for peripherical regions since the estimation of grid point values further away from available station data is favoured by the more structured large-scale circulation during the winter season (Jones et al., 1999). This reconstruction provides an appropriate data base to calculate some atmospheric circulation indices (see section 3).

Temperature data for northern, western and central Europe were preferably taken in an aggregated manner, only the northernmost part is represented by an individual station, the Stockholm record (Moberg and Bergström, 1997). For western Europe the well-known Central England Temperature time series is used (Manley, 1974; Parker et al., 1992), and for Central Europe two (out of eight) thermal regions have been selected from Beck (2000) each comprising several stations out of 72 temperature time series which were drawn from various sources (World Climate Disc, 1992; Schönwiese and Rapp, 1997; several German Archives). Details about the stations are included in Beck (2000) as well as results of Alexandersson-tests for homogeneity and corrections of identified inhomogeneities. Thermal regions finally have been determined by applying clustering techniques to the matrix of pairwise correlations (Beck, 2000). The two selected regions in this context cover parts of southwestern and northeastern Central Europe, respectively, being represented by the following stations:

Central European Region SW (centered around 49.5°N, 8°E):
Aachen, Augsburg, Bamberg, Chalons, Karlsruhe, Langen, Lille, München, Strasbourg, Stuttgart, Trier, Uccle, Würzburg.

Central European Region NE (centered around 54°N, 16°E):
Copenhagen, Elblag, Gdansk, Kaliningrad, Koszalin, Rostock, Szczecin, Stralsund, Sylt, Warszawa.

Thermal indices for these regions are defined by averaging the monthly normalized temperature series from the corresponding stations, respectively.

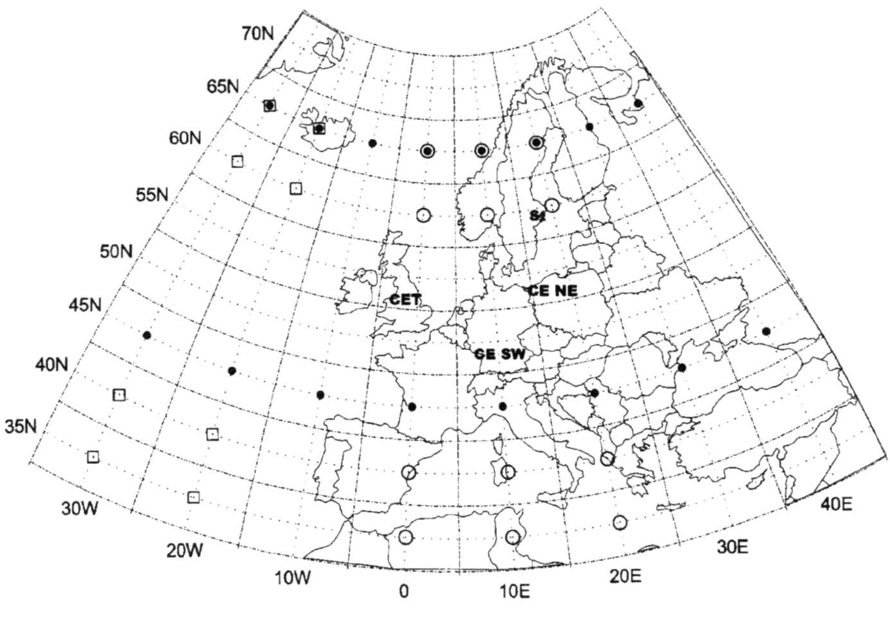

Figure 1. Geographical extension of the historical monthly mean sea-level pressure grid of Jones et al. (1999). Dotted lines show the geographical coordinate system and the grid boxes are indicated as solid lines. Grid boxes used for calculating the various zonal indices are marked as follows:
dots - the overall Zonal Index (ZI);
open squares - the North Atlantic Oscillation Index (NAOI);
open circles - the Central European Zonal Index (CEZI).
Approximate locations of the regional temperature data are indicated with the following abbreviations:
CET - Central England temperature;
ST - Stockholm temperature;
CENE - temperature of Central European Region NE;
CESW - temperature of Central European Region SW.

3. Methods

A simple index method of quantifying the strength of the mid-latitude westerlies is to calculate monthly values of the pressure difference between two latitudinal bands and hereby obtaining an overall zonal index for the whole longitudinal area under consideration. High values of this zonal index correspond to a vigorous westerly flow advecting maritime air masses over widespread regions of Europe whereas low zonal index values coincide with a less maritime climate. Using the historical SLP grid from Jones et al. (1999) we get an overall zonal index ZI for the North-Atlantic-European region (30°W to 40°E); calculating ZI as the pressure difference (hPa) between the latitudes 45°N and 65°N (see Fig. 1) implies that conditions around the main branch of the westerlies will be considered in particular.

Since the whole North-Atlantic-European area includes significant geographical variations with respect to land-sea distribution and to the important centres of circulation it is necessary to consider regional indices, too. First of all, an index approximating the North Atlantic Oscillation (NAOI) is important. Apart from classical approaches to define a difference between normalized station pressure time series in appropriate locations - e.g. Ponta Delgada/Akureyri (Rogers, 1984), Lisbon/Stykkisholmur (Hurrell, 1995) or Gibraltar/Reykjavik (Jones et al., 1997) - there are several approaches based on SLP grid point values: e.g. by determining the moving pressure maxima and minima (Mächel et al., 1998) or by calculating the first grid-based principal component representing the NAO (Wanner et al., 2000). Here we use fixed grid point averages of normalized SLP (see Fig. 1) to remove random variations that would occur within single grid point series. Thus, the NAOI results from differences of normalized SLP averaged over the 4 grid points 35°N/30°W, 35°N/20°W, 40°N/30°W, 40°N/20°W and 60°N/30°W, 60°N/20°W, 65°N/30°W, 65°N/ 20°W, respectively.

Certainly, the North Atlantic Oscillation represents the major mechanism governing the large-scale circulation variability in Europe (especially in winter), but its area of determination covers the peripherical regions of the historical SLP grid where the reconstruction quality is lower than in the central European region, especially for the first decades before nearby station pressure time series become available (Jones et al., 1999). Therefore, another zonal index has been defined (see Fig. 1) as a specific indicator of the conditions above continental Europe, the region of highest quality in SLP reconstruction. Again fixed grid point averages of normalized SLP are used for this Central European Zonal Index (CEZI) related to the same latitudinal bands as for NAOI (i.e. 35°N-40°N and 60°N-65°N) but for the longitudinal section from 0° to 20°E.

Concerning correlations between these circulation indices and the four selected temperature time series (Central England, Stockholm, southwestern and northeastern Central Europe) we focus on the consistency of these relationships during the period 1780–1995. We have used the non-parametric Spearman rank correlation coefficient to counteract incidental deviations from the normal distribution. Correlations have been calculated for running time windows of 21 years with time steps of one year giving time series of 196 correlation coefficients for each case of circulation-temperature relationship. In addition to time series with seasonal resolution (aggregations into the meteorological seasons: DJF, MAM, JJA, SON), running correlations on a monthly basis are calculated for each season comprising samples of 63 months from the corresponding 21 seasons with time steps, in this case, of 1 month resulting in time series of 586 correlation coefficients for each case of circulation-temperature relationship.

4. Results

4.1 ZONAL INDICES

Fig. 2 shows the general structure of variation of the overall zonal index (ZI) for the period 1780-1995 for all months of the year. Since this index is defined by latitudinal pressure differences in absolute terms (hPa), seasonal contrasts are distinctly identified: higher values during the cold season, lower ones within the warmer period with regular minima centered around May when the overall pressure gradient is weak over Europe. Since the 1960s, however, a certain tendency for regular inclusion of early summer into the period of lowest ZI values seems to be indicated. The consistency during summer (July and August) is striking as well as the pronounced decadal-scale variability in January after the 1850s: from a maximum in the late 1850s there is a decrease to a minimum about 1880 and an increase to a second maximum in the 1910s, whereafter there is a decrease until the 1940s and an increase again from the 1960s onwards.

The regional indices, NAOI and CEZI, are shown in Fig. 3 as explicit values with seasonal resolution. In addition to seasonal anomalies and smoothed time series, the normalized cumulative anomalies are included in Fig. 3 calculated as the continuous sum of successive seasonal anomalies. Thus, sustainably declining/rising values in cumulative anomalies indicate the prevalence of negative/ positive seasonal anomalies, and major turning points in cumulative anomalies indicate transitions between periods predominated by contrasting seasonal anomalies. Therefore cumulative anomalies reveal decadal and interdecadal variations particularly clear-cut as shown, for example, by Mächel et al. (1998) analyzing circulation time series since 1881.

The time series in Fig. 3 extended back to 1780 reveal marked differences between the periods before and after the mid-nineteenth century with some particular contrasts between the seasons: thus, NAOI shows a period until the 1850s with accumulating negative anomalies in winter whereas in summer positive anomalies are prevailing until the 1870s. This is in contrast to the most recent evolution of the last few decades with positive anomalies prevailing during winter and negative

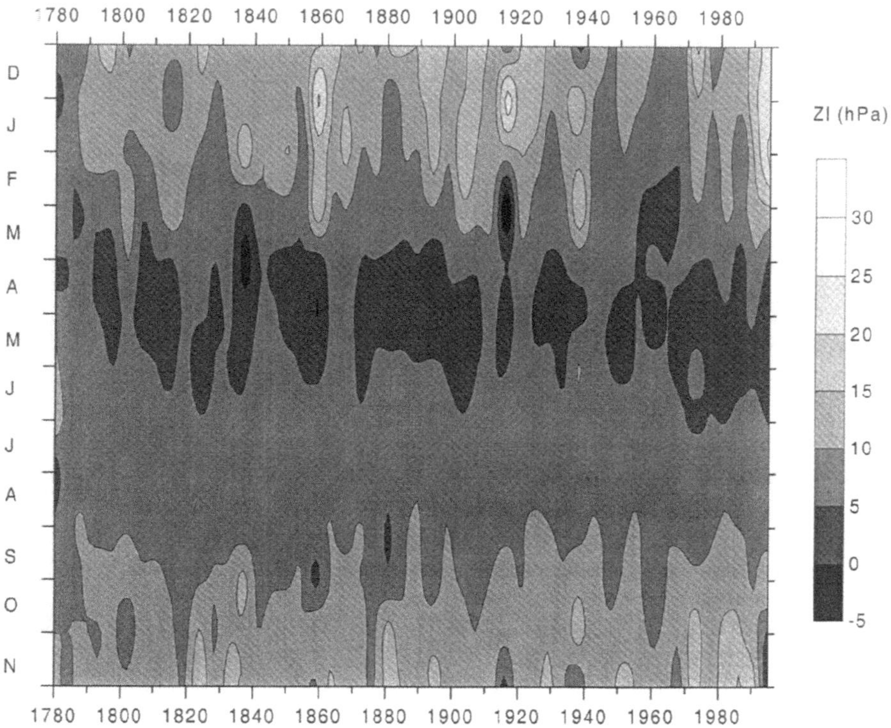

Figure 2. Isopleth diagram showing the evolution over time of the annual course of the overall zonal index (ZI). The contours are smoothed to show decadal scale variations.

ones during summer. Remarkably, CEZI does not reproduce the dominating low values during winter for the early period mentioned above, although the temperatures for Central Europe indicate a significantly more continental climate for this period, i.e. lower winter and higher summer temperatures than during the recent century (Jacobeit et al., 1998). This might be a general indication of instationarities in circulation-temperature relationships that will be discussed in the following chapter.

Further differences between CEZI and NAOI are manifested by an earlier turning point of cumulative anomalies within the twentieth century indicating the transition from prevailing positive to negative anomalies in winter, and by the absence of a period with cumulating negative anomalies in summer around the turn from the nineteenth to the twentieth centuries. This points to the fact that each NAO mode is linked to a variety of different circulation patterns over Europe implying a complex relationship with temperature even during winter time.

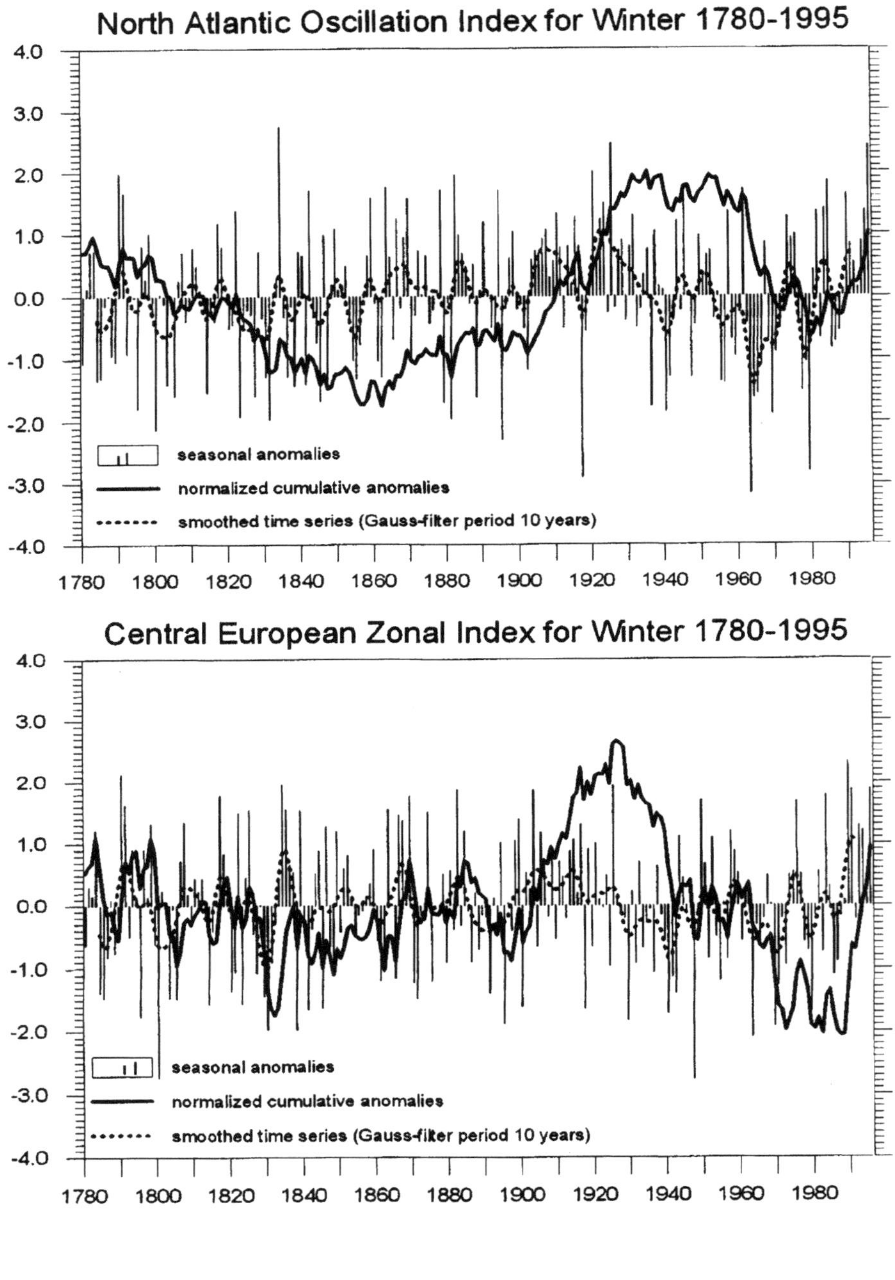

Figure 3. North Atlantic Oscillation Index (NAOI) and Central European Zonal Index (CEZI) derived from gridded North-Atlantic-European sea-level pressure for the meteorological seasons 1780-1995.

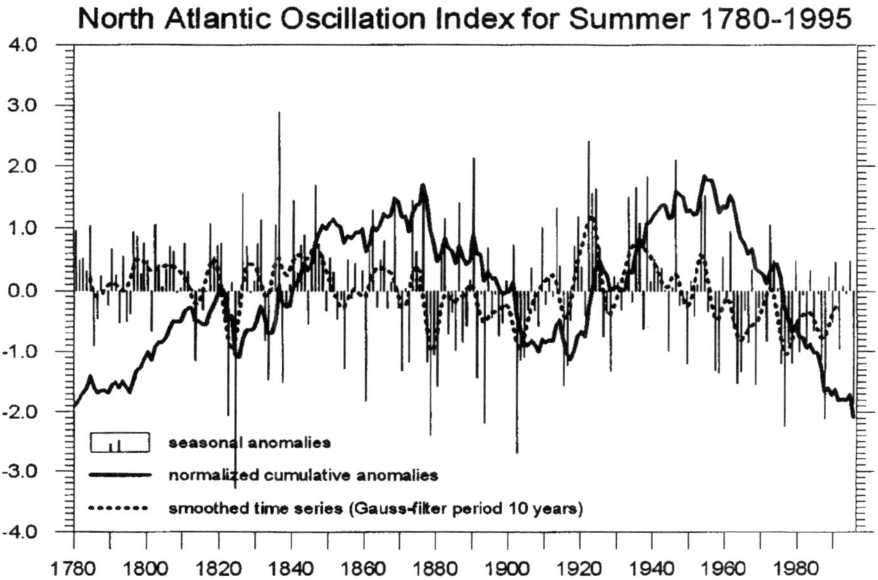

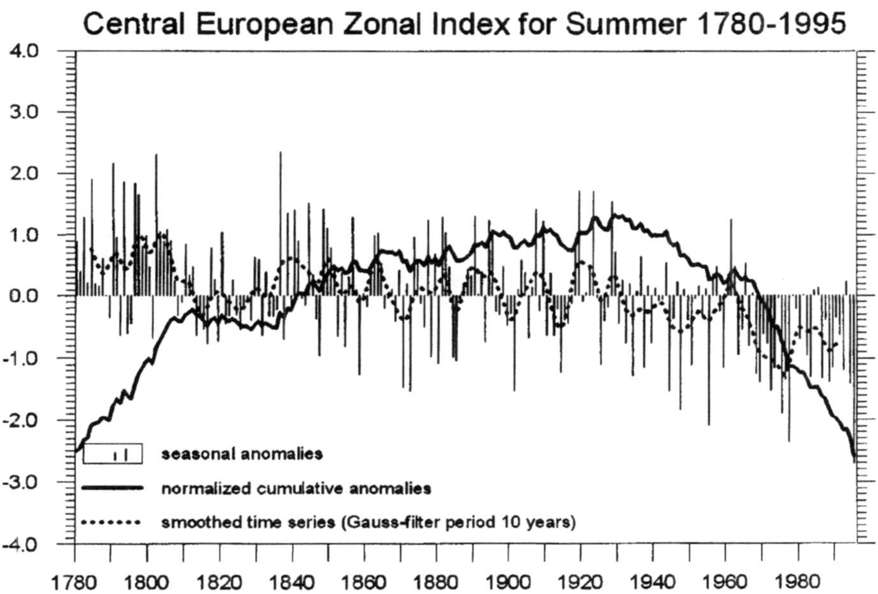

Figure 3. (Continued)

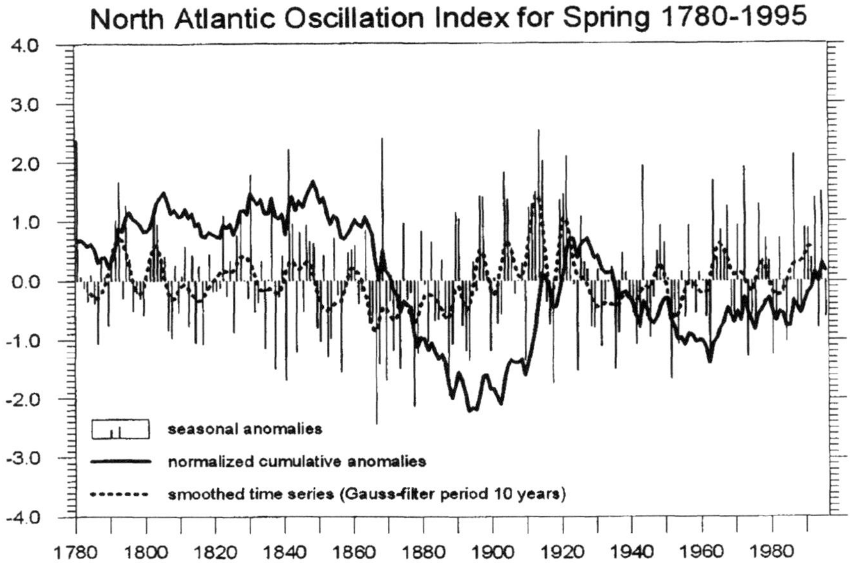

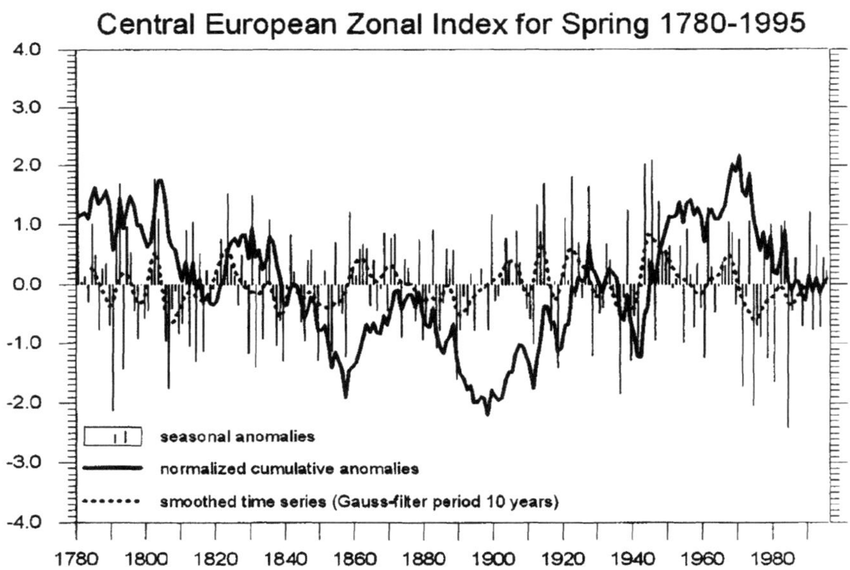

Figure 3. (Continued)

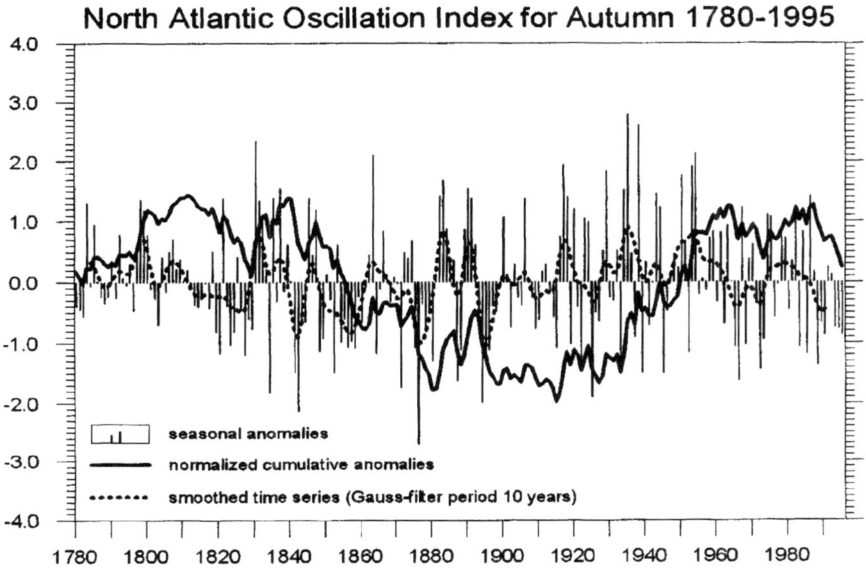

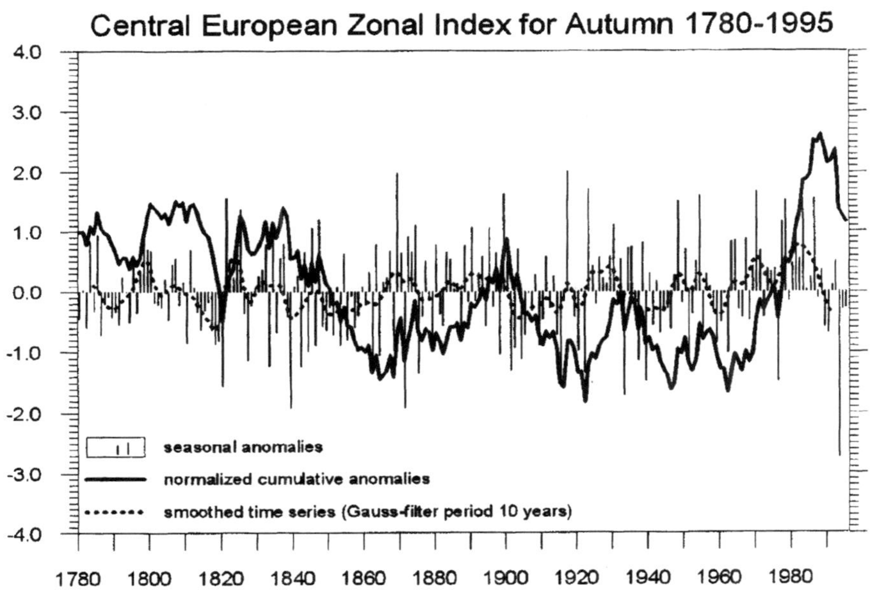

Figure 3. (Continued)

Concerning the extent to which the circulation indices have increased during the most recent period in winter it should be stressed that only for the very short period since 1989 the indices based on gridded data are significantly higher than in the earlier periods with positive seasonal anomalies. Since this short period has not continued during the last years (Jones et al., 1997) it cannot define an extraordinary interval in terms of major periods in cumulative anomalies (Fig. 3).

For the transitional seasons we get declining tendencies in cumulative anomalies at some time during the nineteenth century, i.e. periods of variable length with prevailingly negative seasonal anomalies, most distinctly for NAOI in spring during the second half of the nineteenth century, more oscillatory for CEZI in spring, and generally less marked during autumn. Opposite developments (accumulating positive seasonal anomalies) within the twentieth century again are differing in detail: concentrated to the first decades for NAOI in spring, more elongated but oscillatory for CEZI in spring, lasting until the 1960s for NAOI in autumn, and just starting at this time for CEZI in autumn.

As a main result, however, the hundred years preceding the time when previously available data sets are starting, indicate modified conditions in atmospheric circulation compared to the modern century, affecting all seasons in a different way. This might be important for the transition from the "Little Ice Age" to the subsequent period as will be discussed in chapter 5.

4.2 RUNNING CORRELATIONS WITH TEMPERATURE

Comparisons between different zonal indices in the last chapter have indicated that there might be major instationarities in circulation-climate relationships. This is verified by running correlations between zonal indices and various European temperature time series.

First we look at the overall zonal index, ZI, and its running correlations with temperature for all months of the year (Fig. 4). As a distinct result we get no persistence over time of correlation coefficients within one and the same class, for all included regions (Central England, Stockholm, southwestern and northeastern Central Europe) and at every time during the year (note that each class of coefficients has a range of values of 0.2). The longest interval with variations confined within one class occurs for Stockholm during February extending from the 1830s to the 1940s. But even this example includes differences in correlation coefficients of up to more than 0.6 considering the whole period from 1780 to 1995. Remarkably, even during winter, coefficients may drop to low values. During the decades around 1880 there is a widespread shift of the seasonal correlation structure: the high correlations during winter persist during spring, and the usually low correlations during summer are replaced by higher ones (resembling spring conditions) during these decades.

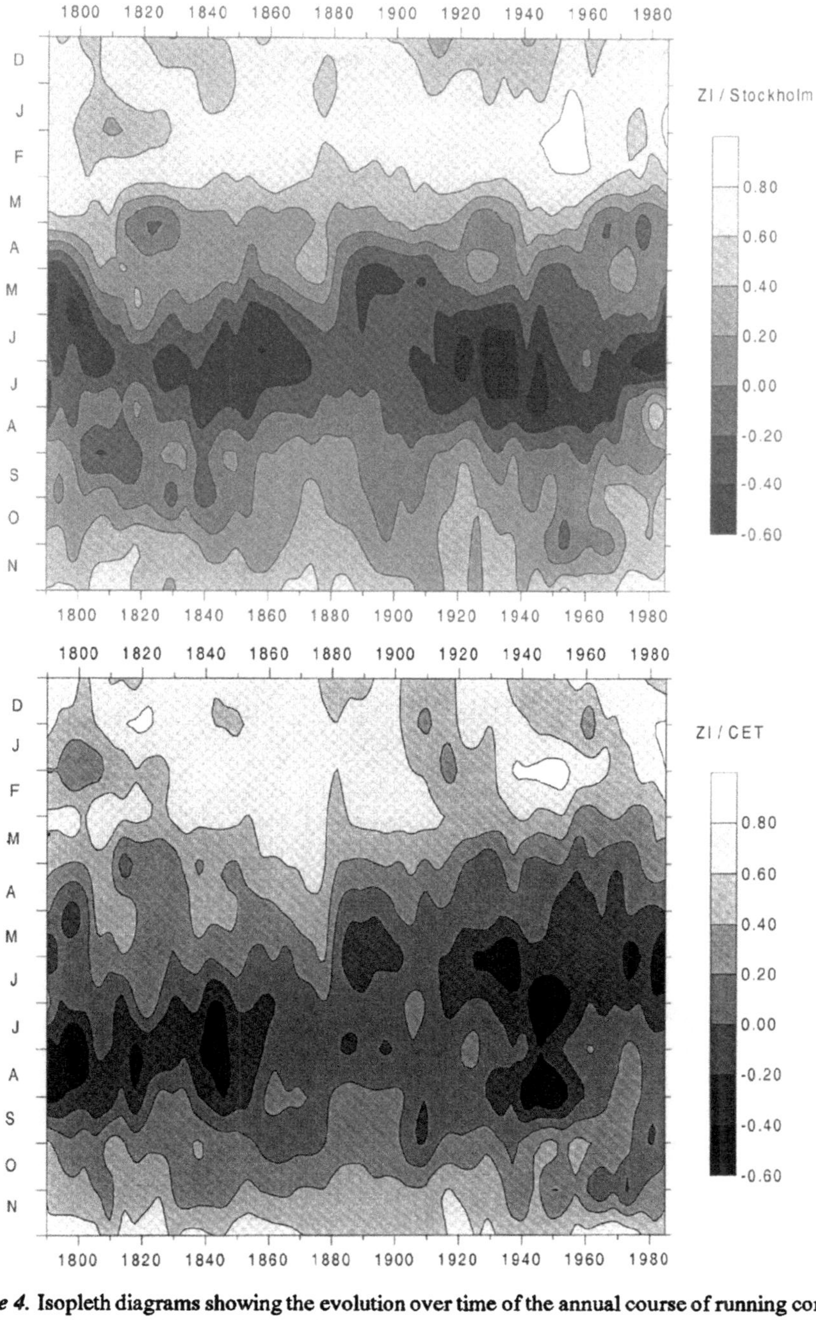

Figure 4. **Isopleth diagrams showing the evolution over time of the annual course of running correlations between the overall zonal index (ZI) and regional temperature time series (Stockholm; Central England - CET). The contours are smoothed to show decadal scale variations.**

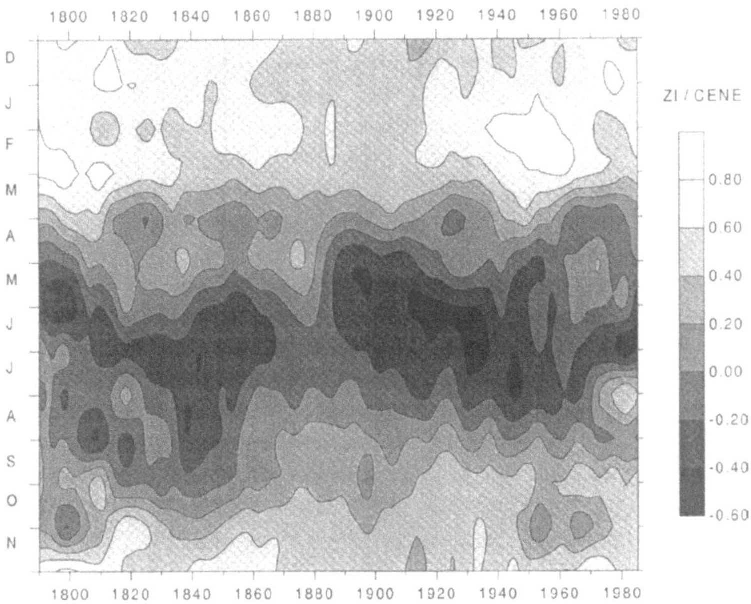

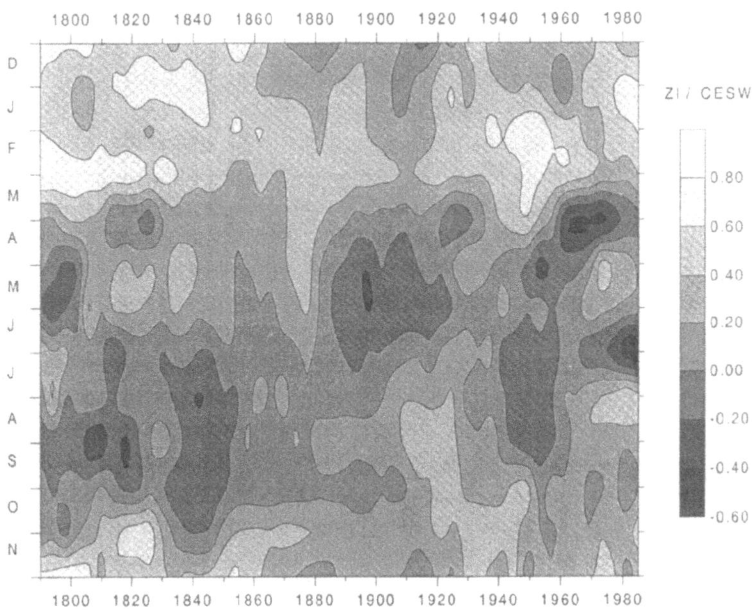

Figure 4. **(Cont.) ZI - overall zonal index; CENE - temperature of Central European Region NE; CESW - temperature of Central European Region SW.**

To resolve variations in circulation-temperature relationships explicitly (i.e. without any classification) Fig. 5 gives selected time series of running correlation coefficients related to the regionally defined indices NAOI and CEZI. Examples are confined to the winter season and to two particular cases from summertime. Due to the effect that seasonal integration may differ between circulation and temperature indices their correlation coefficients sometimes may vary differently comparing monthly and seasonal resolution of the same selected case.

Looking at the four cases relating the temperature time series described above to the NAOI series for winter (Fig. 5) we get considerable ranges of correlation coefficients varying around the mean values for the whole period 1780-1995 (0.57, 0.64, 0.57, 0.58 with monthly resolution and 0.67, 0.66, 0.63, 0.58 with seasonal resolution for Stockholm, Central England, Central European regions NE and SW, respectively). Particularly important is the distinct result of profoundly lower correlations at the beginning of the twentieth century compared with both recent as well as former times during the last two centuries. The drop in correlation coefficients around the turn from the nineteenth to the twentieth centuries is most pronounced for Central England based on seasonal resolution. For southwestern Central Europe the decline in correlation levels already set in from the 1820s onwards whereas the northern regions (NE Central Europe, Stockholm) experienced a first period of lower correlations during the first half of the nineteenth century before recovered values from the second half of the century again dropped to the lower levels mentioned above. With monthly resolution the amplitudes of variation are considerably smaller; thus, declines in correlation coefficients do not reach the lower limit for 95% statistical significance (approximately at $r=0.25$), but still we get major variations reaching absolute differences in correlation coefficients of more than 0.5 for southwestern Central Europe. This region as well as Central England furthermore attain a second minimum around 1950 intervening within the modern period of higher correlation levels.

Correlations with CEZI only are given for the Central European regions since the other regions are located to the west (Central England) or in the northern periphery (Stockholm). For the whole period 1780-1995 correlation coefficients based on monthly and seasonally resolved data are 0.76 and 0.80 for the NE region and 0.66 and 0.67 for the SW region, respectively, thus reaching higher levels than the corresponding cases with NAOI mentioned above. But looking at the running correlations (Fig. 5) we get again distinctly varying values with lowest correlation coefficients at the beginning of the twentieth century. In this case, however, they do not result from a continuous decline because the 1890s - in contrast to the correlations with NAOI - appear as an intervening period of recovered values. In general, correlations with CEZI are varying more oscillatory,

Figure 5. (Next two pages) Running correlations between circulation indices (NAOI, CEZI) and regional temperature time series (Stockholm, Central England, Central European Regions NE and SW). See text for calculation modes of running monthly and seasonal correlations.

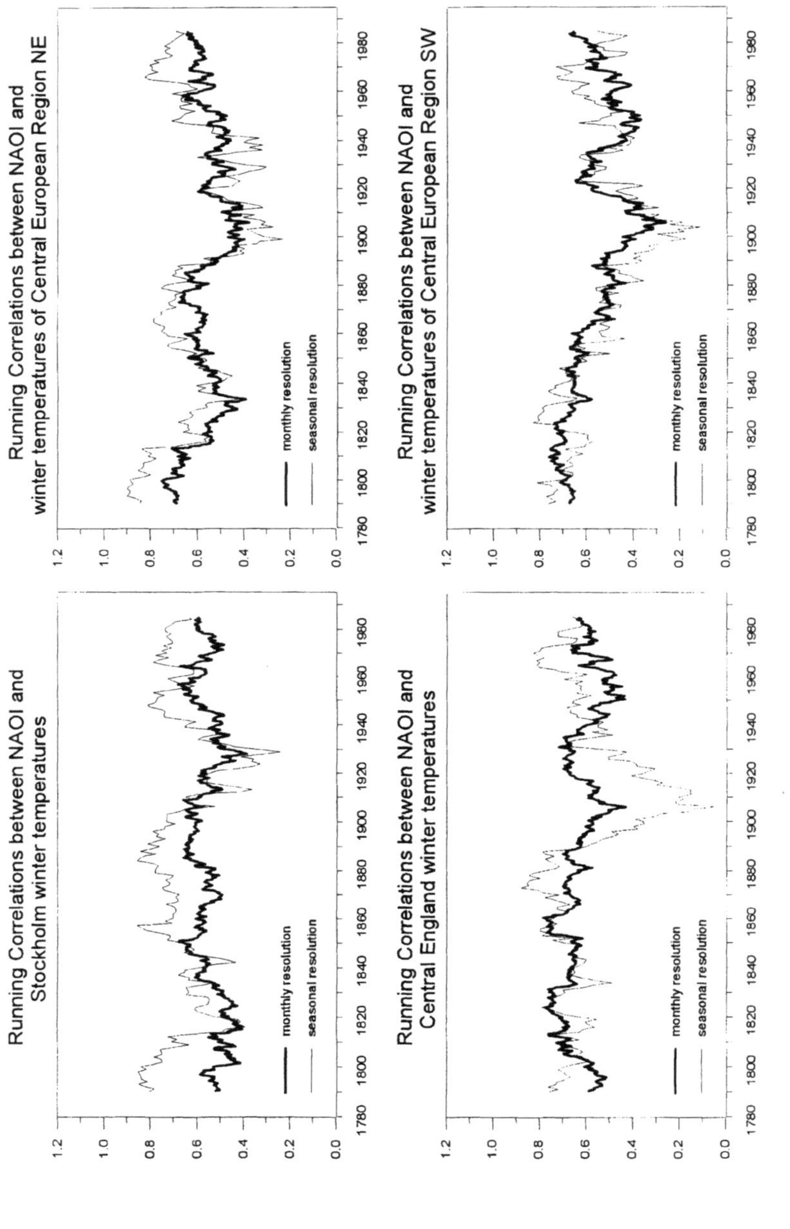
Running Correlations between NAOI and winter temperatures of Central European Region NE
Running Correlations between NAOI and winter temperatures of Central European Region SW
Running Correlations between NAOI and Stockholm winter temperatures
Running Correlations between NAOI and Central England winter temperatures
monthly resolution
seasonal resolution

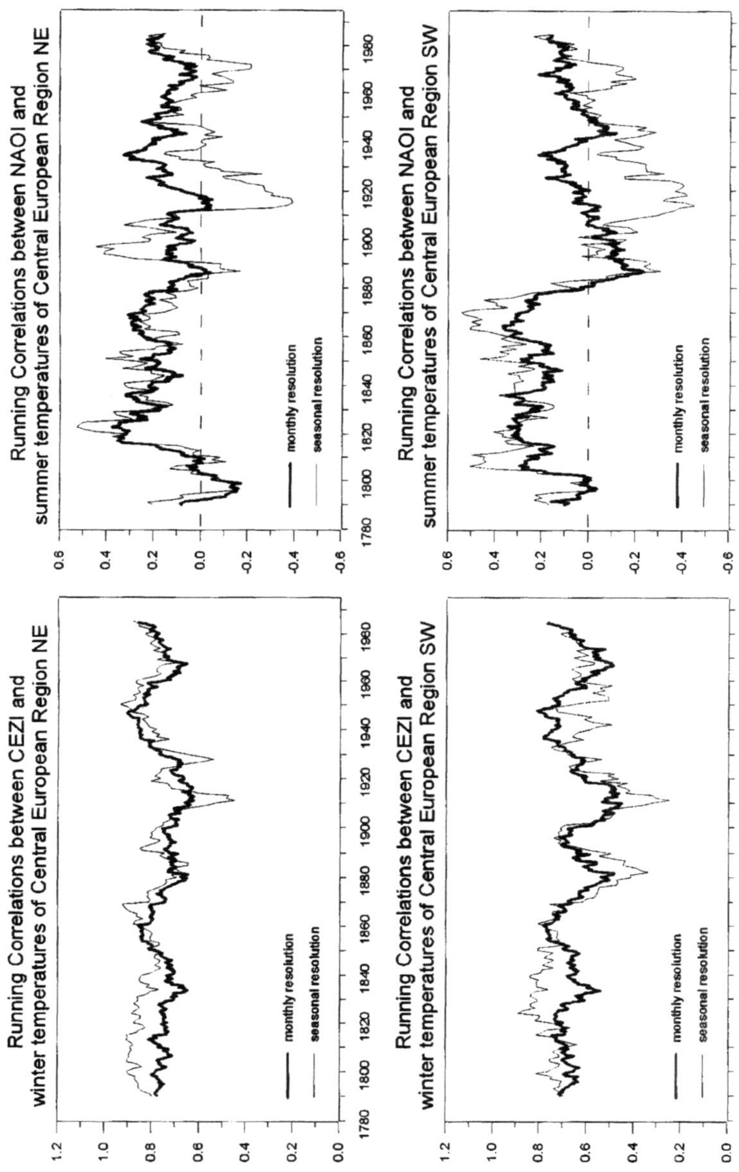
Running Correlations between CEZI and
winter temperatures of Central European Region NE
monthly resolution
seasonal resolution
Running Correlations between NAOI and
summer temperatures of Central European Region NE
monthly resolution
seasonal resolution
Running Correlations between CEZI and
winter temperatures of Central European Region SW
monthly resolution
seasonal resolution
Running Correlations between NAOI and
summer temperatures of Central European Region SW
monthly resolution
seasonal resolution

especially for monthly resolution with several minima during the 1830s, 1880s, 1910s and 1960s and alternating maxima around 1860, during the 1890s, 1940s and 1980s.

Correlations for the summer season generally are weaker and often statistically insignificant. Therefore only two examples implying major variations over time are included in Fig. 5. It is particularly noteworthy that, in contrast to the modern conditions with low correlation coefficients between NAOI and central European temperature, this relationship was positive for the greater part of the nineteenth century (95% statistical significance is reached with $r > 0.25$ on the monthly scale and $r > 0.44$ on the seasonal scale). This was partly due to zonal ridges of high pressure (i.e. high values in NAOI and central European summer temperatures) which were more important for the atmospheric circulation in the former period than during the recent century (Jacobeit et al., 1998). The 'continental' index CEZI actually gives negative correlations with central European summer temperatures (at least for the northern region) but this relationship dropped to insignificant values during the first decades of the analyzed period and between 1870 and 1890. Thus, circulation-climate relationships do not necessarily remain stationary if periods of investigation are extended further into the historical past.

5. Discussion

The mass reduction of many European glaciers since the mid-nineteenth century, often referred to as the end of the "Little Ice Age" (e.g. Flohn, 1993), should be reflected in some atmospheric circulation changes since there exists a general coupling between circulation and the glacier mass balance (Pohjola and Rogers, 1997). For recent boundary conditions, for example, the net mass balance in the European Alps is favoured by westerlies, troughs and depressions increasing accumulation in winter, but reducing ablation in summer (Fitzharris, 1998). Of course, these conditions may not occur simultaneously in winter and summer, and particular large-scale weather types during spring and autumn also influence the mass balance of glaciers as already shown by Hoinkes (1968). Furthermore, the response of indi-vidual glaciers is even more complicated and does not only depend on variables linked to atmospheric circulation (temperature, precipitation and cloudiness) but also on glacier geometry and location (Oerlemans, 1994; Oerlemans et al., 1998) together with topography, altitude and response time (Kuhn et al., 1985).

Based upon this background no immediate relationships between glacier mass balance and zonal circulation indices can be inferred. However, zonal indices are indicating particular states of the atmosphere which might be differing in frequency between different climatic periods. Thus, without a direct link to glacier behaviour, the "Little Ice Age" and the subsequent period of warming might be distinguished by different prevailing circulation modes indicated in the time series of cumulative anomalies of an appropriate zonal circulation index. In fact, the NAOI calculations (Fig. 3) based on gridded SLP reconstructed back to 1780 do reveal distinct circulation changes within the nineteenth century: during winter an extended accumulation of negative anomalies until the 1850s was replaced by prevailing positive anomalies afterwards, in contrast to that, positive anomalies in NAOI accumulated during summer until the 1870s subsequently changing to an opposite evolution. Thus, the different turning points in cumulative anomalies indicate, in terms of circulation modes, the transition period between the "Little Ice Age" and the following warming period, a transition that started in winter during the 1850s and concluded in summer some 20 years later.

No distinct preference for negative anomalies, however, is indicated for the Central European Zonal Index (CEZI) during the historical winter seasons until the 1850s - thus it cannot be excluded that the prevailing low-index mode in NAOI during this time might be due to the lower performance in SLP grid reconstruction for marginal areas, especially during the first decades since 1780. This uncertainty in the early part of the reconstruction highlights the need to elaborate further high quality station time-series covering the earliest part of the period, particularly stations in the periphery of the North-Atlantic-European region. On the other hand, low values in NAOI during winter still are indicated after the 1820s when station pressure time series from Reykjavik and Gibraltar have entered the network for SLP grid reconstruction (Jones et al., 1999). Furthermore, marked differences between NAOI and CEZI also do occur for later periods with enhanced data bases, e.g. the time-lagging turning points to prevailing negative anomalies in winter during the twentieth century or the period with cumulating negative anomalies in NAOI during summer around the turn from the nineteenth to the twentieth centuries which is com-pletely missing within the corresponding CEZI series. Due to different circulation patterns different modes in NAOI and CEZI may in fact be linked with each other, e.g. high-index conditions above the Atlantic with continental blocking patterns decreasing CEZI values to intermediate or even negative levels. Thus, the correlation coefficient between NAOI and CEZI being 0.68 during winter for the whole period 1780-1995 sometimes drops to significantly lower values, e.g. around 0.2 for running 21-year periods at the beginning of the twentieth century when lower values in the circulation-temperature relationships (see below) also hint at intervening changes in circulation patterns. For the end of the "Little Ice Age" we conclude in general that the NAOI most probably preferred below-average values in winter linked with greater variability in CEZI including patterns of mixed circulation type (transitional between zonal and meridional types). For the summer season both indices preferred above-average values even some decades beyond the mid-nineteenth century thus indicating an extended period of transition between the different seasonal circulation modes characterizing the "Little Ice Age" and the subsequent climatic period in Europe.

The second part of the investigation focussed on circulation-temperature relationships. Considering the seasonal cycle in the overall zonal index (ZI)-temperature correlations (Fig. 4) we get a distinct climatic zonation: in the more continental regions of central Sweden (represented by Stockholm) and northeastern Central Europe (CENE, see Fig. 1) there is a large seasonal cycle in the correlations (positive during winter and negative during summer). In the maritime climate of Central England this seasonality is less distinct, even though there is an indication of an increased seasonal cycle after about 1920. The seasonal ZI-temperature cycle of southwestern Central Europe (CESW) has similar characteristics as the maritime Central England region.

This climatic zonation is due to particular geographical circumstances. Thus, during situations with influx of continental air from the east, Central England mostly retains relatively maritime conditions - apart from distinct anomaly periods like the Late Maunder Minimum at the end of the seventeenth century - because of the North Sea. The much smaller Baltic Sea cannot produce a similar transformation of the air mass thus rendering Sweden a more continental climate. This is particularly pronounced during winters when the Baltic sea-ice is extended to the south. These geographical circumstances produce a ZI-temperature correlation pattern that is similar for Stockholm and the northeastern Central European region (CENE). The southwestern Central European region (CESW), however, is more similar to Central England than to the Central European northeasterly region despite the fact that there is no large water body to the east of this region that would modify the continental air mass.

Further investigations have identified significant instationarities in correlations between circulation indices and temperature time series for both monthly and seasonal resolution of the corresponding samples from 1780 onwards (Fig. 5). Similar results (not shown here) are obtained for spring and autumn and for various timescales in running correlations (e.g. for three or four instead of two decades as moving time windows). These instationarities point to the fact that both positive and negative modes of circulation index variations are including different circulation patterns with quite different effects on Central European climate: for example, positive modes of the NAO are not only related to westerly flow types with above-average winter temperatures in Central Europe, but sometimes also to mixed circulation types with Central European high pressure systems leading to below-average winter temperatures due to long-wave radiation losses. On the other hand, even weak zonal flow above the North Atlantic (i.e. negative mode of the NAO) sometimes is related to persistent westerly advection towards Central Europe with quite warm conditions during winter. Schmutz and Wanner (1998) also give examples of circulation patterns with unexpected regional temperatures in terms of different modes of the NAO. Thus, if changes of circulation patterns within a particular index mode do occur, correlations between this circulation index and regional temperatures may not remain persistent over time - the NAOI being an indicator of the state of the climate system in the North Atlantic area must not be used as a stable indicator for surrounding regional climates like in Central Europe as well as these climates are not qualified as proxy indicators for North Atlantic circulation modes on multi-decadal time-scales. In general, the European climate is more complex than can be captured by a simple circulation index like the NAOI.

Major shifts in circulation-temperature relationships have been identified for both winter and summer conditions. During summer changes occurred between the nineteenth and the twentieth centuries from positive to near-zero correlations between NAOI and southwestern Central European temperatures, from near-zero to negative correlations between CEZI and northeastern Central European temperatures. This might be due, at least in part, to the above-mentioned decrease in zonal high pressure ridges for summer. During winter the varying levels of correlations between circulation indices and European temperatures reveal a major event common to all particular cases under consideration: the marked decline in correlations around the turn from the nineteenth to the twentieth centuries leading to significantly lower values in the circulation-temperature relationships at the beginning of this century. Additionally, the northern regions (NE Central Europe, Stockholm) experienced a first period of lower correlations with NAOI during the first half of the nineteenth century, the southwestern regions (Central England, SW Central Europe) a second minimum period around 1950. Most important seems to be the former period around the last turn of centuries, since similar declines in correlations for this time have been reported from other regions, too, e.g. from the eastern Alpine region (Böhm et al., 1998) or from the Greenland area based on oxygen isotope data from ice cores (Steig and Barlow, 1998). Especially the 1890s are particularly striking in terms of circulation dynamics since running correlations with temperature develop differently around this time with increasing values (for an intervening period) in relation to CEZI while correlations with NAOI already decline towards their minimum level.

In order to understand these instationarities in circulation-climate relationships further investigations have to be directed towards the question whether there are significant redistributions among basic circulation patterns within particular index modes between different periods of high and of low correlations with regional European temperatures. Selecting periods of high, of low, and of rapidly changing correlations according to Fig. 5 will allow to investigate whether major NAO modes are dominated by different circulation patterns during these selected periods. Since this kind of variability is operating on decadal to centennial time scales periods covered by objectively reconstructed SLP data are required extending as far as possible into the historical past including sections of the "Little Ice Age" which additionally offer an extended range of natural variability in climate, circulation and their relationships. Thus, historical and dynamic climatology are linked by reconstructing past climates and simultaneously investigating atmospheric circulations over periods of several centuries thereby ensuring results that are valid not only for a few decades with limited range in variability.

Acknowledgements

This work has been supported by the European Commission under grant ENV4-CT95-0129, coordinated by Trevor Davies and Phil Jones (Climatic Research Unit, University of East Anglia, Norwich, UK). We thank the two reviewers, Rudolf Brazdíl and Heinz Wanner, for suggesting useful improvements to the paper.

References

Barnston, A.G. and Livezey, R.E.: 1987, 'Classification, seasonality and persistence of low-frequency atmospheric circulation patterns', *Mon. Wea. Rev.* **115**, 1083-1126.

Bärring, L., Jönsson, P., Achberger, C., Ekström, M. and Alexandersson, H.: 1999, 'The Lund record of meteorological instrument observations: reconstruction of monthly sea-level pressure 1780-1997', *Int. J. Climatol.* **19**, 1427-1443.

Beck, C.: 2000, 'Zirkulationsdynamische Variabilität im Bereich Nordatlantik-Europa seit 1780', *Würzburger Geographische Arbeiten* (in press).

Böhm, R., Auer, I., Hagen, M. and Schöner, W.: 1998, 'Is there a NAO-signal in Alpine climate time series?', *Abstracts of The Second International Climate and History Conference, Norwich* 1998, 36.

Corti, S., Molteni, F. and Palmer, T.N.: 1999, 'Signature of recent climate change in frequencies of natural atmospheric circulation regimes', *Nature* **398**, 799-802.

Dickson, R., Lazier, J., Meincke, J. Rhines, P. and Swift, J.: 1996, 'Long-term coordinated changes in the convective activity of the North Atlantic', *Progr. Oceanogr.* **38**, 241-295.

Fitzharris, B.: 1998, 'A synoptic climatology of glacier mass balance for the European Alps', in *Climate and Environmental Change*, International Geographical Union, Commission on Climatology, Evora 1998, 141-142.

Flohn, H.: 1993, 'Climatic evolution during the last millenium: what can we learn from it?', in Eddy, J.A. and Oeschger, H. (eds.), *Global Changes in the Perspective of the Past*, Wiley & Sons, Chichester et al., pp. 295-316.

Hoinkes, H.C.: 1968, 'Glacier variations and weather', *J. Glaciol.* **7**, 3-19.

Hurrell, J.W.: 1995, 'Decadal trends in the North Atlantic Oscillation: regional temperatures and precipitation', *Science* **269**, 676-679.

Hurrell, J.W.: 1996, 'Influence of variations in extratropical wintertime teleconnections on Northern Hemisphere temperature', *Geophys. Res. Let.* **23**, 665-668.

Hurrell, J.W. and van Loon, H.: 1997, 'Decadal variations in climate associated with the North Atlantic Oscillation', *Clim. Change* **36**, 301-326.

Jacobeit, J., Beck, C. and Philipp, A.: 1998, 'Annual to decadal variability in climate in Europe - objectives and results of the German contribution to the European climate research project ADVICE', *Würzburger Geographische Manuskripte* **43**, 163 pp.

Jones, P.D., Jonsson, T. and Wheeler, D.: 1997, 'Extension of the North Atlantic Oscillation using early instrumental pressure observations from Gibraltar and south-west Iceland', *Int. J. Climatol.* **17**, 1433-1450.

Jones, P. D., Davies, T. D., Lister, D. H., Slonosky, V., Jonsson, T., Bärring, L., Jönsson, P., Maheras, P., Kolyva-Machera, F., Barriendos, M., Martin-Vide, J., Alcoforado, M. J., Wanner, H., Pfister, C., Schuepbach, E., Kaas, E., Schmith, T., Jacobeit, J. and Beck, C.: 1999: 'Monthly mean pressure reconstructions for Europe for the 1780-1995 period', *Int. J. Climatol.* **19**, 347-364.

Jönsson, P. and Bärring, L.: 1994, 'Zonal index variations, 1899-1992, and links to air temperature in southern Scandinavia', *Geogr. Ann.* **76A**, 207-219.
Kuhn, M., Markl, G., Kaser, G., Nickus, U., Obleitner, F. and Schneider, H.: 1985, 'Fluctuations of climate and mass balance: different responses of two adjacent glaciers', *Z. Gletscherkunde u. Glazialgeologie* **21**, 409–416.
Lamb, P.J. and Peppler, R.A.: 1987, 'North Atlantic Oscillation: Concept and application', *Bull. Am. Met. Soc.* **68**, 1218-1225.
Mächel, H., Kapala, A. and Flohn, H.: 1998, 'Behaviour of the centres of action above the Atlantic since 1881. Part I: Characteristics of seasonal and interannual variability', *Int. J. Climatol.* **18**, 1-22.
Malberg, H. and Bökens, G.: 1997, 'Die Winter- und Sommertemperaturen in Berlin seit 1929 und ihr Zusammenhang mit der Nordatlantischen Oszillation (NAO)', *Meteorol. Z. N.F.* **6**, 230-234.
Manley G.: 1974, 'Central England Temperatures: monthly means 1659 to 1973', *Quart. J. Roy. Met. Soc.* **100**, 389-405.
Moberg, A. and Bergström, H.: 1997, 'Homogenization of Swedish temperature data. Part III: The temperature records from Uppsala and Stockholm', *Int. J. Climatol.* **17**, 667-699.
Moses, T., Kiladis, G. N., Diaz, H. F. and Barry, R. G.: 1987, 'Characteristics and frequency of reversals in mean sea level pressure in the North Atlantic sector and their relationship to long-term temperature trends', *J. Climatol.* **7**, 12-30.
Oerlemans, J., Anderson, B., Hubbard, A., Huybrechts, Ph., Jóhannesson, T., Knap, W. H., Schmeits, M., Stroeven, A. P., van de Wal, R. S. W., Wallinga, J. and Zou, Z.: 1998, 'Modelling the response of glaciers to climate warming', *Clim. Dyn.* **14**, 267–274.
Oerlemans, J.: 1994, 'Quantifying global warming from the retreat of glaciers', *Science* **264**, 243–245.
Parker D. E., Legg T. P. and Folland C. K.: 1992, 'A new daily Central England Temperature series, 1772-1991', *Int. J. Climatol.* **12**, 317-42.
Pohjola, V. A. and Rogers, J. C.: 1997; 'Coupling between the atmospheric circulation and extremes of the mass balance of Storglaciären, northern Scandinavia', *Ann. Glaciol.* **24**, 229–233.
Rodríguez, R., et al: (subm.), 'The long pressure series of Barcelona (Spain), daily reconstruction and monthly homogenization', *Int. J. Climatol.* (submitted).
Rogers, J.C.: 1984, 'The association between the North Atlantic Oscillation and the Southern Oscillation in the Northern Hemisphere', *Mon. Wea. Rev.* **112**, 1999-2015.
Rogers, J. C.: 1985, 'Atmospheric circulation changes associated with the warming over the northern North Atlantic in the 1920s', *J. Clim. Appl. Met.* **24**, 1303-1310.
Rogers, J.C.: 1990, 'Patterns of low-frequency monthly sea-level pressure variability (1899-1986) and associated wave cyclone frequencies', *J. Clim.* **3**, 1364-1379.
Schmutz, C. and Wanner, H.: 1998, 'Low frequency variability of atmospheric circulation over Europe between 1785 and 1994', *Erdkunde* **52**, 81-94.
Schönwiese, C.-D. and Rapp, J.: 1997, '*Climate trend atlas of Europe, based on observations 1891-1990*', Kluwer, Dordrecht.
Steig, E. and Barlow, L.: 1998, 'High-resolution proxy climate data from ice cores - an overview', *Abstracts of The Second International Climate and History Conference, Norwich* 1998, 29.
Steinrücke, J.: 1999, 'Changes in the Northern Hemispheric Zonal Circulation and their relationship to precipitation frequencies in Europe', *Bochumer Geographische Arbeiten* **65**, 135 pp.
Wanner, H., Gyalistras, D., Luterbacher, J., Rickli, R., Salvisberg, E. and Schmutz, C.: 2000, '*Klimawandel im Schweizer Alpenraum*', Vdf Hochschulverlag AG, Zürich.
World Climate Disc: 1992, '*Global Climate Change Data*', Climatic Research Unit, University of East Anglia.

(Received 26 March 1999; in revised form 28 March 2000)

A QUANTITATIVE ASSESSMENT OF BUFFERS AMONG TEMPERATURE VARIATIONS, LIVESTOCK, AND THE HUMAN POPULATION OF ICELAND, 1784 TO 1900

DANIEL E. VASEY

Program in Cross-Cultural Studies, Divine Word College, Epworth, Iowa 52045

Abstract. In this study two statistical techniques were used to assess buffers within the farm food chain in late preindustrial Iceland. Both methods arrived at similar quantitative assessments. The chain's tangible links were grass, livestock, and the human population, but it began with temperature, the primary determinant of grass yield and a proxy for grass yield in this study. The buffers were large in spite of the marginal nature of cattle, sheep, and horse farming in the subarctic climate, and the poverty of much of the population. One method of study involved comparison of livestock reductions and human population decrease in seven crises between 1784 and 1888, with attention to temperature data for crises after 1823, when continuous systematic meteorological observations began. The other method is time-series analysis using temperature observations, livestock counts which began in 1817, and several human demographic variables. The discussion focuses on likely buffers, notably alternative sources of winter fodder for livestock and of food for the human population.

Key Words: temperature variation, livestock, human population, Iceland 1784-1900

1. Introduction

The focus of the present study is the assessment of buffers between short-term temperature fluctuations and a population which relied on farming in a difficult climate. During the period under examination, 1784 to 1900, Icelanders, the subject, lived in households which were largely self-sufficient in food. Farming consisted almost entirely of grazing sheep, cattle, and horses. Fishing was an important secondary occupation, a somewhat undeveloped one in 1784, the start of the study period, and a more productive one at its end, in 1900. Gathering, gardening, and grain imports supplemented the food supply.

The goal is a quantitative assessment of buffers. These are described qualitatively in most published accounts. The framework of the study is an ecological one. Buffers are assessed by comparing temperature, livestock counts, and human population size and vital rates. Data on grass yields were unavailable, so that temperature was used as a proxy. Commodity prices, a proxy for food production in many other studies, are not considered here. They are well along the chain from field to consumer and thus would not be a good starting point even if continuous annual series were available for the study period, which they are not. Some studies of historical European populations have dealt with the effects of grain prices and temperature [1], in particular with the more direct effects of temperature, those which act by producing physiological stress

Climatic Change **48**: 243–263, 2001.

or affecting the transmission of infectious diseases. This paper touches on these effects, but they are not a focus.

Discourse on the role of climate in Icelandic history goes back more than a century (see Ogilvie and Jónsson, this volume). It is part of a larger debate concerning the respective roles of the natural environment and culture in shaping population and well-being [2]. Farming was near its northern limit in Iceland. The "Little Ice Age" has often been cited as a period of stress on farming, although it is often unclear whether the blame is being placed on general cooling or a greater frequency of particularly cold years. Few would deny that natural phenomena, including unfavorable weather, frequently resulted in reductions in livestock and subsequent hunger and human misery. Arguments within the debate consider the respective roles of environment and culture in capping production, and to what extent the political economy made the population more or less vulnerable. It has previously been argued that both the environment and the political economy set limits [3]. Regarding vulnerability, the present author has no argument in principle with a critique of climatic determinism made with reference to Iceland:

> The climatic interpretation of history is rather like restating the obvious fact that it was the weather which destroyed the hay without asking the question why no sheds were built. Instead a history of hay destruction during winter is written, which then will be used for weather reconstruction in the past, followed by fine diagrams showing the correlation between hay destruction and stormy and rainy weather during winter [4].

Diagrams that follow, whether or not they are fine, present sundry correlations involving climate. It may appear contradictory to proceed after agreeing with the quotation given above, but acknowledging the likely existence of non-climatic factors does not preclude studying climate's impact. The goal here is to contribute to better accounting of climatic effects, not to impose climate on historiography.

Portrayals of well-being in preindustrial populations range from primitive affluence to chronic dearth that made lives nasty and short, if not necessarily brutish, and a connection has often been suggested between well-being in normal times and vulnerability to climatic fluctuations and other exogenous changes. In the anthropological literature one can find the claim that large reserves of normally little used foods give hunter-gatherers security from famine even in climatically harsh years [5], or, conversely, that preindustrial populations tend to reach recurring states of "demographic saturation," rendering them highly vulnerable to exogenous shocks [6]. Several historical studies of early modern European peasantries find them highly vulnerable [7], but others either downplay the demographic significance of exogenous shocks and resulting famine [8] or emphasize the vulnerability of only disadvantaged classes [9]. No one case study will resolve these issues, but a quantitative estimate of buffers in Iceland should help.

2. Sources

2.1. TEMPERATURE AND GRASS YIELD

Temperature is the main determinant of grass growth or yield in Iceland, for which only qualitative descriptions are available from the study period. Temperature is therefore a proxy for yield, assuming the relationship found in Bergþórsson's comparison of Icelandic temperature observations and hay yields from 1901 to 1975 [10]. Since he included a control for added combined nitrogen, the findings presumably apply to pasture as well. A decrease of one degree in October-April mean temperature results in a 30% decline in hay yields. Bergþórsson added without quantification that May-September temperature has less impact [11].

In the text and tables that follow, October-April is abbreviated as winter, May-September as summer. Winters are dated the year in which they ended. There are advantages besides brevity to dividing the year into two seasons. They approximate to the normal periods of dormancy and growth. The two-season division is common in Icelandic, particularly where farming is concerned, although words do exist for spring and autumn. Other seasons, months, or series of months are distinguished here as needed.

Official temperature observations began at Stykkishólmur, western Iceland, in November, 1845. Here they have been combined with private observations previously taken at Reykjavík, beginning in 1823, which were previously calibrated with the Stykkishólmur observations [12, 13]. Also available was a compilation of sea-ice observations from 1601 to 1850 [14]. An association between sea-ice frequency and cold in Iceland is well known, though it is by no means perfect [15].

Some climatic factors have been wholly or largely ignored here, not to discount them, but because of insufficiently precise information. Precipitation sometimes limits grass growth in Iceland, but mainly in interior sections of the north, not near the stations where the nineteenth century temperature observations used here were taken. Wet weather at haymaking time was a frequent problem and occasional calamity. In winter snow cover protects soil from deep freezing, but it delays the resumption of growth if it lingers too long. Prolonged icing, either from ice storms or from alternate thawing and freezing of snow, may badly damage a sward.

2.2. LIVESTOCK COUNTS

Livestock statistics cited here are from *Hagskinna*, a collection of historical statistics compiled by the Icelandic Statistical Office [16]. They will be referred to as counts. Farmers counted their own herds and declared the numbers to district officials at annual meetings, in early autumn through 1852, and about June thereafter. The change in meeting times did not cause a discernible shift in the counts, since all were taken after the birthing season and before the main culling, which was in autumn before animals were put on hay. Animal numbers were here entered as the year in which they were declared to district officials, presumably the year after the autumn culling which largely determined them.

Hagskinna lists seven counts of cattle, sheep, and horses in the eighteenth century, six between 1801 and 1815, then every year beginning in 1817, except 1860, 1870, and 1887. Counts after 1790 generally included categories of each species, but they are not entirely consistent. For example, yearling cattle, bulls, and steers were lumped through 1852, after which yearlings were counted separately. The consistent categories after 1790 are calves, cows and heifers, ewes with lambs, shearlings, and foals, except that from 1881 to 1888 all sheep were counted together. Any analysis involving ewes or shearlings thus ends in 1880.

Two measures of aggregate livestock were used, based upon grass consumption. *Hay units*, short for hay-consuming livestock units, are based on a 20:1:2.5 ratio of cattle, sheep, and horses [17], *pasture units*, a like shorthand, on a 6:1:2.5 ratio [18]. Of the two measures, pasture units probably better correspond to the supply of food from livestock, and they were used in analysis as the predictor of human population and vital rates, but substitution of hay units does not greatly alter the results. Horses were kept primarily for work and transport. Their meat was not well regarded but it was eaten, particularly in hard times.

2.3. POPULATION DATA

National civil censuses were taken in 1703, 1762 (heads of households), 1769 (head count, without names), 1785 (lost), 1801, and at five-year intervals beginning in 1835. A 1729 civil census covers three counties. Parish censuses were taken at fairly frequent intervals. A collection of them from 1816-18 covers nearly the whole country.

Registration of births, marriages, and deaths began in some places as early as 1668. Surviving registers cover very little of the population before 1750, more than two-thirds by 1785, and nearly all of it by 1819, but there are gaps in places throughout the nineteenth century.

The Icelandic Statistical Office has reconstructed Iceland's population history several times, and the latest version is in *Hagskinna*. The method used was projection of population totals forward and backward from the censuses, using vital rates extrapolated from surviving parish registers and occasional supplementary sources. The accuracy of the resulting estimates of population and vital events is quite good from 1800 on, because of the quality and frequency of censuses and the completeness of surviving parish registers. Before that the potential error of the published data increases further back in time, because extrapolation must be done from increasingly thin parish registration data. The Statistical Office has not attempted to estimate annual population series or vital rates before 1734.

All monthly and seasonal vital rates used here were extracted from the machine-readable files of the Iceland Population Database (IPDB), which ends in 1870 [19].

Only non-infant seasonal mortality series were used, to control for the effects of seasonal variation in births.

The greatest risk of hunger was typically from March through June, at which season mortality and conceptions (December-March fertility) were of particular interest as indicators of hunger. This was the time before new grass revived milk production, when winter food stores had dwindled. The effects may be seen in the crisis years of 1802-05. Population declined in all four years, though slightly in 1805. The crude death rate, which had been 2.1 and 3.0 percent in 1800 and 1801, rose to 3.7, 4.9, 4.1, and 3.3 percent over the next four years. The proportion of deaths in March-June went from 0.31 in 1800 to a peak of 0.50 in 1804, while March-June conceptions fell by nearly half. Similar but more extreme responses occurred during the great famine of 1784-1785 [20].

3. Methods

Two methodological approaches were used here. One was to compare known livestock crises between 1784 and 1900 with temperature observations and human population data. The other was analysis of temperature, livestock, and human demographic time-series. In order to discern trends, reliance must be put on the crisis study, because the time-series cover too short a span to allow appropriate statistical controls and breaking down into shorter periods.

3.1. CRISES

Cold-induced crises are of obvious interest. An extreme crisis with another cause, the one in 1784-85, was also taken to be pertinent where the task is to assess buffers between livestock and humans. Livestock crises are defined as one or more consecutive years of reduction, totaling five percent or more in all three species (sheep, cattle, and horses) and ten percent or more in either sheep or cattle. The earliest such event that is evident in before and after counts was in 1784. Others followed in 1801-03/04, 1811/13-1813/14, 1835-37, 1867-68, 1882-83, and 1886-87/88. The lack of counts in 1803, 1811, 1812, 1814, and 1887 results in ambiguous beginning and ending dates. For example, the 1813 count was down from that in 1810, but, because of the lack of counts in either 1811 or 1812, it cannot be determined from this source alone whether the crisis began in 1811, 1812, or 1813. Because the counts preceded culling, each crisis would have actually begun as early as the autumn before the first year listed and would have ended during the last year. It is essential to base the definition on all three species, as epidemics sometimes caused major losses in one of them, even though the restriction leaves out some periods of possible interest. For example, cattle counts went down 0.111 from 1859 to 1861, for which two quite cold winters, in 1859 and 1860, were a plausible cause, but sheep herds recovered a little over the same period, following an 0.346 reduction from 1857 to 1859.

Human population declines or changes in vital rates are of interest if they occurred along with or immediately following livestock crises. Such declines occurred in 1784-85 [21], 1802-05, 1811-14, 1869, 1882-83, and 1887-88. Other population declines occurred during the study period but were disconnected from livestock reductions.

Runs of two cold winters in 1801-02, 1811-12, 1866-67, 1881-82, and 1885-86, and three in 1835-37, were the reported proximate causes of all the nineteenth century livestock crises. In 1801-02 and 1811-12 sea-ice observations and qualitative testimony indicate cold. The other runs were consecutive years in which the mean winter temperature was more than 0.2° below the 1824-1900 mean, henceforth referred to as the norm. Some runs included cold summers, most notably in 1882, which was 2.3 degrees below normal. As stated above, the primary proximate cause of the 1784-85 crisis was an immense volcanic eruption [22].

Buffers between grass and livestock during the crises were assessed by comparing the predicted decrease in grass yield, using Bergþórsson's formula, with actual livestock reductions. The temperature statistic used was the deviation from normal of the mean of the winters in a run. Substituting the coldest winter of the run for the mean would be preferable if there were no buffers to be depleted, as each winter's effect would be independent of the previous winter. However, substituting this method for the one which was used would always enlarge the projected decreases in grass yield, thereby boosting the estimated buffers, contrary to the method's premise. A sign that consecutive cold winters did deplete buffers is that only small livestock reductions followed single cold winters, in 1825, 1874, and 1892, years bracketed by warm or near normal winters. Three year runs may have had more impact than two year runs, but the single case is no basis for comparison.

Differences in wealth and status would be expected to affect the risk of mortality in hard times. Most people were either members of householders' families or servants. Servants, both men and women, made up 23.0% of the population in 1801 and 25.6% in 1850 [23]. They were especially numerous among young adults. Paupers were possibly a more disadvantaged group. Most were children or elderly persons, especially the latter. Censuses counted 3.0% in 1801, 6.3% in 1816-18 (IPDB), 3.4% in 1840, and 2.1% in 1850 [24].

Study of differential mortality in 1803 and 1804 was done by matching the dead in March-June of both years with their status in the 1801 census. There were 1170 deaths in those months in 1803 and 1804 in the IPDB. After excluding records with missing ages or names, verbal ages, ages under five, and names too common for probable linkage, 321 death records remained, of which 315 were successfully linked with census records. A previous study, which used parish censuses and death registers to break down mortality in the great 1784-85 famine by age, sex, and status [25], was available for comparison.

Only differences within households were studied. Servants generally lived in their employers' households. Although the poor relief laws provided for the support of destitute households *in situ*, civil district officials placed the great majority in households which were reimbursed for their expenses. Although there have been many assertions that servants and paupers suffered unduly in hard times, more claims have been made for differences among households, something which was not studied, since the records thus far entered into the IPDB do not reliably indicate household wealth or status [26].

3.2. TIME-SERIES ANALYSIS

A large number of time-series regressions have been run, in the following categories:

- annual livestock counts on annual or seasonal mean temperatures,
- human population size on annual livestock counts,
- annual or seasonal death rates or birth rates on annual livestock counts, and
- annual, seasonal, or monthly death rates or birth rates on annual, seasonal or monthly mean temperatures.

Breakdowns of death rates by age and sex were tried but were found unpromising. They made little difference in regression coefficients and reduced significance.

The purpose of the time-series analysis was to assess buffers against short-term temperature variation, that stemming from year to year changes in summer and winter temperatures. There was, however, a problem of possible confounding of the effects of temperature over a few years with longer-term trends in livestock and human populations (Figure 1).

A secular upwards trend is evident in pasture units and especially in the human population. In addition, pasture units followed apparent cycles of about 20 years. There is a possible association between these cycles and winter temperature (Figure 2), but the period of observation is too short to rigorously test the relationship.

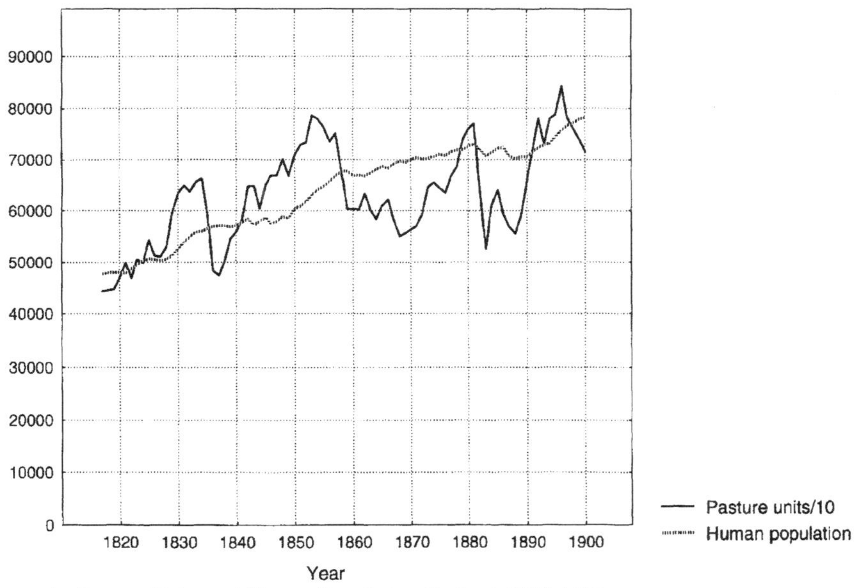

Figure 1. Pasture Units and Human Population, Iceland, 1817-1900.

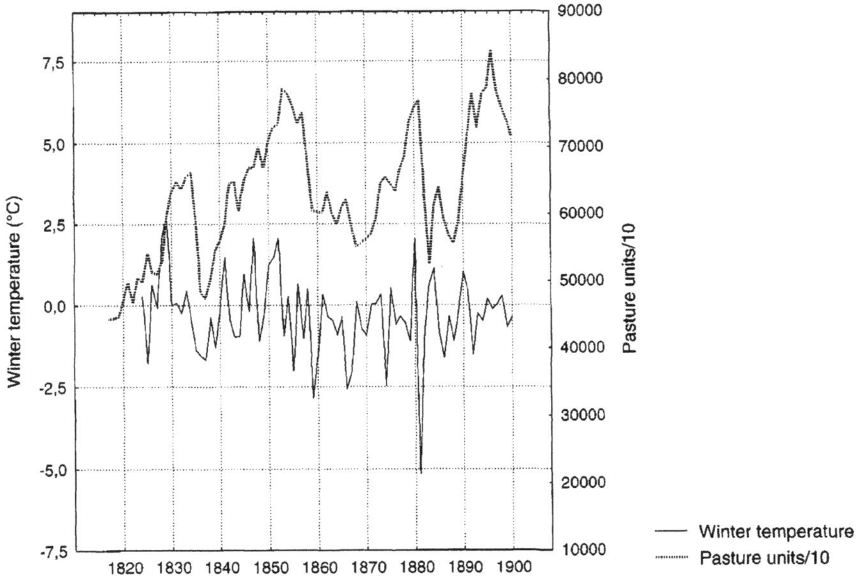

Figure 2. Pasture Units, 1817-1900, and Winter Mean Temperatures, 1824-1900.

Two different methods were used to control for the effects of secular trends and long-term cyclical changes. Each method used moving averages of the eleven years of which each observation is the mid-points. In regressions of human population size or vital rates on pasture units, each observed value was divided by the moving average [27]. Each regression coefficient is the ratio of predicted demographic change to change in pasture units. Because the detrending method uses the ratio of the moving average to the series mean, it is essential that the variables are on ratio scales. Adjustments made by this method are sensitive to the size of observations in the series in relation to variation among them. The same method could not be used to detrend the temperature series. If those observations are expressed on a ratio (Kelvin) scale, variation in the moving average is trivial. In regressions of livestock counts on temperatures, each observation was reduced by the positive deviation of the moving average from the series mean or increased by the negative deviation. Adjustments in the variable are thus in proportion to absolute temperature deviations, and each regression coefficient is the predicted change in livestock numbers (expressed in the results below as a proportion of the series mean) for each degree change in temperature.

Corrections were made for first-order autoregressive disturbance [28]. Linearly interpolated values were substituted for the missing years 1860, 1870, and 1886 in the livestock series, as missing values could not be included if corrections were to be made for autoregressive disturbance.

Lags up to eight years were initially tried in regressions. Those which indicated slight and inconsistent effects were dropped and are not reported below.

4. Results

4.1. CRISES

Table I shows actual livestock reductions and the decreases in grass yield projected from observed temperature deviations in runs of cold winters. In each case the former are a fraction of the latter. For the sake of simplicity and clarity the table displays only the ratio of reduction of hay units to projected decrease in grass yield ($\Delta H/\Delta G$), but the reduction in pasture units is not very different from the reduction in hay units.

TABLE I

October-April Temperature Deviations and Predicted and Actual Livestock Reductions in Three Crises.

Period	Mean Temperature Deviation (°)	Predicted Grass Reduction (ΔG)	Actual Livestock Reduction Period	Pasture Units	Hay Units (ΔH)	$\Delta H/\Delta G$
1835-37	-1.54	0.423	1835-37	0. 285	0. 277	0.65
1866-67	-2.00	0.510	1867-68	0. 114	0. 123	0.24
1881-82	-2.93	0.648	1882-83	0. 317	0..288	0.44
1885-86	-0.98	0.294	1886-87/88	0. 132	0. 135	0.46

Table II displays the impacts of the livestock crises on human population size. The ratio of Icelandic population decrease to livestock reduction ($\Delta I/\Delta P$) is at most just over a fourth. Lest the ratio of pasture units per Icelander (P/I) on the eve of the crisis give a misleading impression of abundance, owing to the multipliers used in calculating pasture units, a sample breakdown might be helpful. Ratios of specific livestock to people in 1834, the highest P/I value in the table, were 0.35 cows, 0.07 other cattle, 3.43 ewes with lambs, 1.47 shearlings, 1.55 other sheep older than one year, and 0.49 horses.

TABLE II

Livestock Reductions and Human Population Declines in Major Crises, 1783-1888.

		Change in Livestock				Human Pop. Change			
Years	P/I*	Cattle	Sheep	Horses	Pasture Units	Years	Size	Natural Increase	$\Delta I/\Delta P$†
1784	9.0	-0.511	-0.832	-0.814	-0.709	1784-85	-0.189	- 0.189	0.267
1801-03/04	11.0	-0.141	-0.562	-0.185	-0.429	1802-05	-0.034	-0.034	0.079
1811/13-13/14	8.7	-0.061	-0.115	-0.066	-0.091	1811-14	-0.022	-0.022	0.242
1835-37	11.8	-0.263	-0.322	-0.172	-0.285	NONE			0.000
1867-68	9.1	-0.142	-0.107	-0.103	-0.114	1869	-0.003	-0.003	0.263
1882-83	10.5	-0.218	-0.357	-0.248	-0.317	1882-83	-0.034	-0.013	0.107
1885-87/88	8.8	-0.141	-0.123	-0.164	-0.132	1887-88	-0.032	NONE	0.242

* Pasture units per Icelander, one year before crisis.

† (Icelandic) population change / pasture units change.

The study of differential mortality in March-June, 1803 and 1804, found significantly higher mortality among paupers in all adult age intervals, but among servants only in the oldest interval, 68-87. Table III displays the results.

TABLE III

Frequencies of Servants and Paupers, 1801 Census and the Dead, March-June, 1803 and 1804 (age intervals two and three years older, respectively, than in 1801 census).

Age in 1801	Number in Census	# of Linkages	Servants Census	Servants Dead	difference p(Z)<	Paupers Census	Paupers Dead	difference p(Z)<
3-17	14317	54	0			.055	.074	0.28
18-42	17355	67	.321	.284	0.27	.005	.090	0.01
43-67	9471	99	.197	.222	0.27	.021	.071	0.01
68-87	2883	95	.001	.074	0.01	.115	.179	0.03

4.2. TIME-SERIES ANALYSIS

Selection was necessary among the results of the large number of regressions. Details of results summarized in the text are available from the author.

The results of regressions of the three livestock species and of hay and pasture units on winter and summer mean temperatures are in Table IV. Winter mean temperature was weakly correlated with the other independent variable used, that is, means in the preceding ($r = 0.27$) or following ($r = 0.26$) summers. Substituting mean temperatures of other seasons -- February-April, March-June, and July-September -- in the analysis or including them along with winter and summer means led to less significant associations.

TABLE IV

Coefficients of Regressions of Animal Numbers on Winter and Summer Mean Temperatures, 1823-1900.

Temperature Period	Cattle	Sheep	Horses	Pasture Units (6:1:2.5)	Hay Units (20:1:2.5)
Winter	0.0035	0.0026	0.0002	0.0023	0.0024
Winter, 1 yr lag	0.0228	0.0261	0.0119	0.0233	0.0240
Winter, 2 yr lag	0.0140	0.0206	0.0091	0.0175	0.0180
Winter Lag Sum	0.0403	0.0493	0.0212	0.0431	0.0444
Summer, 1 yr lag	0.0105	0.0176	0.0082	0.0147	0.0151
One-Tail Significance, p(t)					
Winter	0.211	0.339	0.472	0.316	0.306
Winter, 1 yr lag	0.000	0.001	0.002	0.001	0.000
Winter, 2 yr lag	0.001	0.001	0.003	0.002	0.000
Summer, 1 yr lag	0.050	0.025	0.038	0.014	0.144
p(F)	0.000	0.000	0.000	0.000	0.000
R^2	0.625	0.604	0.601	0.625	0.625
ρ	0.403	0.304	0.403	0.383	0.383

Among livestock categories, those whose counts were most significantly associated with temperature variation were calves, shearlings, and foals, indicating preferential culling of young animals. When regressions were run on the same independent variables and with the same lags as in Table IV, the largest predicted cumulative effect (lag sum) from a one degree change in mean winter temperature, 0.107, was in shearlings (1817-1880), the least, 0.019, in cows (1817-1900).

Table V reports the results of regressions of some human demographic variables on pasture units. At the lags the large swings between positive and negative regression coefficients and the mostly low significance are reason to regard as undemonstrated any carryover effect from livestock reductions into subsequent years. When the regressions were run only on pasture units in the same year, the coefficients were 0.051 for population size, -0.552 for annual crude death rate, -1.799 for March-June non-infant death rate, 0.405 for annual crude birth rate, and 0.818 for December-April birth rate. Significance (one-tail p(t)) was 0.01 or better in each of these regressions, except that it was < 0.12 when March-June non-infant death rate was the dependent variable.

TABLE V

Coefficients of Regressions (Means of All Variables = 1.00) of Human Population Size and Vital Rates on Pasture Units, 1817-1900.

	Population	Crude Death Rate		Crude Birth Rate	
Lag	Size	Annual	Mar-Jun*	Annual	Dec-Apr*
Same Year	0.0533	-1.1046	-1.6354	0.3166	0.5571
1	0.0107	1.2632	0.3694	0.2104	-0.0053
2	0.0342	-0.9587	-1.4667	-0.1443	0.9839
3	0.0006	0.9019	1.0580	0.0923	-0.2187
Lag Sum	0.1028	0.1018	-1.6747	0.3827	1.3170
One-Tail Signifance, p(t) <					
Same Year	0.004	0.046	0.042	0.004	0.132
1	0.294	0.076	0.378	0.063	0.497
2	0.044	0.134	0.116	0.145	0.079
3	0.486	0.075	0.120	0.209	0.327
p(F) <	0.001	0.364	0.051	0.001	0.039
R^2	0.479	0.080	0.268	0.371	0.282
ρ	0.101	- 0.001	0.082	0.039	0.039

* 1817-1870.

Results from regressions of these same demographic variables on temperature were inconclusive, except that the coefficient of regression of annual crude death rate on annual mean temperature was 0.505 (one-tail p(t) < 0.084). Additional analysis involving monthly non-infant death rates and temperature indicated several significant associations. Most of the year, and especially from April to June, elevated mortality was associated with below-normal means, but in August and September it was associated with above-normal means.

5. Discussion

The results from the crisis study and the time-series analysis were consistent. They showed effective buffers, the most so in minor setbacks, but appreciable even under catastrophic shocks. The crisis study provided some signs that buffers increased over the course of the nineteenth century.

5.1. TEMPERATURE, PROJECTED GRASS YIELD, AND LIVESTOCK COUNTS

The results of the time-series analysis corroborate three of Bergþórsson's points [29]. The effect of winter temperature on livestock counts is more than twice that of summer temperature, in line with his nonquantified comparison in the paper cited. An October-April and May-September division of the year gave more significant results than did analysis involving other seasons. Finally, hay yield was the main determinant of herd size, since the main effect of either winter or summer temperature variation was at a one year lag; ergo, insufficient hay mandated severe culling that autumn.

A further comparison suggests an explanation for the difference in apparent buffers among the last four crises, those for which temperature data are available. The higher the ratio of livestock counts to human population, the more vulnerable the herds were to reduction during the crises. The effect was found in all classes of livestock, but is probably best illustrated using hay units, since hay yield determined culling. One variable is the ratio of hay units to humans the year before the crisis, the other an inverse indicator of buffers, i.e., the ratio of reduction in hay units to the projected decrease in grass yield ($\Delta H/\Delta G$ in table I). The correlation between them is fairly high ($r = 0.67$). Thus, the greater the prior ratio of livestock to humans, the closer the livestock reduction during the crisis came to the projected decrease in grass yield.

In at least one important case, the 1801-1803/04 livestock crisis, another crucial climatic factor is known. The winters of 1801 and 1802 were cold, but probably not extreme. Sea ice was fairly frequent in both years, yet not as much as in 1807, 1808, 1811, and 1812 [30]. Contemporary accounts indict very heavy snowfall during the winter of 1802, especially in the north and northwest. In many places snow refroze after partial thaws, producing icing, and lingered into summer [31].

Time-series analysis found greater differences between actual livestock reductions and those predicted from temperature variation than did the crisis study. Put simply, farmers more readily compensated for small reductions in grass yield than the large ones of the crises [32].

That the association between livestock numbers and mean temperatures, either winter or summer, was insignificant and inconsistent at long lags suggests fairly rapid recovery of herds. The preference for culling young animals, rather than breeding stock, is a good explanation. Recovery could take a little longer after livestock crises, when there were some reductions in breeding stock. Livestock counts reached pre-crisis levels four years after the 1835-37 crisis and three years after the less severe 1867-68 crisis. After the shocks of 1882-83 and 1886-87/88 they took until 1892.

5.2. LIVESTOCK AND THE HUMAN POPULATION

For the most part the findings on livestock reduction and human demographic impact parallel those on temperature and livestock, in that time-series analysis mostly found greater buffers than did the crisis study. The difference in findings is not completely consistent, however, especially since the impact of the 1835-37 livestock crisis on human population size was nil. An explanation for the variation is found by again taking into account the adequacy of herds on the eve of the crises. Figure 3 is a scattergram comparing two statistics. The independent variable is the ratio of pasture units to Icelanders in the last year before the livestock crisis (P/I in Table II), the dependent variable the ratio of Icelandic population decrease to pasture unit reduction ($\Delta I/\Delta P$ in Table II).

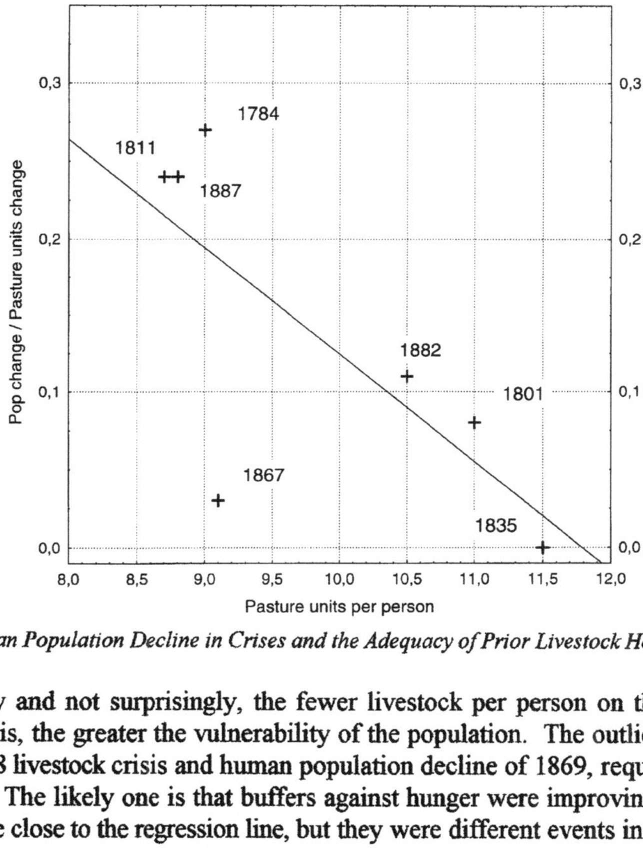

Figure 3. Human Population Decline in Crises and the Adequacy of Prior Livestock Holdings.

Evidently and not surprisingly, the fewer livestock per person on the eve of a livestock crisis, the greater the vulnerability of the population. The outlier in Figure 3, the 1867-68 livestock crisis and human population decline of 1869, requires further explanation. The likely one is that buffers against hunger were improving. The two later crises are close to the regression line, but they were different events in that excess mortality was not the main cause of human population decrease. Iceland's first large-scale emigration began in the 1880s. It was 60% as large as population decline in 1882-83 and exceeded it in 1887-88. Had the emigrants stayed and mostly survived, the 1880s population crises would join that of 1869 near the bottom of Figure 3.

The extent of hunger-related mortality is unclear after 1814. Starvation was a listed cause of deaths in 1784-85, 1803-04, and 1811-14, afterwards disappearing from registers [33], yet anecdotal reports continued, and in 1882 hunger prompted an international relief effort. Certainly some people continued to experience undernutrition. Numerous accounts of badly impoverished households continued throughout the nineteenth century. So did seasonal hunger. *Independent People*, Halldór Laxness' novel of rural life in early twentieth century Iceland, reflects conventional wisdom of the time. People could "feel reasonably sure of surviving the winter" if they maintained good credit and "restricted themselves to the customary one meal a day in the spring" [34]. Synergy between hunger and disease may have elevated mortality even where outright starvation was absent. It is a complex and controversial topic where many diseases are concerned, but undernutrition is well known to increase tuberculosis morbidity and mortality and is implicated in pneumonia mortality as well. In 1911-15, when official statistics on causes of death began, tuberculosis accounted for 21% of deaths, pneumonia and influenza 13% [35].

Time-series analysis of livestock counts and death rates indicated lingering hunger-related mortality. There are some grounds for doubt but also plausible rejoinders. Livestock reductions were significantly associated with elevated mortality in the same year, especially in the March-June lean season, which is what would be expected if hunger mortality were present, since the culling would have occurred months before. A possible explanation why the registers did not report starvation is that it may have occurred in socially marginal individuals, whose deaths tended to prompt minimal entries in parish registers. Another possibility is that already sick individuals, particularly those with tuberculosis, succumbed from hunger which was widely experienced but not lethal in most of the population.

The case for continuing hunger-related mortality would be stronger if analysis involving birth rates gave significant and parallel results, but they were only sometimes significant and on the whole not parallel. Pasture units were most significantly associated with December-March birth rates at zero lag, yet the women who accounted for those births would already have been pregnant during the autumn culling, which typically made ample food available for a time. Undernutrition is known to lower fecundity, as it did in several Icelandic famines before 1815; controversy is mostly over how severe it needs to be to have an effect. Unless extreme, it does not substantially affect the likelihood of existing pregnancies going to term. If March-June hunger affected fecundity, the most significant associations of annual or December-March crude birth rates with pasture units should have been at a one year lag, but they were not.

The inability to demonstrate an effect of livestock numbers on fertility does not disallow hunger-related mortality. Hunger may have exacerbated mortality without being severe enough to suppress fecundity, or the particularly vulnerable parts of the population which suffered detectable hunger-related mortality may not have been the fertile ones. In 1820 nearly all births were to householders, and household formation was restricted mostly to couples with rights to a farm or one of a tightly controlled number of cottages. Curbs broke down over the century, as households increased in towns and fishing villages, more servants married, and illegitimate births increased, but the change was gradual. Direct effects best explain the results from the analysis of temperature and mortality. If food chain effects were at work cold should have mostly increased deaths at a one year lag, but the main effect was in the same year, indeed the same month [36].

5.3. LIKELY BUFFERS

A buffer common to humans and their livestock lies in the physiology of large mammals, who survive for some time without food by metabolizing body fat and muscle mass. Other buffers must be sought in specific strategies of management of Icelandic farms and households. One was simply to defer the renewal of herds by culling mostly young animals. The farmer who accurately anticipates winter hay requirements is a figure of Icelandic folklore and literature, and that of Scandinavia more generally, as is the improvident one who underestimates them. It was no easy task, since one determinant, how soon new grass would allow grazing in the spring, was unknown.

One of the more pertinent stories, though it comes not fom Iceland but from the 17th century Faroe Islands, concerns a farmer who wants to keep a yearling ram over the winter and an experienced hired hand who thinks it should be culled to ensure enough hay is available for the farm's cows. The hand disobeys orders, secretly killing and eating the yearling. The spring is cold, delaying grass growth, and the farmer resigns himself to killing the cows. To the hand he says the hay consumed by the ram would not have made a difference. Oli, the hand, replies, asking the farmer:

> ...to come out to the shed and look at the yearling ram. When the door is opened, he tells the farmer that now he will tell him the truth: he killed the yearling ram in the fall, but he had kept in the shed the hay that was intended for it; he now asks him to tell him if it made no difference. This hay was so much, that it saved all the cows of Lamba that year [37].

When yields were down in their usual hay fields, farmers had recourse to alternative sources of winter fodder, but only by assuming increased costs or risks. Farmers could cut grass over a wider area, which in an often rugged and stony landscape often involved cutting thin or short stands and hauling the hay long distances. Poor growth over the summer often meant there was one less crop of hay, so that the work of scything, raking, and hauling had to be crowded into a shorter time span than in a good year. The added costs were a liability as well as a burden. Under the best conditions Icelandic haymaking was frenetic. Long hours were spent to make enough to get animals through the long winter, often in poor condition by its end. Seaweed was an alternative to hay, but one that came with added risks if used too much. It was in some places dried and stored, even hauled well inland, operations which could be carried out after the haymaking season. Sheep were turned onto beaches during the winter to graze it. Both practices were reportedly expanded when fodder was scant, so that each constituted a buffer, but within limits imposed by the poor nutritional quality of seaweed. If ewes are fed too much of it, many of the lambs born to them in the spring die of what Icelandic farmers call *þaraskjögur* [seaweed stagger] [38]. Poor judgment or inexperience could carry a heavy penalty. As one farmer recalled: "...because his father was then an inexperienced farmer who let his ewes eat too much seaweed -- seventy of Svalbarð's two hundred newborn lambs died of *þaraskjögur*" [39].

One alternative, imported grain, was added over the course of the nineteenth century, although nothing suggests its use became common. As early as 1836, during a livestock crisis, a sheriff in northern Iceland advocated with some success its use in emergencies, to get animals through the winter when hay was short [40]. The end of the Napoleonic wars, improving trade terms, falling shipping costs, and increasing exports of wool and fish had made grain, previously a luxury even for human consumption, more affordable.

Buffers between livestock and humans included non-livestock sources of human food, recourse to normally eschewed foods, food stores, and cash or barter reserves which could be used to purchase imported foods. Both potato cultivation and imports of food slowly expanded over the course of the century. They played important roles feeding a growing population after 1820. In many locations the potato crop is highly vulnerable to cold summers. The cold summer of 1882 followed the poor hay crop of 1881, adding to hardship, but potato crop failures should not as a rule have occurred the same year as low livestock counts, which resulted mainly from cold winters ending the previous year. Imports became more affordable over the nineteenth century. Improvements in farming, resulting in increased hay yields, mostly waited until the closing years of the century. Some controversy has therefore surrounded the upward trend in livestock numbers over the century, its relation to climatic change, reoccupation of highland pastures, and favorable prices for sheep products [41]. It should be noted, however, that only in the 1890s, when improvements in hay yields were beginning to be felt, did pasture units or hay units exceed levels at the beginning of the eighteenth century.

Average food intake was sufficient in good times, at least in energy content. In one reconstruction of national food supply, adult "male equivalents" averaged somewhat less than 3000 kcal/day at various times between 1770 and the closing years of the nineteenth century, when a decidedly upward trend began [42]. In a physically active population of fairly large stature these intake levels, a few hundred kilocalories above modern reference standards, would not have supported much excess body fat, and judging from contemporary accounts and photographs they did not, but neither did most people enter hard times undernourished. To the degree that food from sources other than livestock was normally consumed, the effect would have been to dilute the impact of a livestock reduction. Dependence on food from livestock declined during the nineteenth century, so the dilution should have become greater. Food energy from livestock sources was about 78% of total food intake in 1819, 64% in 1855, and 57% in 1880 [43]. The buffers found in this study exceed what would be expected from dilution alone, suggesting a considerable capacity to expand non-livestock sources of food when milk and meat fell short. How much this entailed more purchases of imported foods and how much expanded fishing and gathering is beyond the scope of this paper.

Sharing of resources helped buffer the population against hunger. Sharing within households was effective until conditions became extreme. The poor relief system worked up to a point, having its intended buffering effect. Adult paupers died disproportionately, but people often became paupers because of ill health, which may in turn have increased their vulnerability. In this vein it is interesting that paupers aged 3-17, who would most likely have been on the poor rolls because of their parents' ill health, not their own, died in 1803 and 1804 at a rate that was not significantly disproportionate to their representation in the 1801 census. The 1784-85 famine was deadly to adult paupers, yet paupers up to 19 years old died at a higher rate than householders' children only in the hardest-hit parishes [44].

These findings should not be interpreted to mean there were no social differentials in the risk of hunger-related mortality, however. This study did not test the conventional wisdom, supported by much Icelandic historiography, that hunger-related mortality mostly fell on the poorer households. One piece of evidence is especially compelling. A Danish physician, Peter A. Schleisner, described the average daily diets in seven households around Iceland in 1847, during a favorable period in which livestock to person ratios were fairly high. From that account energy intake per male equivalent has been calculated, ranging from 1437 to 4370 kilocalories [45]. Calculation involved some extrapolation, while some foods, such as fresh fish and fish liver oil, may have been omitted, but the demonstration of wide differences among households is probably valid.

6. Conclusions

Buffers are not a novel finding, and they are an old topic of Icelandic agricultural and economic history, if not always discussed in this term. A good rhetorical question is what quantification adds, to which the answer is, in laconic terms, a lot. It illustrates in certain terms how effective buffers were, throughout the study period and especially towards its end. The Icelanders were acutely aware of weather and frequently suffered its manifestations, but they were not altogether at its mercy.

The well-being of the Icelanders did not fit well the opposed portrayals of preindustrial populations, either primitive affluence or demographic saturation. Foreign visitors commented on the widespread poverty, yet even the poor must have enjoyed some protection, or the effects of bad years would have been much worse.

It is difficult to make comparisons with studies of other historical populations, in Europe or elsewhere, since quantitative studies of the sensitivity of populations to changes in food supply have not generally traced the effects of weather or other exogenous factors through the food chain to human consumers. For future reference let it be noted that the Icelanders likely had both disadvantages and advantages in comparison with most European populations. They were singularly impoverished in the view of many compatriots and visitors from Europe, even as late as the middle decades of the nineteenth century, when real incomes had improved in much of northern and western Europe. An advantage of the Icelanders was the ready access of many households to fisheries, another their position well along the farm food chain. Most European farmers practiced mixed farming. Poor households typically had few livestock and relied heavily on grain for subsistence, rent payments, and cash sales. If grain yields fell by some sizeable fraction they were quite likely worse off than Icelandic farmers were when grass yields fell by as much.

The findings of this study apply to the effects of short-term variation in temperature, up to a few years, not longer-term variation as in some other studies [46]. These findings underline the significance of runs of cold winters, which could have significance for the interpretation of the human dimesnsions of the "Little Ice Age" in Iceland. They do not have any direct bearing on the consequences of variation in mean temperatures over long periods, if that occurred.

Acknowledgments

The Iceland Population Database was made possible by grants BNS-8820264 and BNS-8820264 from the (U.S.) National Science Foundation, support in data entry from the University of Iceland, and an exchange of data with the Genealogical Society of Utah. Additional support for work by the author in Iceland was provided by a grant from the Fulbright Commission of Iceland in summer, 1999, and the hospitality of the host institution, the Stefansson Arctic Institute, its Director, Níels Einarsson, and its Senior Scientist and my long time colleague Jón Haukur Ingimundarsonn. Travel at other times was supported by Divine Word College and by NSF through the North Atlantic Biocultural Organization (NABO) cooperative. The author thanks his Icelandic Database Project workers Rhonda Gotto, Linda Mathewson, and Greer Vasey. Trausti Jónsson supplied invaluable data, and he and Astrid Ogilvie provided helpful comments and references to literature. Guidance and assistance regarding data sources and their quality came from Guðmundur Jónsson, Magnús Magnússon, Ólafur Ásgeirsson, Stefán F. Hjartarson, and Unnur Dís Skaptadóttir. Patrick Galloway and Eugene Hammel sparked and aided my use of time-series analysis. Participants at the 1999 NABO meeting in Akureyri offered helpful comments on a preliminary version of this paper, as did the anonymous reviewers of the submitted version. All these individuals are responsible for what I got right and none of them for what I got wrong.

References and Notes

1. Lee, R. D.: 1981, 'Short-Term Variation: Vital Rates, Prices, and Weather', in *The Population History of England 1541-1871: A Reconstruction*, E.A. Wrigley and R. Schofield, (eds), Harvard University Press, Cambridge, Massachusetts; Eckstein, Z., T. P. Schultz, K. I. Wolpin, 1985: 'Short-Run Fluctuations in Fertility and Mortality in Pre-Industrial Sweden.', *Eur. Econ. Rev.* **26**, 295-317; Galloway, P..: 1988, 'Basic Patterns in Annual Variations in Fertility, Nuptiality, Mortality, and Prices in Pre-Industrial Europe', *Pop. Stud.* **42**, 275-303; Galloway, P.: 1994, 'Secular Changes in the Short-Term Preventive, Positive, and Temperature Checks to Population Growth in Europe, 1460 to 1909', *Clim. Change* **26**, 3-63.
2. Gunnarsson, G.: 1980, *A Study of Causal Relations in Climate and History, with an Emphasis on the Icelandic Experience*, Meddelande från Ekonomisk Historiska Institutioner, Lunds Universitet No. 17; Gunnarsson, G.: 1983, *Monopoly Trade and Economic Stagnation: Studies in Foreign Trade of Iceland 1602-1787*, Skrifter Utgivna av Ekonomisk-Historiska Föreningen i Lund No. 38; Vasey, D.E.: 1996, 'Icelandic Population Regimes: Premodern, Modern, and Modern Constructions Thereof', in *Images of Iceland*, E. P. Durrenberger and Gísli Pálsson (eds), University of Iowa Press, Iowa City, Iowa.
3. Vasey, D.E.,: 1996, 'Population Regulation, Ecology, and Political Economy in Preindustrial Iceland', *Am. Ethnol.* **23(2)**, 366-392.
4. Gunnarsson: 1980, op. cit, p. 5,. note 2.
5. Lee, R.B.:1968, 'What Hunters Do for a Living, or, How to Make Out on Scarce Resources', in *Man the Hunter*, R.B. Lee and I. DeVore (eds), Aldine, Chicago, Illinois.
6. Wood, J.W.: 1998, 'A Theory of Preindustrial Population Dynamics: Demography, Economy, and Well-Being in Malthusian Systems', *Curr. Anthro.* **39(1)**, 99-135.
7. Particularly good examples include McKeown, T.: 1976, *The Modern Rise of Population*, Arnold, London, and Ladurie, E.L.: 1977, *The Peasants of Languedoc*, J. Day, trans., University of Illinois Press, Champaign, Illinois.
8. Livi-Bacci, L.: 1991, *Population and Nutrition: An Essay on European Demographic History*, Cambridge University Press, Cambridge, United Kingdom.
9. Summarized in Galloway: 1988, op. cit., and Galloway: 1994, op. cit.
10. Bergthorsson[Bergþórsson], P.: 1985, 'Sensitivity of Icelandic Agriculture to Climatic Variations', in *The Sensitivity of Natural Ecosystems and Agriculture to Climatic Change*, *Clim. Change*, special issue, M.L. Parry (ed.).
11. A partly differing conclusion based on contemporary qualitative testimony on cold and yields of grass was that spring temperature is generally the best predictor of grass yield (Ogilvie, A. E. J.: 1984, 'The Impact of Climate on Grass Growth and Hay Yield in Iceland: A.D. 1601 to 1780', in *Climatic Changes on a Yearly to Millennial Basis*, N.-A. Mörner, W. Karlén, (eds), D. Reidel, Norwell, Massachusetts).
12. Diskette supplied by Trausti Jónsson, Icelandic Meteorological Office.
13. Both Reykjavík and Stykkishólmur are along the west coast and best represent temperature in the south and west of Iceland, where about 60% of the population dwelled during the study period. The rest lived to the north and east of the central highlands. The two regions frequently, but not always, share unusually warm or cold seasons. The published livestock counts used in this study do not differentiate them, and published temperature series for the north and east began later in the study period, not permitting a sufficient length of time for separate analysis. Regressions of human vital rates on temperature series were tried breaking down the dependent variable by climatic region. Little difference was found in the results.
14. Ogilvie, A. E. J.: 1996, 'Sea-Ice Conditions Off the Coasts of Iceland A.D. 1601-1850 with Special Reference to Part of the Maunder Minimum Period (1675-1715)', in *North European Climate Data in the Latter Part of the Maunder Minimum Period A.D. 1675-1715*, E. S. Pedersen (ed.), Arkeologisk Museum I Stavanger, Stavanger, Norway.

15. Some other proxies were available or reliable only in decadal series, e.g., Ogilvie's ordinal level reconstructions of temperature from contemporary descriptions (14). Good correlations were found between Stykkishólmur observations and either various northern European means or southern Greenland ice-core oxygen-18 readings supplied by Paul Mayewski, but only when decadal series, not annual or seasonal ones, were used.
16. Hagstofa Íslands: 1997, *Hagskinna: sögulegar hagtölur um Ísland*, G. Jónsson and M. Magnússon (eds), Hagstofa Íslands, Reykjavík, Iceland.
17. Bergthorsson: 1985, op. cit. note 5.
18. Fridriksson, S.: 1972, 'Grass and Grass Utilization in Iceland', *Ecology* **53**, 785-797.
19. Iceland Population Database files and manual are available from the author or by file transfer protocol from the Divine Word College Web site, at http://www.dwci.edu. Births were entered in the IPDB by the Genealogical Society of Utah, deaths and census records by workers of the Iceland Database Project from microfilm copies of the originals provided by the Genealogical Society. Comparisons of the IPDB with official statistics, where available, have indicated a less than one percent error in data entry in censuses and death records, discrepancies being attributable to differences in the reporting of stillbirths, missing registers, or gaps in registers. The IPDB ends in 1870, by agreement with the Genealogical Society.
20. Vasey, D.E.: 1991, 'Population, Agriculture, and Famine: Iceland, 1784-85', *Hum. Ecology* **19**, 1-28.
21. There was also a decline in 1786, but a mild smallpox epidemic was the cause and it followed a decided improvement in mortality during the second half of 1785.
22. Cold may have contributed to the 1784-85 crisis. A detailed contemporary account (Finnsson, H.: 1796, *Um mannfækun af hallærum á Íslandi*. Reykjavík: Rit fless Konunglega Íslenzka Lærdomslista-Félags 14:30-226) indicted it. Sea ice was heavy in 1782, 1783, and 1784, and the eruption itself may have cooled much of the northern hemisphere in the winter of 1784. Nonetheless, the distribution of excess mortality largely reflected proximity to and direction from the eruption. Extreme northern, coastal locations which should have been worst affected by sea ice but were distant from the eruption experienced the least mortality (Vasey: 1991, op. cit. note 15).
23. Guttormsson, L.: 1983, *Bernska,ungdómur og uppeldi á einveldisöld: tilraun til félagslegrar og lýðfræðilegrargreiningar*, Sagnfræðistofnun Háskola Íslands, Reykjavík, Iceland, Table 19, p. 120, Table 20, p. 122
24. Gunnlaugsson, G. Á.: 1982, *Family and Household in Iceland 1801-1930*, Almqvist-Wiksell, Sweden, 1982, p. 102.
25. Vasey: 1991, op. cit. note 20.
26. Censuses often identify main farms, dependent farms, and cottages, but two obstacles encountered in using this information in a test of social differentials among households were turnover of householders, most of whom were tenants, and the discovery that some cottages or dependent farms acquired larger herds than did many main farms.
27. Lee: 1981, op. cit. note 1.
28. Hall, R. E., J. Johnston, D. M. Lillien: 1990, *MicroTSP User's Manual: Version 7.0*, Quantitative Micro Software, Irvine, California. Without correction for first-order autoregressive disturbance, serial correlation was often problematical, the Durbin-Watson statistic sometinmes below 1.00.
29 Bergthorsson: 1985, op. cit. note 5.
30. Ogilvie: 1996, op. cit. note 9.
31. Thoroddsen, Þ.: 1916-17, *Árferði á Íslandi í þúsund ar*, Hið Íslenzka Fræðafjelagi, København, pp. 199-202.
32. The method of regression makes a considerable difference. Bergthorsson (1985, op. cit. note 5) arrived at predicted effects of winter temperature variation on livestock greater than those obtained here. His study period, 1846 to 1900, differed in part, but differing statistical methods are the probable main reason. He performed a linear regression of hay units, "ewe-values" in his terminology, on weighted annual temperature means. He found that the maximum effect was at a one year lag, a predicted 0.087 reduction per degree decrease. For the best comparison with his results,

I ran a time-series regression of hay units on 1846-1900 annual mean temperatures. By this method the maximum effect was also at a one year lag, but the predicted reduction was 0.032. An even greater difference was found in predicted cumulative effects, an 0.29 decrease in livestock from a one degree temperature decrease in his version, less than an 0.07 decrease in mine. He estimated cumulative effects by deriving a weight of one at a one year lag and a weight in each year thereafter 5/7 as much as the previous year. I tried lags up to eight years, and found that the lag sum increases to 6.5% with lags zero to three included, then oscillates slightly, always below seven percent as more lags are added.

33. Steffensen, J.: 1975, *Menning og meinsemdir*, Sögufélag, Reykjavík, Iceland, p. 396.
34. Laxness, H.: 1997, *Independent People*, tr. by J. A. Thompson. Vintage Books, New York, New York, p. 17.
35. Hagskinna: 1997, op. cit. note 11.
36. Other studies have found an association between higher late summer temperature and elevated mortality (Galloway: 1984, op. cit. note 1; Sakamoto-Momiyama, M.: 1977, *Seasonality in Human Mortality*, University of Tokyo Press, Tokyo), even at high latitudes (Turpeinen, O.: 1978, 'Infectious Diseases and Regional Differences in Finnish Death Rates, 1749-1773', *Pop. Stud.* **32(3)**, 523-533; Lithell, U.-B.: 1981, ' Breast-Feeding Habits and Their Relation to Infant Mortality and Marital Fertility', *J. Fam. Hist.* **6**, 182-194; Brändström, A.: 1984, *De Kärlekslösa Mödrarna: Spädbarnsdödligheten i Sverige under 1800-Talet med Särskild Hänsyn till Nedertorneå*, Umeå University and Almqvist & Wiksell International, Stockholm, Sweden). Several causal connections have been suggested, none of them involving nutritional sufficiency.
37. Wylie, J.: 1987, *The Faroe Islands: Interpretations of History*, University of Kentucky Press, . Lexington, Kentucky Press, pp. 35-36.
38. Ingimundarson, J. H..: 1995, *Of Sagas and Sheep: Toward a Historical Anthropology of Social Change and Production for Market, Subsistence and Tribute in Early Iceland (10th to the 13th Century)*, Ph.D. dissertation, Department of Anthropology , University of Arizona, p. 95.
39. Ibid, p. 95.
40. Bjarnason, B.: 1941, *Brandsstaðaannáll*, Sögufélagi Húnvetningur og Húnvetningafélagi í Reykjavík, Reykjavík, Iceland, 1941, p. 121.
41. Ingimundarson, J.H..: 1995, op. cit. pp. 67-69; Kristjánsson, A., and G.Á. Gunnlaugsson: 1990, 'Félags- og hagþróun á Íslandi á fyrri hluta 19. aldar', Saga **27**, 7-62.
42. Jonsson [Jónsson], G.: 1997, 'Changes in Food Consumption in Iceland Ca. 1770-1940', in *Kultur och Konsumption i Norden 1750-1950*, J. Söderberg and L. Magnusson (eds), Finska Historiska Samfundet, Helsinki, Finland. Jónsson's estimates by design reflect general trends in food consumption and are probably too high in bad years, because they are in part based upon assumed constant ratios of livestock counts to food from livestock sources, whereas badly underfed cows and ewes produce little milk.
43. Ibid, Table 5, p. 52
44. Vasey: 1991, op. cit. note 15.
45. Jonsson: 1997, op. cit. note 34, pp. 44-45
46. Lamb, H. H.: 1977, *Climate Present, Past, and Future*, vol. 2, Methuen, London, United Kingdom; Lamb, H. H.: 1982, *Climate, History, and the Modern World*, Methuen, London, United Kingdom; Galloway, P.: 1986, 'Long-Term Fluctuations in Climate and Population in the Preindustrial Era', *Pop. Dev. Rev.* **12(1)**, 1-24.

(Received 3 March 2000; in revised form 25 September 2000)

MIX
Papier aus verantwortungsvollen Quellen
Paper from responsible sources
FSC® C105338

FSC
www.fsc.org

If you have any concerns about our products,
you can contact us on
ProductSafety@springernature.com

In case Publisher is established outside the EU,
the EU authorized representative is:
Springer Nature Customer Service Center GmbH
Europaplatz 3, 69115 Heidelberg, Germany

Printed by Libri Plureos GmbH
in Hamburg, Germany